1 MONTH OF
FREE
READING

at

www.ForgottenBooks.com

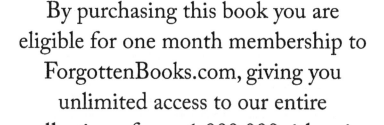

By purchasing this book you are
eligible for one month membership to
ForgottenBooks.com, giving you
unlimited access to our entire
collection of over 1,000,000 titles via
our web site and mobile apps.

To claim your free month visit:
www.forgottenbooks.com/free60367

ISBN 978-0-365-20785-6
PIBN 10060367

HAND-BOOK

OF

PHYSIOLOGY

BY

W. MORRANT BAKER, F.R.C.S.

SURGEON TO ST. BARTHOLOMEW'S HOSPITAL AND CONSULTING SURGEON TO THE EVELINA HOSPITAL
FOR SICK CHILDREN; LECTURER ON PHYSIOLOGY AT ST. BARTHOLOMEW'S HOSPITAL,
AND LATE MEMBER OF THE BOARD OF EXAMINERS OF THE ROYAL COLLEGE
OF SURGEONS OF ENGLAND.

AND

VINCENT DORMER HARRIS, M.D., LOND.

DEMONSTRATOR OF PHYSIOLOGY AT ST. BARTHOLOMEW'S HOSPITAL.

ELEVENTH EDITION

WITH NEARLY 500 ILLUSTRATIONS

VOLUME I

NEW YORK

WILLIAM WOOD & COMPANY

56 & 58 LAFAYETTE PLACE

1885

THE PUBLISHERS'
BOOK COMPOSITION AND ELECTROTYPING CO.,
39 AND 41 PARK PLACE,
NEW YORK.

PREFACE TO THE ELEVENTH EDITION.

In the preparation of the present edition of Kirkes' Physiology, we have endeavored to maintain its character as a guide for students, especially at an early period of their career; and, while incorporating new facts and observations which are fairly established, we have as far as possible omitted the controvertible matters which should only find a place in a complete treatise or in a work of reference.

A large number of new illustrations have been added, for many of which we are indebted to the courtesy of Dr. Klein, Professor Michael Foster, Professor Schaefer, Dr. Mahomed, Mr. Gant, and Messrs. McMillan, who have been so good as to allow various figures to be copied. Our thanks are also due to Mr. Wm. Lapraik, F.C.S., who has kindly prepared a table of the absorption spectra of the blood and bile, based upon his own observations; as well as to Mr. S. K. Alcock for several careful drawings of microscopical preparations, and for reading several sheets in their passage through the press.

Mr. Danielsson, of the firm of Lebon & Co., has executed all the new figures to our entire satisfaction; and for the skill and labor he has expended upon them we are much indebted to him.

We are desirous also of acknowledging the help we have derived from the following works: Klein's Histology; M. Foster's Text-Book of Physiology; Pavy's Food and Dietetics; Quain's Anatomy, Vol. II., Ed. ix.; Wickham Legg's Bile, Jaundice, and Bilious Diseases;

218839

Watney's Minute Anatomy of the Thymus; Rosenthal's Muscles and
Nerves; Cadiat's Traité D'Anatomie Générale; Ranvier's Traité
Technique D'Histologie; Landois' Lehrbuch der Physiologie des
Menschen, and the Journal of Physiology.

W. MORRANT BAKER.

V. D. HARRIS.

WIMPOLE STREET,
    August, 1884.

CONTENTS TO VOLUME I.

CHAPTER I.

CHAPTER II.

CHAPTER III.

CHAPTER IV.

CHAPTER V.

CHAPTER VI.

CHAPTER VII.

CHAPTER VIII.

CHAPTER IX.

CHAPTER X.

CHAPTER XI.

CHAPTER XII.

CHAPTER XIII.

HAND-BOOK OF PHYSIOLOGY.

CHAPTER I.

THE GENERAL AND DISTINCTIVE CHARACTERS OF LIVING BEINGS.

HUMAN PHYSIOLOGY is the science which treats of the life of man—of the way in which he lives, and moves, and has his being. It teaches how man is begotten and born; how he attains maturity; and how he dies.

Having, then, man as the object of its study, it is unnecessary to speak here of the laws of life in general, and the means by which they are carried out, further than is requisite for the more clear understanding of those of the life of man in particular. Yet it would be impossible to understand rightly the working of a complex machine without some knowledge of its motive power in the simplest form; and it may be well to see first what are the so-called *essentials* of life—those, namely, which are manifested by all living beings alike, by the lowest vegetable and the highest animal—before proceeding to the consideration of the structure and endowments of the organs and tissue belonging to man.

The essentials of life are these,—*Birth, Growth and Development, Decline and Death.*

The term **birth,** when employed in this general sense of one of the conditions essential to life, without reference to any particular kind of living being, may be taken to mean, separation from a parent, with a greater or less power of independent life. Taken thus, the term, although not defining any particular stage in development, serves well enough for the expression of the fact, to which no exception has yet been proved to exist, that the capacity for life in all living beings is obtained by inheritance.

Growth, or inherent power of increasing in size, although essential to our idea of life, is not confined to living beings. A crystal of common salt, or of any other similar substance, if placed under appropriate conditions for obtaining fresh material, will grow in a fashion as definitely characteristic and as easily to be foretold as that of a living creature. It is, therefore, necessary to explain the distinctions which exist in this respect between living and lifeless structures; for the manner of growth in the two cases is widely different.

Differences between Living and Lifeless Growth.—(1.) The growth of a crystal, to use the same example as before, takes place merely by additions to its outside; the new matter is laid on particle by particle, and layer by layer, and, when once laid on, it remains unchanged. The growth is here said to be *superficial.* In a living structure, on the other hand, as, for example, a brain or a muscle, where growth occurs, it is by addition of new matter, not to the surface only, but throughout every part of the mass; the growth is not superficial, but *interstitial.*

(2.) All living structures are subject to constant decay; and life consists not, as once supposed, in the power of preventing this never-ceasing decay, but rather in making up for the loss attendant on it by never-ceasing repair. Thus, a man's body is not composed of exactly the same particles day after day, although to all intents he remains the same individual. Almost every part is changed by degrees; but the change is so gradual, and the renewal of that which is lost so exact, that no difference may be noticed, except at long intervals of time. A lifeless structure, as a crystal, is subject to no such laws; neither decay nor repair is a necessary condition of its existence. That which is true of structures which never had to do with life is true also with respect to those which, though they are formed by living parts, are not themselves alive. Thus, an oyster-shell is formed by the living animal which it encloses, but it is as lifeless as any other mass of inorganic matter; and in accordance with this circumstance its growth takes place, not *interstitially,* but layer by layer, and it is not subject to the constant decay and reconstruction which belong to the living. The hair and nails are examples of the same fact.

(3.) In connection with the growth of lifeless masses there is no alteration in the chemical constitution of the material which is taken up and added to the previously existing mass. For example, when a crystal of common salt grows on being placed in a fluid which contains the same material, the properties of the salt are not changed by being taken out of the liquid by the crystal and added to its surface in a solid form. But the case is essentially different in living beings, both animal and vegetable. A plant, like a crystal, can only grow when fresh material is presented to it; and this is absorbed by its leaves and roots; and animals, for the same purpose of getting new matter for growth and nutrition, take food into their stomachs. But in both these cases the materials are much altered before they are finally *assimilated* by the structures they are destined to nourish.

(4.) The growth of all living things has a definite limit, and the law which governs this limitation of increase in size is so invariable that we should be as much astonished to find an individual plant or animal without limit as to growth as without limit to life.

·Development is as constant an accompaniment of life as growth. The term is used to indicate that change to which, before maturity, all living parts are constantly subject, and by which they are made more and more capable of performing their several functions.   For example, a full-grown man is not merely a magnified child; his tissues and organs have not only grown, or increased in size, they have also *developed*, or become better in quality.

No very accurate limit can be drawn between the end of development and the beginning of decline; and the two processes may be often seen together in the same individual.   But after a time all parts alike share in the tendency to degeneration, and this is at length succeeded by death.

Differences between Plants and Animals.—It has been already said that the essential features of life are the same in all living things; in other words, in the members of both the animal and vegetable kingdoms.   It may be well to notice briefly the distinctions which exist between the members of these two kingdoms.   It may seem, indeed, a strange notion that it is possible to confound vegetables with animals, but it is true with respect to the lowest of them, in which but little is manifested beyond the essentials of life, which are the same in both.

(1.) Perhaps the most essential distinction is the presence or absence of power to live upon *inorganic* material.   By means of their green coloring matter, *chlorophyl*—a substance almost exclusively confined to the vegetable kingdom, plants are capable of decomposing the carbonic acid, ammonia, and water, which they absorb by their leaves and roots, and thus utilizing them as food.   The result of this chemical action, which occurs only under the influence of light, is, so far as the carbonic acid is concerned, the fixation of carbon in the plant structures and the exhalation of oxygen.   Animals are incapable of thus using inorganic matter, and never exhale oxygen as a product of decomposition.

The power of living upon organic as well as inorganic matter is less decisive of an animal nature; inasmuch as fungi and some other plants derive their nourishment in part from the former source.

(2.) There is, commonly, a marked difference in general chemical composition between vegetables and animals, even in their lowest forms; for while the former consist mainly of *cellulose*, a substance closely allied to starch and containing carbon, hydrogen, and oxygen only, the latter are composed in great part of the three elements just named, together with a fourth, nitrogen; the chief proximate principles formed from these being identical, or nearly so, with *albumen*.   It must not be supposed, however, that either of these typical compounds alone, with its allies, is confined to one kingdom of nature.   Nitrogenous compounds are freely produced in vegetable structures, although they form a very much smaller proportion of the whole organism than cellulose or starch. And while the presence of the latter in animals is much more rare than is that of the former in vegetables, there are many animals in which traces of it may be discovered, and some, the Ascidians, in which it is found in considerable quantity.

(3.) Inherent power of movement is a quality which we so commonly consider an essential indication of animal nature, that it is difficult at first to conceive it existing in any other. The capability of simple motion is now known, however, to exist in so many vegetable forms, that it can no longer be held as an essential distinction between them and animals, and ceases to be a mark by which the one can be distinguished from the other. Thus the zoospores of many of the Cryptogamia exhibit ciliary or amoeboid movements (p. 8) of a like kind to those seen in animalcules; and even among the higher orders of plants, many, e. g., *Dionæa Muscipula* (Venus's fly-trap), and *Mimosa Sensitiva* (Sensitive plant), exhibit such motion, either at regular times, or on the application of external irritation, as might lead one, were this fact taken by itself, to regard them as sentient beings. Inherent power of movement, then, although especially characteristic of animal nature, is, when taken by itself, no proof of it.

(4.) The presence of a digestive canal is a very general mark by which an animal can be distinguished from a vegetable. But the lowest animals are surrounded by material that they can take as food, as a plant is surrounded by an atmosphere that it can use in like manner. And every part of their body being adapted to absorb and digest, they have no need of a special receptacle for nutrient matter, and accordingly have no digestive canal. This distinction then is not a cardinal one.

It would be tedious as well as unnecessary to enumerate the chief distinctions between the more highly developed animals and vegetables. They are sufficiently apparent. It is necessary to compare, side by side, the lowest members of the two kingdoms, in order to understand rightly how faint are the boundaries between them.

CHAPTER II.

By dissection, the human body can be proved to consist of various dissimilar parts, bones, muscles, brain, heart, lungs, intestines, etc., while, on more minute examination, these are found to be composed of different tissues, such as the connective, epithelial, nervous, muscular, and the like.

Cells.—Embryology teaches us that all this complex organization has been developed from a microscopic body about $\frac{1}{120}$ in. in diameter (ovum), which consists of a spherical mass of jelly-like matter enclosing a smaller spherical body (germinal vesicle). Further, each individual tissue can be shown largely to consist of bodies essentially similar to an ovum, though often differing from it very widely in external form. They are termed *cells :* and it must be at once evident that a correct knowledge of the nature and activities of the cell forms the very foundation of physiology.

Cells are, in fact, physiological no less than histological units.

The prime importance of the cell as an element of structure was first established by the researches of Schleiden, and his conclusions, drawn from the study of vegetable histology, were at once extended by Schwann to the animal kingdom. The earlier observers defined a cell as a more or less spherical body limited by a membrane, and containing a smaller body termed a *nucleus,* which in its turn encloses one or more *nucleoli.* Such a definition applied admirably to most vegetable cells, but the more extended investigation of animal tissues soon showed that in many cases no limiting membrane or cell-wall could be demonstrated.

The presence or absence of a cell-wall, therefore, was now regarded as quite a secondary matter, while at the same time the cell-substance came gradually to be recognized as of primary importance. Many of the lower forms of animal life, *e.g.*, the Rhizopoda, were found to consist almost entirely of matter very similar in appearance and chemical composition to the cell-substance of higher forms: and this from its chemical resemblance to flesh was termed *Sarcode* by Dujardin. When recognized in vegetable cells it was called *Protoplasm* by Mulder, while Remak applied the same name to the substance of animal cells. As the presumed formative matter in animal tissues it was termed *Blastema,* and in the belief that, wherever found, it alone of all substances has to do with generation and nutrition, Beale has named it *Germinal matter* or *Bioplasm*. Of these terms the one most in vogue at the present day is Protoplasm, and inasmuch as all life, both in the animal and vegetable kingdoms, is associated with protoplasm, we are justified in describing it, with Huxley, as the "physical basis of life."

A cell may now be defined as a nucleated mass of protoplasm,[1] of microscopic size, which possesses sufficient individuality to have a life-history of its own. Each cell goes through the same cycle of changes as the whole organism, though doubtless in a much shorter time. Beginning with its origin from some pre-existing cell, it grows, produces other cells, and finally dies. It is true that several lower forms of life consist of non-nucleated protoplasm, but the above definition holds good for all the higher plants and animals.

Hence a summary of the manifestations of cell-life is really an account of the vital activities of protoplasm.

Protoplasm.—*Physical* characters.—Physically, protoplasm is viscid, varying in consistency from semi-fluid to stronglyc oherent. *Chemical* characters.—Chemically, living protoplasm is an extremely unstable albuminoid substance, insoluble in water. It is neutral or weakly alkaline in reaction. It undergoes heat stiffening or coagulation at about 130°F. (54·5°C.), and hence no organism can live when its own temperature is raised beyond this point, though, of course, many can exist for a time in a much hotter atmosphere, since they possess the means of regulating their own temperature. Besides the coagulation produced by heat, protoplasm is coagulated by all the reagents which produce this change in albumen. If not-living protoplasm be subjected to chemical analysis it is found to be made up of numerous bodies[2] besides albumen, *e.g.*, of glycogen, lecithin, salts and water, so that if living protoplasm be, as some believe, an independent chemical body, when it no longer possesses life, it undergoes a disintegration which is accompanied by the appearance of these new chemical substances. When it is examined under the microscope two varieties of protoplasm are recognized—the *hyaline,* and the *granular.* Both are alike transparent, but the former is perfectly homogeneous, while the latter (the more common variety) contains small granules or molecules of various sizes and shapes. Globules of watery fluid are also sometimes found in protoplasm; they look like clear spaces in it, and are hence called *vacuoles.*

Vital or Physiological characters.—These may be conveniently treated under the three heads of—I. **Motion;** II. **Nutrition;** and III. **Reproduction.**

---

[1] In the human body the cells range from the red blood-cell ($\frac{1}{3000}$ in.) to the ganglion-cell ($\frac{1}{300}$) in.

[2] For an account of which, reference should be made to the Appendix.

I. **Motion.**—It is probable that the protoplasm of all cells is capable at some time of exhibiting movement; at any rate this phenomenon, which not long ago was regarded as quite a curiosity, has been recently observed in cells of many different kinds. It may be readily studied in the Amœbæ, in the colorless blood-cells of all vertebrata, in the branched cornea-cells of the frog, in the hairs of the stinging-nettle and Tradescantia, and the cells of Vallisneria and Chara.

These motions may be divided into two classes—(*a*) Fluent and (*b*) Ciliary.

Another variety—the molecular or vibratory—has also been classed by some observers as vital, but it seems exceedingly probable that it is nothing more than the well-known "Brownian" molecular movement, a purely mechanical phenomenon which may be observed in any minute particles, *e.g.*, of gamboge, suspended in a fluid of suitable density, such as water.

Such particles are seen to oscillate rapidly to and fro, and not to progress in any definite direction.

(*a*.) *Fluent.*—This movement of protoplasm is rendered perceptible (1) by the motion of the granules, which are nearly always imbedded in it, and (2) by changes in the outline of its mass.

If part of a hair of Tradescantia (Fig. 1) be viewed under a high magnifying power, streams of protoplasm containing crowds of granules hurrying along, like the foot passengers in a busy street, are seen flowing steadily in definite directions, some coursing round the film which lines the interior of the cell-wall, and others flowing toward or away from the irregular mass in the centre of the cell-cavity. Many of these

FIG. 1.—Cell of Tradescantia drawn at successive intervals of two minutes. The cell-contents consist of a central mass connected by many irregular processes to a peripheral film: the whole forms a vacuolated mass of protoplasm, which is continually changing its shape. (Schofield.)

streams of protoplasm run together into larger ones, and are lost in the central mass, and thus ceaseless variations of form are produced.

In the Amœba, a minute animal consisting of a shapeless and structureless mass of sarcode, an irregular mass of protoplasm is gradually thrust out from the main body and retracted: a second mass is then protruded in another direction, and gradually the whole protoplasmic substance is, as it were, drawn into it. The Amœba thus comes to occupy a new position, and when this is repeated several times we have locomotion in a definite direction, together with a continual change of form. These movements when observed in other cells, such as the colorless blood-corpuscles of higher animals (Fig. 2) are hence termed *amœboid.*

Colorless blood-corpuscles were first observed to migrate, *i.e.*, pass through the walls of the blood-vessels (p. 159), by Waller, whose observations were confirmed and extended to connective tissue corpuscles by the researches of Recklinghausen, Cohnheim, and others, and thus the phenomenon of migration has been proved to play an important part in many normal, and pathological processes, especially in that of inflammation.

This amœboid movement enables many of the lower animals to capture their prey, which they accomplish by simply flowing round and enclosing it. •

The remarkable motions of pigment-granules observed in the branched pigment-cells of the frog's skin by Lister are probably due to amœboid movement. These granules are seen at one time distributed uniformly through the body and branched processes of the cell, while under the action of various stimuli (*e.g.*, light and electricity) they collect in the central mass, leaving the branches quite colorless.

(b.) *Ciliary* action must be regarded as only a special variety of the general motion with which all protoplasm is endowed.

The grounds for this view are the following: In the case of the Infusoria, which move by the vibration of cilia (microscopic hair-like processes projecting from the surface of their bodies) it has been proved that these are simply processes of their protoplasm protruding through pores of the

Fig. 2.—Human colorless blood-corpuscle, showing its successive changes of outline within ten minutes when kept moist on a warm stage. (Schofield.)

investing membrane, like the oars of a galley, or the head and legs of a tortoise from its shell: certain reagents cause them to be partially retracted. Moreover, in some cases cilia have been observed to develop from, and in others to be transformed into, amœboid processes.

The movements of protoplasm can be very largely modified or even suspended by external conditions, of which the following are the most important.

1. *Changes of temperature.*—Moderate heat acts as a stimulan.: this is readily observed in the activity of the movements of a human colorless blood-corpuscle when placed under conditions in which its normal temperature and moisture are preserved. Extremes of heat and cold stop the motions entirely.

2. *Mechanical stimuli.*—When gently squeezed between a cover and object glass under proper conditions, a colorless blood-corpuscle is stimulated to active amœboid movement.

3. *Nerve influence.*—By stimulation of the nerves of the frog's cornea, contraction of certain of its branched cells has been produced.

4. *Chemical stimuli.*—Water generally stops amœboid movement, and by imbibition causes great swelling and finally bursting of the cells.

In some cases, however, (myxomycetes) protoplasm can be almost entirely dried up, and is yet capable of renewing its motions when again moistened.

Dilute salt-solution and many dilute acids and alkalies, stimulate the movements temporarily.

Ciliary movement is suspended in an atmosphere of hydrogen or carbonic acid, and resumed on the admission of air or oxygen.

5. *Electrical.*—Weak currents stimulate the movement, while strong currents cause the corpuscles to assume a spherical form and to become motionless.

II. **Nutrition.**—The nutrition of cells will be more appropriately described in the chapters on Secretion and Nutrition.

Before describing the Reproduction of cells it will be necessary to consider their structure more at length.

Minute Structure of Cells.—(a.) *Cell-wall.*—We have seen (p. 5) that the presence of a limiting-membrane is no essential part of the definition of a cell.

In nearly all cells the outer layer of the protoplasm attains a firmer consistency than the deeper portions: the individuality of the cell becoming more and more clearly marked as this cortical layer becomes more and more differentiated from the deeper portions of cell-substance.   Side by side with this physical, there is a gradual chemical differentiation, till at length, as in the case of the fat-cells, we have a definite limiting-membrane differing chemically as well as physically from the cell-contents, and remaining as a shriveled-up bladder when they have been removed. Such a membrane is transparent and structureless, flexible, and permeable to fluids.

The cell-substance can, therefore, still be nourished by imbibition through the cell-wall.   In many cases (especially in fat) a membrane of some toughness is absolutely necessary to give to the tissue the requisite consistency.   When these membranes attain a certain degree of thickness and independence they are termed capsules: as examples, we may cite the capsules of cartilage-cells, and the thick, tough envelope of the ovum termed the "primitive chorion."

(b.) *Cell contents.*—In accordance with their respective ages, positions, and functions, the contents of cells are very varied.

The original protoplasmic substance may undergo many transformations; thus, in fat-cells we may have oil, or fatty crystals, occupying nearly the whole cell-cavity: in pigment-cells we find granules of pigment; in the various gland-cells the elements of their secretions. Moreover, the original protoplasmic contents of the cell may undergo a gradual chemical change with advancing age; thus the protoplasmic cell-substance of the deeper layers of the epidermis becomes gradually converted into keratin as the cell approaches the surface.   So, too, the original protoplasm of the embryonic blood-cells is replaced by the bæmo-globin of the mature colored blood-corpuscle.

The minute structure of cells has lately been made the subject of careful investigation, and what was once regarded as homogeneous protoplasm with a few scattered granules, has been stated to be an exceedingly complex structure. In colorless blood-corpuscles, epithelial cells, connective tissue corpuscles, nerve-cells, and many other varieties of cells, an *intracellular network* of very fine fibrils, the meshes of which are occupied by a hyaline interstitial substance, has been demonstrated (Heitzmann's network) (Fig. 3). At the nodes, where the fibrils cross, are little swellings, and these are the objects described as granules by the older observers: but in some cells, *e.g.*, colorless blood corpuscles, there are real granules, which appear to be quite free and unconnected with the intra-cellular network.

(c.) *Nucleus.*—Nuclei (Fig. 3) were first pointed out in the year 1833, by Robert Brown, who observed them in vegetable cells. They are either

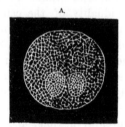

FIG. 3. — (A). Colorless blood-corpuscle showing intra-cellular network of Heitzmann, and two nuclei with intra-nuclear network. (Klein and Noble Smith.)
(B.) Colored blood-corpuscle of newt showing intra-cellular network of fibrils (Heitzmann). Also oval nucleus composed of limiting-membrane and fine intra-nuclear network of fibrils. × 800. (Klein and Noble Smith.)

small transparent vesicular bodies containing one or more smaller particles (nucleoli), or they are semi-solid masses of protoplasm always in the resting condition bounded by a well-defined envelope. In their relation to the life of the cell they are certainly hardly second in importance to the protoplasm itself, and thus Beale is fully justified in comprising both under the term "germinal matter." They exhibit their vitality by initiating the process of division of the cell into two or more cells (fission) by first themselves dividing. Distinct observations have been made showing that spontaneous changes of form may occur in nuclei as also in nucleoli.

Histologists have long recognized nuclei by two important characters:—

(1.) Their power of resisting the action of various acids and alkalies, particularly acetic acid, by which their outline is more clearly defined, and they are rendered more easily visible. This indicates some chemical difference between the protoplasm of the cell and nuclei, as the former is destroyed by these reagents.

(2.) Their quality of staining in solutions of carmine, hæmatoxylin, etc. Nuclei are most commonly oval or round, and do not generally conform to the diverse shapes of the cells; they are altogether less variable elements than cells, even in regard to size, of which fact one may see a good example in the uniformity of the nuclei in cells so multiform as those of epithelium. But sometimes nuclei appear to occupy the whole of the cell, as is the case in the lymph corpuscles of lymphatic glands and in some small nerve cells.

Their position in the cell is very variable. In many cells, especially where active growth is progressing, two or more nuclei are present.

The nuclei of many cells have been shown to contain a fine *intra-nuclear network* in every respect similar to that described above as intra-cellular (Fig. 3), the interstices of which are occupied by semi-fluid protoplasm.

III. **Reproduction.**—The life of individual cells is probably very short in comparison with that of the organism they compose: and their constant decay and death necessitate constant reproduction. The mode in which this takes place has long been the subject of great controversy.

In the case of plants, all of whose tissues are either cellular or composed of cells which are modified or have coalesced in various ways, the theory that all new cells are derived from pre-existing ones was early advanced and very generally accepted. But in the case of animal tissues Schwann and others maintained a theory of spontaneous or free cell formation.

According to this view a minute corpuscle (the future nucleolus) springs up spontaneously in a structureless substance (blastema) very much as a crystal is formed in a solution. This nucleolus attracts the surrounding molecules of matter to form the nucleus, and by a repetition of the process the substance and wall are produced.

This theory, once almost universally current, was first disputed and finally overthrown by Remak and Virchow, whose researches established the truth expressed in the words "Omnis cellula e cellulâ."

It will be seen that this view is in strict accordance with the truth established much earlier in Vegetable Histology that every cell is descended from some pre-existing (mother-) cell. This derivation of cells from cells takes place by (1) *gemmation*, or (2) *fission* or *division*.

(1.) *Gemmation.*—This method has not been observed in the human body or the higher animals, and therefore requires but a passing notice. It consists essentially in the budding off and separating of a portion of the parent cell.

(2.) *Fission* or *Division.*—As examples of reproduction by fission, we may select the ovum, the blood cell, and cartilage cells.

In the frog's ovum (in which the process can be most readily observed) after fertilization has taken place, there is first some amœboid movement, the oscillation gradually increasing until a permanent dimple appears, which gradually extends into a furrow running completely round the spherical ovum, and deepening until the entire yelk-mass is divided into two hemispheres of protoplasm each containing a nucleus (Fig. 4, *b*). This process being repeated by the formation of a second furrow at right angles to the first, we have four cells produced (*c*): this subdivision is

FIG. 4.—Diagram of an ovum (*a*) undergoing segmentation. In (*b*) it has divided into two; in (*c*) into four; in (*d*) the process has ended in the production of the so-called "mulberry mass." (Frey.)

carried on till the ovum has been divided by segmentation into a mass of cells (mulberry-mass) (*d*) out of which the embryo is developed.

Segmentation is the first step in the development of most animals, and doubtless takes place in man.

Multiplication by fission has been observed in the colorless blood-cells of many animals. In some cases (Fig. 5), the process has been seen to commence with the nucleolus which divides within the nucleus. The nucleus then elongates, and soon a well-marked constriction occurs, rendering it hour-glass shaped, till finally it is separated into two parts, which gradually recede from each other: the same process is repeated in the cell-substance, and at length we have two cells produced which by

FIG. 5.—Blood-corpuscle from a young deer embryo, multiplying by fission. (Frey.)

rapid growth soon attain the size of the parent cell (*direct division*). In some cases there is a primary fission into three instead of the usual two cells.

In cartilage (Fig. 6), a process essentially similar occurs, with the exception that (as in the ovum) the cells produced by fission remain in the original capsule, and in their turn undergo division, so that a large number of cells are sometimes observed within a common envelope. This process of fission within a capsule has been by some described as a separate method, under the title "endogenous fission," but there seems to be no sufficient reason for drawing such a distinction.

It is important to observe that fission is often accomplished with great rapidity, the whole process occupying but a few minutes, hence the comparative rarity with which cells are seen in the act of dividing.

Indirect cell division.—In certain and numerous cases the division of cells does not take place by the simple constriction of their nuclei and surrounding protoplasm into two parts as above described (direct division), but is preceded by complicated changes in their nuclei (karyokinesis).

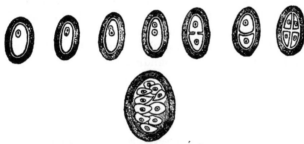

FIG. 6.—Diagram of a cartilage cell undergoing fission within its capsule. The process of division is represented as commencing in the nucleolus, extending to the nucleus, and at length involving the body of the cell. (Frey.)

These changes consist in a gradual re-arrangement of the intranuclear network of each nucleus, until two nuclei are formed similar in all respects to the original one. The nucleus in a resting condition, *i.e.*, before any changes preceding division occur, consists of a very close meshwork of fibrils, which stain deeply in carmine, imbedded in protoplasm, which does not possess this property, the whole nucleus being contained in an envelope. The first change consists of a slight enlargement, the disappearance of the envelope, and the increased definition and thickness of

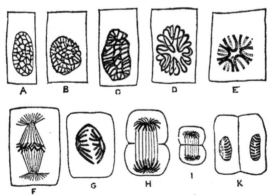

FIG. 7.—Karyokinesis. A, ordinary nucleus of a columnar epithelial cell; B, C, the same nucleus in the stage of *convolution;* D, the *wreath or rosette* form; E, the *aster* or *single star;* F, a nuclear spindle from the Descemet's endothelium of the frog's cornea; G, H, I, diaster; K, two daughter nuclei. (Klein.)

the nuclear fibrils, which are also more separated than they were and stain better. This is the stage of *convolution* (Fig. 7, B, C). The next step in the process is the arrangement of the fibrils into some definite figure by an alternate looping in and out around a central space, by which means the *rosette* or *wreath* stage (Fig. 7, D) is reached. The loops of the rosette next become divided at the periphery, and their central points become more angular, so that the fibrils, divided into portions of about equal length, are, as it were, doubled at an acute angle, and radiate V-shaped from the centre, forming a *star* (aster) or wheel (Fig. 7, E), or perhaps from two centres, in which case a double star (diaster) results (Fig. 7, G, H, and I). After remaining almost unchanged for some time, the V-shaped fibres being first re-arranged in the centre, side by side (angle outward), tend to separate into two bundles, which gradually assume position at either pole. From these groups of fibrils the two nuclei of the new cells are formed (daughter nuclei) (Fig. 7, K), and the changes they pass through before reaching the resting condition are exactly those through which the original nucleus (mother nucleus) has gone, but in a reverse order, viz., the star, the rosette, and the convolution. During or shortly after the formation of the daughter nuclei the cell itself becomes constricted, and then divides in a line about midway between them.

Functions of Cells.—The functions of cells are almost infinitely varied and make up nearly the whole of Physiology. They will be more appropriately considered in the chapters treating of the several organs and systems of organs which the cells compose.

Decay and Death of Cells.—There are two chief ways in which the comparatively brief existence of cells is brought to an end. (1) Mechanical abrasion, (2) Chemical transformation.

1. The various epithelia furnish abundant examples of mechanical abrasion. As it approaches the free surface the cell becomes more and more flattened and scaly in form and more horny in consistence, till at length it is simply rubbed off. Hence we find epithelial cells in the mucus of the mouth, intestine, and genito-urinary tract.

2. In the case of chemical transformation the cell-contents undergo a degeneration which, though it may be pathological, is very often a normal process.

Thus we have (a.) *fatty* metamorphosis producing oil-globules in the secretion of milk, fatty degeneration of the muscular fibres of the uterus after the birth of the fœtus, and of the cells of the Graafian follicle giving rise to the "corpus luteum." (See chapter on Generation.)

(b.) *Pigmentary* degeneration from deposit of pigment, as in the epithelium of the air-vesicles of the lungs.

(c.) *Calcareous* degeneration which is common in the cells of many cartilages.

Having thus reviewed the life-history of cells in general, we may now discuss the leading varieties of form which they present.

In passing, it may be well to point out the main *distinctions between animal and vegetable cells.*

It has been already mentioned that in animal cells an envelope or cell-wall is by no means always present. In adult vegetable cells, on the other hand, a well-defined cellulose wall is highly characteristic; this, it should be observed, is non-nitrogenous, and thus differs chemically as well as structurally from the contained mass.

Moreover, in vegetable cells (Fig. 8, B), the protoplastic contents of the cell fall into two subdivisions: (1) a continuous film which lines the interior of the cellulose wall; and (2) a reticulate mass containing the

A  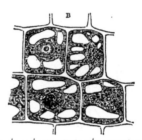 B

FIG. 8.—(A). Young vegetable cells, showing cell-cavity entirely filled with granular protoplasm enclosing a large oval nucleus, with one or more nucleoli.
(B.) Older cells from the same plant, showing distinct cellulose-wall and vacuolation of protoplasm.

nucleus and occupying the cell-cavity; its interstices are filled with fluid. In young vegetable cells such a distinction does not exist; a finely granular protoplasm occupies the whole cell-cavity (Fig. 8, A).

Another striking difference is the frequent presence of a large quantity of intercellular substance in animal tissues, while in vegetables it is comparatively rare, the requisite consistency being given to their tissues by the tough cellulose walls, often thickened by deposits of lignin. In animal cells this end is attained by the deposition of lime-salts in a matrix of intercellular substance, as in the process of ossification.

Forms of Cells.—Starting with the spherical or spheroidal (Fig. 9, *a*) as the typical form assumed by a free cell, we find this altered to a polyhedral shape when the pressure on the cells in all directions is nearly the same (Fig. 9, *b*).

Of this, the primitive segmentation-cells may afford an example.

The discoid shape is seen in blood-cells (Fig. 9, *c*), and the scale-like

FIG. 9.—Various forms of cells. *a.* Spheroidal, showing nucleus and nucleolus. *b.* Polyhedral. *c.* Discoidal (blood-cells). *d.* Scaly or squamous (epithelial cells).

form in superficial epithelial cells (Fig. 9, *d*). Some cells have a jagged outline (prickle-cells) (Fig. 13).

Cylindrical, conical, or prismatic cells occur in the deeper layers of laminated epithelium, and the simple cylindrical epithelium of the intestine and many gland ducts. Such cells may taper off at one or both ends into fine processes, in the former case being caudate, in the latter fusiform (Fig. 10). They may be greatly elongated so as to become fibres. Ciliated cells (Fig. 10, *d*) must be noticed as a distinct variety: they possess, but only on their free surfaces, hair-like processes (cilia). These vary immensely in size, and may even exceed in length the cell itself. Finally we have the branched or stellate cells, of which the large

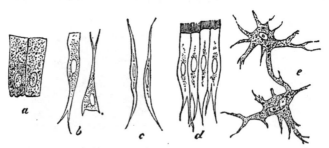

FIG. 10.—Various forms of cells. *a.* Cylindrical or columnar. *b.* Caudate. *c.* Fusiform. *d.* Ciliated (from trachea). *e.* Branched, stellate.

nerve-cells of the spinal cord, and the connective tissue corpuscle are typical examples (Fig. 10, *e*). In these cells the primitive branches by secondary branching may give rise to an intricate network of processes.

Classification of Cells.—Cells may be classified in many ways. According to:—

(a.) *Form:* They may be classified into spheroidal or polyhedral, discoidal, flat or scaly, cylindrical, caudate, fusiform, ciliated and stellate.

(b.) *Situation:*—we may divide them into blood cells, gland cells, connective tissue cells, etc.

(c.) *Contents:*—fat and pigment cells and the like.

(d.) *Function:*—secreting, protective, contractile, etc.

(e.) *Origin:*—hypoblastic, mesoblastic, and epiblastic cells. (See chapter on Generation.)

It remains only to consider the various ways in which cells are connected together to form tissues, and the transformations by which intercellular substance, fibres and tubules are produced.

Modes of connection.—Cells are connected:—

(1) By a cementing intercellular substance. This is probably always present as a transparent, colorless, viscid, albuminous substance, even between the closely apposed cells of cylindrical epithelium, while in the case of cartilage it forms the main bulk of the tissue, and the cells only appear as imbedded in, not as cemented by, the intercellular substance.

This intercellular substance may be either homogeneous or fibrillated.

In many cases (*e.g.* the cornea) it can be shown to contain a number of irregular branched cavities, which communicate with each other, and in which the branched cells lie: through these branching spaces nutritive fluids can find their way into the very remotest parts of a non-vascular tissue.

As a special variety of intercellular substance must be mentioned the basement membrane (*membrana propria*) which is found at the base of the epithelial cells in most mucous membranes, and especially as an investing tunic of gland follicles which determines their shape, and which may persist as a hyaline saccule after the gland-cells have all been discharged.

(2) By anastomosis of their processes.

This is the usual way in which stellate cells, *e.g.*, of the cornea, are united: the individuality of each cell is thus to a great extent lost by its connection with its neighbors to form a reticulum: as an example of a network so produced, we may cite the stroma of lymphatic glands.

Sometimes the branched processes breaking up into a maze of minute fibrils, adjoining cells are connected by an intermediate reticulum: this is the case in the nerve-cells of the spinal cord.

Besides the Cell, which may be termed the primary tissue-element, there are materials which may be termed secondary or derived tissue-elements. Such are Intercellular substance, Fibres and Tubules.

Intercellular substance is probably in all cases directly derived from the cells themselves. In some cases (*e.g.* cartilage), by the use of re-agents the cementing intercellular substance is, as it were, analyzed into various masses, each arranged in concentric layers around a cell or group of cells, from which it was probably derived (Fig. 6).

Fibres.—In the case of the crystalline lens, and of muscle both striated and non-striated, each fibre is simply a metamorphosed cell: in the case of striped fibre the elongation being accompanied by a multiplication of the nuclei.

The various fibres and fibrillæ of connective tissue result from a gradual transformation of an originally homogeneous intercellular substance. Fibres thus formed may undergo great chemical as well as physical transformation: this is notably the case with yellow elastic tissue, in which the sharply defined elastic fibres, possessing great power of resistance to re-agents, contrast strikingly with the homogeneous matter from which they are derived.

Tubules which were originally supposed to consist of structureless membrane, have now been proved to be composed of flat, thin cells, cohering along their edges. (See Capillaries.)

With these simple materials the various parts of the body are built up; the more elementary tissues being, so to speak, first compounded of them; while these again are variously mixed and interwoven to form more intricate combinations.

Thus are constructed epithelium and its modifications, connective tissue, fat, cartilage, bone, the fibres of muscle and nerve, etc.; and these, again, with the more simple structures before mentioned, are used as materials wherewith to form arteries, veins, and lymphatics, secreting and vascular glands, lungs, heart, liver, and other parts of the body.

CHAPTER III.

STRUCTURE OF THE ELEMENTARY TISSUES.

IN this chapter the leading characters and chief modifications of two great groups of tissues—the Epithelial and Connective—will be briefly described; while the Nervous and Muscular, together with several other more highly specialized tissues, will be appropriately considered in the chapters treating of their physiology.

EPITHELIUM.

Epithelium is composed of cells of various shapes held together by a small quantity of cementing intercellular substance.

Epithelium clothes the whole exterior surface of the body, forming the *epidermis* with its appendages—nails and hairs; becoming continuous at the chief orifices of the body—nose, mouth, anus, and urethra—with the epithelium which lines the whole length of the alimentary and genito-urinary tracts, together with the ducts of their various glands. Epithelium also lines the cavities of the brain, and the central canal of the spinal cord, the serous and synovial membranes, and the interior of all blood-vessels and lymphatics.

The cells composing it may be arranged in either one or more layers, and thus it may be subdivided into (*a*) *Simple* and (*b*) *Stratified or laminated Epithelium*. A simple epithelium, for example, lines the whole intestinal mucous membrane from the stomach to the anus: the epidermis on the other hand is laminated throughout its entire extent.

Epithelial cells possess an intracellular and an intranuclear network (p. 10). They are held together by a clear, albuminous, cement substance. The viscid semi-fluid consistency both of cells and intercellular substance permits such changes of shape and arrangement in the individual cells as are necessary if the epithelium is to maintain its integrity in organs the area of whose free surface is so constantly changing, as the stomach, lungs, etc. Thus, if there be but a single layer of cells, as in the epithelium lining the air vesicles of the lungs, the stretching of this membrane causes such a thinning out of the cells that they change their shape from spheroidal or short columnar, to squamous, and *vice versâ*, when the membrane shrinks.

Classification of Epithelial Cells.

Epithelial cells may be conveniently classified as:
1. *Squamous, scaly, pavement, or tessellated.*
2. *Spheroidal, glandular, or polyhedral.*
3. *Columnar, cylindrical, conical, or goblet-shaped.*
4. *Ciliated.*
5. *Transitional.*

Although, for convenience, epithelial cells are thus classified, yet the first three forms of cells are sometimes met with at different depths in

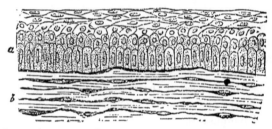

FIG. 11.—Vertical section of Rabbit's cornea. *a.* Anterior epithelium, showing the different shapes of the cells at various depths from the free surface. *b.* Portion of the substance of cornea. (Klein.)

the same membrane. As an example of such a laminated epithelium showing these different cell-forms at various depths, we may select the anterior epithelium of the cornea (Fig. 11).

1. *Squamous Epithelium* (Fig. 12).—Arranged (A) in several super-posed layers (*stratified or laminated*), this form of epithelium covers (*a*) the skin, where it is called the Epidermis, and lines (*b*) the mouth, pharynx, and œsophagus, (*c*) the conjunc-tiva, (*d*) the vagina, and entrance of the urethra in both sexes; while, as (B) a single layer, the same kind of epithelium forms (*a*) the pigmentary layer of the retina, and lines (*b*) the interior of the serous and synovial sacs, and (*c*) of the heart, blood and lymph-vessels (Endothelium). It con-sists of cells, which are flattened and scaly,

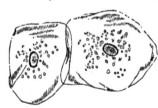

FIG. 12.—Squamous epithelium scales from the inside of the mouth. × 260. (Henle.)

with an irregular outline: and, when laminated, may form a dense horny investment, as on parts of the palms of the hands and soles of the feet. The nucleus is often not apparent. The really cellular nature of even the dry and shriveled scales cast off from the surface of the epidermis, can be proved by the application of caustic potash, which causes them rapidly to swell and assume their original form.

Squamous cells are generally united by an intercellular substance; but in many of the deeper layers of epithelium in the mouth and skin, the outline of the cells is very irregular.

Such cells (Fig. 13) are termed "ridge and furrow," "cogged" or "prickle" cells. These "prickles" are prolongations of the intra-cellular network which run across from cell to cell, thus joining them together, the interstices being filled by the transparent intercellular cement substance. When this increases in quantity in inflammation, the cells are pushed further apart and the connecting fibrils or "prickles" elongated, and therefore more clearly visible.

Squamous epithelium, *e.g.* the pigment cells of the retina, may have a deposit of pigment in the cell-substance. This pigment consists of minute molecules of *melanin*, imbedded in the cell-substance and almost concealing the nucleus, which is itself transparent (Fig. 14).

In white rabbits and other albino animals, in which the pigment of

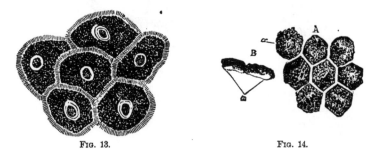

FIG. 13. FIG. 14.

FIG. 13.—Jagged cells of the middle layers of pavement epithelium, from a vertical section of the gum of a new-born infant. (Klein.)
FIG. 14.—Pigment cells from the retina. A, cells still cohering, seen on their surface; *a*, nucleus indistinctly seen. In the other cells the nucleus is concealed by the pigment granules. B, two cells seen in profile; *a*, the outer or posterior part containing scarcely any pigment. × 370. (Henle.)

the eye is absent, this layer is found to consist of colorless pavement epithelial cells.

Endothelium.—The squamous epithelium lining the serous membranes, and the interior of blood-vessels, presents so many special features as to demand a special description; it is called by a distinct name—Endothelium.

The main points of distinction above alluded to are, 1. the very flattened form of these cells; 2. their constant occurrence in only a single layer; 3. the fact that they are developed from the "mesoblast," while all other epithelial cells are derived from the "epiblast," or "hypoblast;" 4. they line closed cavities not communicating with the exterior of the body. Endothelial cells form an important and well-defined subdivision of squamous epithelial cells, which has been especially studied during the last few years. Their examination has been much facilitated by the adoption of the method of staining serous membranes with silver nitrate.

When a small portion of a perfectly fresh serous membrane, as the mesentery or omentum (Fig. 15), is immersed for a few minutes in a quarter per cent. solution of this re-agent, washed with water and exposed to the action of light, the silver oxide is precipitated along the bounda-

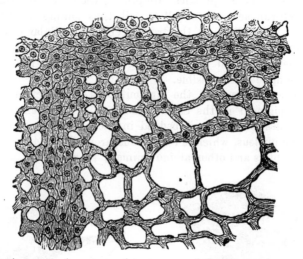

FIG. 15.—Part of the omentum of a cat, stained in silver nitrate, × 100. The tissue forms a "*fenestrated membrane*," that is to say, one which is studded with holes or windows. In the figure these are of various shapes and sizes, leaving trabeculæ, the basis of which is fibrous tissue. The trabeculæ are of various sizes, and are covered with endothelial cells, the nuclei of which have been made evident by staining with hæmatoxylin after the silver nitrate has outlined the cells by staining the intercellular substance. (V. D. Harris.)

ries of the cells, and the whole surface is found to be marked out with exquisite delicacy, by fine dark lines, into a number of polygonal spaces (endothelial cells) (Figs. 15 and 16).

Endothelium lines, as before mentioned, all the serous cavities of the

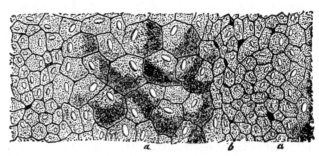

FIG. 16.—Abdominal surface of centrum tendineum of diaphragm of rabbit, showing the general polygonal shape of the endothelial cells; each is nucleated. (Klein.) × 300.

body, including the anterior chamber of the eye, also the synovial membranes of joints, and the interior of the heart and of all blood-vessels and lymphatics. It forms also a delicate investing sheath for nerve-fibres and peripheral ganglion-cells. The cells are scaly in form, and irregular in outline; those lining the interior of blood-vessels and lymphatics having a spindle-shape with a very wavy outline. They enclose a clear, oval nucleus, which, when the cell is viewed in profile, is seen to project from its surface.

Endothelial cells may be ciliated, *e.g.*, those in the mesentery of frogs, especially about the breeding season.

Besides the ordinary endothelial cells above described, there are found on the omentum and parts of the pleura of many animals, little bud-like processes or nodules, consisting of small polyhedral granular cells, rounded on their free surface, which multiply very rapidly by division (Fig. 17). These constitute what is known as "germinating endothelium."

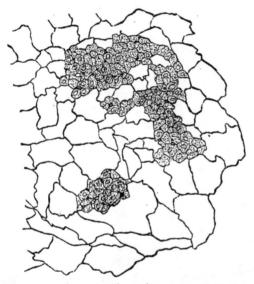

Fig. 17.—Silver-stained preparation of great omentum of dog, which shows, amongst the flat endothelium of the surface, small and large groups of germinating endothelium, between which numbers of stomata are to be seen. (Klein.) × 300.

The process of germination doubtless goes on in health, and the small cells which are thrown off in succession are carried into the lymphatics, and contribute to the number of the lymph corpuscles. The buds may be enormously increased both in number and size in certain diseased conditions.

On those portions of the peritoneum and other serous membranes where lymphatics abound, there are numerous small orifices—*stomata*—(Fig.18) between the endothelial cells: these are really the open mouths of lymphatic vessels, and through them lymph-corpuscles, and the serous fluid from the serous cavity, pass into the lymphatic system.

2. *Spheroidal* epithelial cells are the active secreting agents in most

secreting glands, and hence are often termed glandular; they are generally more or less rounded in outline: often polygonal from mutual pressure.

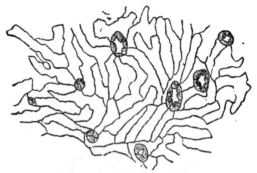

FIG. 18.—Peritoneal surface of septum cisternæ lymphaticæ magnæ of frog. The stomata, some
of which are open, some collapsed, are surrounded by germinating endothelium. (Klein.) × 160.

Excellent examples are to be found in the liver, the secreting tubes of
the kidney, and in the salivary and peptic glands (Fig. 19).

3. *Columnar* epithelium (Fig. 20, A and B) lines (a.) the mucous membrane of the stomach and intestines, from the cardiac orifice of the stomach
to the anus, and (b.) wholly or in part the ducts of the glands opening on

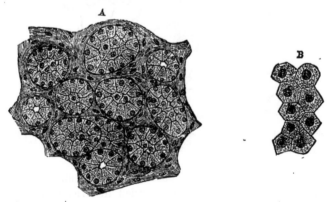

FIG. 19.—Glandular epithelium. A, small lobule of a mucous gland of the tongue, showing nucleated glandular spheroidal cells. B, Liver cells. × 200. (V. D. Harris.)

its free surface; also (c.) many gland-ducts in other regions of the body,
e.g., mammary, salivary, etc.; (d.) the cells which form the deeper layers
of the epithelial lining of the trachea are approximately columnar.

It consists of cells which are cylindrical or prismatic in form, and contain a large oval nucleus. When evenly packed side by side as a single
layer, the cells are uniformly columnar; but when occurring in several
layers as in the deeper strata of the epithelial lining of the trachea, their

shape is very variable, and often departs very widely from the typical columnar form.

Goblet-cells.—Many cylindrical epithelial cells undergo a curious transformation, and from the alteration in their shape are termed goblet-cells (Fig. 20, A, *c*, and B).

These are never seen in a perfectly fresh specimen: but if such a specimen be watched for some time, little knobs are seen gradually appear-

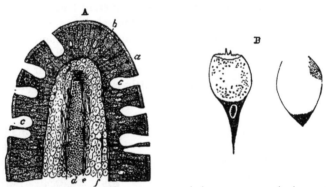

FIG. 20.—A. Vertical section of a villus of the small intestine of a cat. *a.* Striated basilar border of the epithelium. *b.* Columnar epithelium. *c.* Goblet cells. *d.* Central lymph-vessel, *e.* Smooth muscular fibres. *f.* Adenoid stroma of the villus in which lymph corpuscles lie. B. Goblet cells. (Klein.)

ing on the free surface of the epithelium, and are finally detached; these consist of the cell-contents which are discharged by the open mouth of the goblet, leaving the nucleus surrounded by the remains of the protoplasm in its narrow stem.

Some regard this transformation as a normal process which is continually going on during life, the discharged cell-contents contributing to form *mucus*, the cells being supposed in many cases to recover their original shape.

The columnar epithelial cells of the alimentary canal possess a structureless layer on their free surface: such a layer, appearing striated when viewed in section, is termed the "striated basilar border" (Fig. 20, A, *a*).

4. *Ciliated* cells are generally cylindrical (Fig. 21, B), but may be spheroidal or even almost squamous in shape (Fig. 21, A).

This form of epithelium lines (a.) the whole of the respiratory tract from the larynx to the finest subdivisions of the bronchi, also the lower parts of the nasal passages, and some portions of the generative apparatus —in the male (b.) lining the "vasa efferentia" of the testicle, and their prolongations as far as the lower end of the epididymis; in the female (c.) commencing about the middle of the neck of the uterus, and extending throughout the uterus and Fallopian tubes to their fimbriated extremities, and even for a short distance on the peritoneal surface of the latter.   (d.) The ventricles of the brain and the central canal of the spinal cord are clothed with ciliated epithelium in the child, but in the adult it is limited to the central canal of the cord.

The *Cilia*, or fine hair-like processes which give the name to this variety of epithelium, vary a good deal in size in different classes of animals, being very much smaller in the higher than among the lower orders, in which they sometimes exceed in length the cell itself.

The number of cilia on any one cell ranges from ten to thirty, and those attached to the same cell are often of different lengths. When living ciliated epithelium, *e.g.*, the gill of a mussel, is examined under the microscope, the cilia are seen to be in constant rapid motion; each cilium being fixed at one end, and swinging or lashing to and fro. The general impression given to the eye of the observer is very similar to that produced by waves in a field of corn, or swiftly running and rippling water,

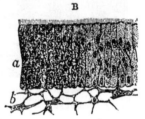

FIG. 21.—A. Spheroidal ciliated cells from the mouth of the frog. × 300 diameters. (Sharpey.) B. *a.* Ciliated columnar epithelium lining a bronchus. *b.* Branched connective-tissue corpuscles. (Klein and Noble Smith.)

and the result of their movement is to produce a continuous current in a definite direction, and this direction is invariably the same on the same surface, being always, in the case of a cavity, toward its external orifice.

5. *Transitional Epithelium.*—This term has been applied to cells which are neither arranged in a single layer, as is the case with simple epithelium, nor yet in many superimposed strata as in laminated; in other words, the term is employed when epithelial cells are found in two, three, or four superimposed layers. The upper layer may be either columnar, ciliated, or squamous. When the upper layer is columnar or ciliated, the second layer consists of smaller cells fitted into the inequalities of the cells above them, as in the trachea (Fig. 21, B). The epithelium which is met with lining the urinary bladder and ureters is, however, the transitional *par excellence*. In this variety there are two or three layers of cells, the upper being more or less flattened according to the full or collapsed condition of the organ, their under surface being marked with one or more depressions, into which the heads of the next layer of club-shaped cells fit. Between the lower and narrower parts of the second row of cells, are fixed the irregular cells which constitute the third row, and in like manner sometimes a fourth row (Fig. 22). It can be easily understood, therefore, that if a scraping of the mucous membrane of the blad-

der be teazed, and examined under the microscope, cells of a great variety of forms may be made out (Fig. 23). Each cell contains a large nucleus, and the larger and superficial cells often possess two.

. **Special. Epithelium in Organs of Special Sense.**—In addition to the above kinds of epithelium, certain highly specialized forms of epithelial cells are found in the organs of smell, sight, and hearing, viz.,

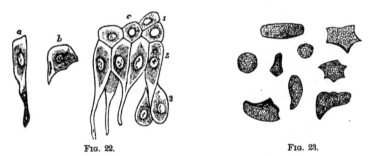

Fig. 22. Fig. 23.

Fig. 22.—Epithelium of the bladder; *a*, one of the cells of the first row; *b*, a cell of the second row; *c*, cells *in situ*, of first, second, and deepest layers. (Obersteiner.)
Fig. 23.—Transitional epithelial cells from a scraping of the mucous membrane of the bladder of the rabbit. (V. D. Harris.)

olfactory cells, retinal rods and cones, auditory cells; they will be described in the chapters which deal with their functions.

Functions of Epithelium.—According to function, epithelial cells may be classified as:—

(1.) *Protective, e.g.*, in the skin, mouth, blood-vessels, etc.

(2.) *Protective and moving*—ciliated epithelium.

(3.) *Secreting*—glandular epithelium; or, *Secreting formed elements* —epithelium of testicle secreting spermatozoa.

(4.) *Protective and secreting, e.g.*, epithelium of intestine.

(5.) *Sensorial, e.g.*, olfactory cells, rods and cones of retina, organ of Corti.

Epithelium forms a continuous smooth investment over the whole body, being thickened into a hard, horny tissue at the points most exposed to pressure, and developing various appendages, such as hairs and nails, whose structure and functions will be considered in a future chapter. Epithelium lines also the sensorial surfaces of the eye, ear, nose, and mouth, and thus serves as the medium through which all impressions from the external world—touch, smell, taste, sight, hearing—reach the delicate nerve-endings, whence they are conveyed to the brain.

The ciliated epithelium which lines the air-passages serves not only as a protective investment, but also by the movements of its cilia is enabled to propel fluids and minute particles of solid matter so as to aid their expulsion from the body. In the case of the Fallopian tube, this agency assists the progress of the ovum toward the cavity of the uterus. Of the purposes served by cilia in the ventricles of the brain, nothing is known. (For an account of the nature and conditions of ciliary motion, see chapter on Motion.)

The epithelium of the various glands, and of the whole intestinal tract, has the power of *secretion, i.e.,* of chemically transforming certain materials of the blood; in the case of mucus and saliva this has been proved to involve the transformation of the epithelial cells themselves; the cell-substance of the epithelial cells of the intestine being discharged by the rupture of their envelopes, as mucus.

Epithelium is likewise concerned in the processes of transudation, diffusion, and absorption.

It is constantly being shed at the free surface, and reproduced in the deeper layers. The various stages of its growth and development can be well seen in a section of any laminated epithelium, such as the epidermis.

The Connective Tissues.

This group of tissues forms the Skeleton with its various connections —bones, cartilages, and ligaments—and also affords a supporting framework and investment to various organs composed of nervous, muscular, and glandular tissue. Its chief function is the mechancial one of support, and for this purpose it is so intimately interwoven with nearly all the textures of the body, that if all other tissues could be removed, and the connective tissues left, we should have a wonderfully exact model of almost every organ and tissue in the body, correct even to the smallest minutiæ of structure.

Classification of Connective Tissues.—The chief varieties of connective tissues may be thus classified:—

I. The Fibrous Connective Tissues.

A.—*Chief Forms.*   B.—*Special Varieties.*

a. Areolar.   a. Gelatinous.
b. White fibrous.   b. Adenoid or Retiform.
c. Elastic.   c. Neuroglia.
                            d. Adipose.

II. Cartilage.
III. Bone.

All of the varieties of connective tissue are made up of two parts, namely, *cells* and *intercellular substance.*

Cells.—The cells are of two kinds.

(*a.*) *Fixed.*—These are cells of a flattened shape, with branched processes, which are often united together to form a network: they can be most readily observed in the cornea in which they are arranged, layer above layer, parallel to the free surface. They lie in spaces, in the intercellular or ground substance, which are of the same shape as the cells they contain but rather larger, and which form by anastomosis a system of branching canals freely communicating (Fig. 24).

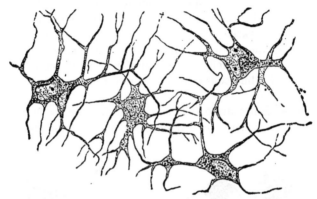

Fig. 24.—Horizontal preparation of cornea of frog, stained in gold chloride; showing the network of branched cornea corpuscles. The ground-substance is completely colorless. × 400. (Klein.)

To this class of cells belong the flattened tendon corpuscles which are arranged in long lines or rows parallel to the fibres (Fig. 29).

These branched cells, in certain situations, contain a number of pigment-granules, giving them a dark appearance: they form one variety of pigment-cells. Branched pigment-cells of this kind are found in the outer layers of the choroid (Fig. 25). In many lower animals, such as the frog, they are found widely distributed, not only in the skin, but also in many internal parts, *e.g.*, the mesentery and sheaths of blood-vessels. In the web of the frog's foot such pigment-cells may be seen, with pigment evenly distributed through the body of the cell and its processes; but under the action of light, electricity, and other stimuli, the pigment-granules become massed in the body of the cell, leaving the processes quite hyaline; if the stimulus be removed, they will gradually be distributed again all over the processes. Thus the skin in the frog is sometimes uniformly dusky, and sometimes quite light-colored, with isolated dark spots. In the choroid and retina the pigment-cells absorb light.

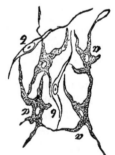

Fig. 25.—Ramified pigment-cells from the tissue of the choroid coat of the eye. × 350. *a*, cell with pigment; *b*, colorless fusiform cells. (Kölliker.)

(*b.*) *Amœboid cells*, of an approximately spherical shape: they have a great general resemblance to colorless blood corpuscles (Fig. 2), with

which some of them are probably identical. · They consist of finely granular nucleated protoplasm, and have the property, not only of changing
their form, but also of moving about, whence they are termed migra-
tory.   They are readily distinguished from the branched connective-tissue
corpuscles by their free condition, and the absence of processes.   Some
are much larger than others, and are found especially in the sublingual
gland of the dog and guinea pig and in the mucous membrane of the
intestine.   A second variety of these cells called *plasma cells* (Waldeyer)
are larger than the amœboid cells, apparently granular, less active in their
movements.   They are chiefly to be found in the inter-muscular septa,
in the mucous and submucous coats of the intestine, in lymphatic glands,
and in the omentum.

Intercellular Substance.—This may be *fibrillar*, as in the fibrous
tissues and certain varieties of cartilage; or *homogeneous*, as in hyaline
cartilage.

Fɪɢ. 26.   Fɪɢ. 27.

Fɪɢ. 26.—Flat, pigmented, branched, connective-tissue cells from the sheath of a large blood-vessel of frog's mesentery: the pigment is not distributed uniformly through the substance of the larger
cell, consequently some parts of the cell look blacker than others (uncontracted state).   In the two
smaller cells most of the pigment is withdrawn into the cell-body, so that they appear smaller, blacker, and less branched.   × 350.   (Klein and Noble Smith.)

Fɪɢ. 27.—Fibrous tissue of cornea, showing bundles of fibres with a few scattered fusiform cells
lying in the inter-fascicular spaces.   × 400.   (Klein and Noble Smith.)

The fibres composing the former are of two kinds—(*a.*) White fibres.
(*b.*) Yellow elastic fibres.

(*a.*) *White Fibres.*—These are arranged parallel to each other in wavy
bundles of various sizes: such bundles may either have a parallel arrangement (Fig. 27), or may produce quite a felted texture by their interlacement.   The individual fibres composing these fasciculi are homogeneous,
unbranched, and of the same diameter throughout.   They can readily
be isolated by macerating a portion of white fibrous tissue (*e.g.*, a small
piece of tendon) for a short time in lime, or baryta-water, or in a solution
of common salt, or potassium permanganate: these reagents possessing
the power of dissolving the cementing interfibrillar substance (which is
nearly allied to syntonin), and thus separating the fibres from each other.

(*b.*) *Yellow Elastic Fibres* (Fig. 28) are of all sizes, from excessively
fine fibrils up to fibres of considerable thickness: they are distinguished

from white fibres by the following characters:—(1.) Their great power of resistance even to the prolonged action of chemical reagents, *e.g.*, Caustic Soda, Acetic Acid, etc. (2.) Their well-defined outlines. (3.) Their great. tendency to branch and form networks by anastomosis. (4.) They very often have a twisted corkscrewlike appearance, and their free ends usually curl up. (5.) They are of a yellowish tint and very elastic.

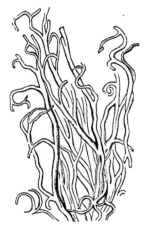

FIG. 28.—Elastic fibres from the ligamenta subflava. × 200. (Sharpey.)

VARIETIES OF CONNECTIVE TISSUE.

I. FIBROUS CONNECTIVE TISSUES.

A.—*Chief Forms.*—(*a.*) *Areolar Tissue.*

Distribution.—This variety has a very wide distribution, and constitutes the subcutaneous, subserous and submucous tissue. It is found in the mucous membranes, in the true skin, in the outer sheaths of the blood-vessels. It forms sheaths for muscles, nerves, glands, and the internal organs, and, penetrating into their interior, supports and connects the finest parts.

Structure.—To the naked eye it appears, when stretched out, as a fleecy, white, and soft meshwork of fine fibrils, with here and there wider films joining in it, the whole tissue being evidently elastic. The openness of the meshwork varies with the locality from which the specimen is taken. On the addition of acetic acid the tissue swells up, and becomes gelatinous in appearance. Under the microscope it is found to be made up of fine white fibres, which interlace in a most irregular manner, together with a variable number of elastic fibres. These latter resist the action of acetic acid as above mentioned, so that when this reagent is added to a specimen of areolar tissue, although the white fibres swell up and become homogeneous, certain elastic fibres may still be seen arranged in various directions, sometimes even appearing to pass in a more or less circular or in a spiral manner round a small mass of the gelatinous mass of changed white fibres. The cells of the tissue are arranged in no very regular manner, being contained in the spaces (areolæ) between the fibres. They communicate, however, with one another by their branched processes, and also apparently with the cells forming the walls of the capillary blood-vessels in their neighborhood, connecting together the fibrils in a certain amount of albuminous *cement* substance.

(*b.*) *White Fibrous Tissue.*

Distribution.—Typically in tendon; in ligaments, in the periosteum and perichondrium, the dura mater, the pericardium, the sclerotic coat of the eye, the fibrous sheath of the testicle; in the fasciæ and aponeurosis of muscles, and in the sheaths of lymphatic glands.

Structure.—To the naked eye, tendons and many of the fibrous membranes, when in a fresh state, present an appearance as of watered silk. This is due to the arrangement of the fibres in wavy parallel bundles. Under the microscope, the tissue appears to consist of long, often parallel, wavy bundles of fibres of different sizes. Sometimes the fibres intersect each other. The cells in tendons are arranged in long chains in the ground substance separating the bundles of fibres, and are more or less regularly quadrilateral with large round nuclei containing nucleoli, which are generally placed so as to be contiguous in two cells. The cells consist of a body, which is thick, from which processes pass in various directions into, and partially filling up the spaces between the bundles of fibres.

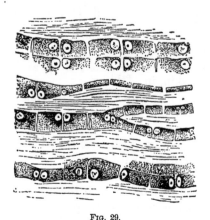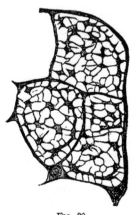

Fig. 29.　　　　　　　　　　　　Fig. 30.

Fig. 29.—Caudal tendon of young rat, showing the arrangement, form, and structure of the tendon cells. × 300. (Klein.)

Fig. 30.—Transverse section of tendon from a cross-section of the tail of a rabbit, showing sheath, fibrous septa, and branched connective-tissue corpuscles. The spaces left white in the drawing represent the tendinous fibres in transverse section. × 250. (Klein.)

The rows of cells are separated from one another by lines of cement substance. The cell spaces can be brought into view by silver nitrate. The cells are generally marked by one or more lines or stripes when viewed longitudinally. This appearance is really produced by the laminar extension either projecting upward or downward.

(*c.*) *Yellow Elastic Tissue.*

Distribution.—In the ligamentum nuchæ of the ox, horse, and many other animals; in the ligamenta subflava of man; in the arteries, constituting the fenestrated coat of Henle; in veins; in the lungs and trachea; in the stylo-hyoid, thyro-hyoid, and crico-thyroid ligaments; in the true vocal cords.

Structure.—Elastic tissue occurs in various forms, from a structureless, elastic membrane to a tissue whose chief constituents are bundles of elastic fibres crossing each other at different angles: these varieties may be classified as follows:—

(*a.*) Fine elastic fibrils, which branch and anastomose to form a network: this variety of elastic tissue occurs chiefly in the skin and mucous membranes, in subcutaneous and submucous tissue, in the lungs and true vocal cords.

(*b.*) Thick fibres, sometimes cylindrical, sometimes flattened like tape, which branch and form a network: these are seen most typically in the ligamenta subflava and also in the ligamentum nuchæ of such animals as the ox and horse, in which it is largely developed.

(*c.*) Elastic membranes with perforations, *e.g.*, Henle's fenestrated membrane: this variety is found chiefly in the arteries and veins.

(*d.*) Continuous, homogeneous elastic membranes, *e.g.*, Bowman's

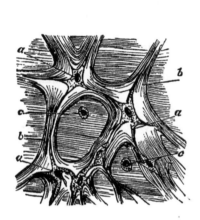
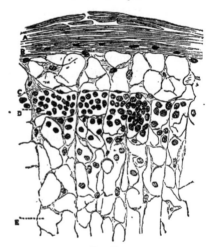

FIG. 31.   FIG. 32.

FIG. 31.—Tissue of the jelly of Wharton from umbilical cord. *a.* connective-tissue corpuscles; *b.* fasciculi of connective tissue; *c.* spherical formative cells. (Frey.)

FIG. 32.—Part of a section of a lymphatic gland, from which the corpuscles have been for the most part removed, showing the adenoid recticulum. (Klein and Noble Smith.)

anterior elastic lamina, and Descemet's posterior elastic lamina, both in the cornea.

A certain number of flat connective tissue cells are found in the ground substance between the elastic fibres constituting this variety of connective tissue.

B.—*Special Forms.*—(*a.*) *Gelatinous Tissue.*

Distribution.—Gelatinous connective tissue forms the chief part of the bodies of jelly fish; it is found in many parts of the human embryo, but remains in the adult only in the vitreous humor of the eye. It may be best seen in the last-named situation, in the "Whartonian jelly" of the umbilical cord, and in the enamel organ of developing teeth.

Structure.—It consists of cells, which in the vitreous humor are rounded, and in the jelly of the enamel organ are stellate, imbedded in a soft jelly-like intercellular substance which forms the bulk of the tissue, and which contains a considerable quantity of mucin. In the umbilical cord, that part of the jelly immediately surrounding the stellate cells shows marks of obscure fibrillation.

(*b.*) *Adenoid or Retiform.*

Distribution.—It composes the stroma of the spleen and lymphatic glands, and is found also in the thymus, in the tonsils, in the follicular glands of the tongue, in Peyer's patches and in the solitary glands of the intestines, and in the mucous membranes generally.

Structure.—Adenoid or retiform tissue consists of a very delicate network of minute fibrils, formed originally by the union of processes of branched connective-tissue corpuscles the nuclei of which, however, are visible only during the early periods of development of the tissue (Fig. 32).

The nuclei found on the fibrillar meshwork do not form an essential part of it. The fibrils are neither white fibrous nor elastic tissue, as they are insoluble in boiling water, although readily soluble in hot alkaline solutions.

(*c.*) *Neuroglia.*—This tissue forms the support of the Nervous elements in the Brain and Spinal cord. It consists of a very fine meshwork of fibrils, said to be elastic, and with nucleated plates which constitute the connective-tissue corpuscles imbedded in it.

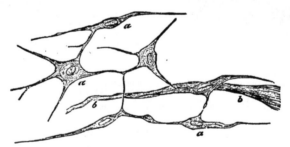

Fig. 33.—Portion of the submucous tissue of gravid uterus of sow. *a*, branched cells, more or less spindle-shaped; *b*. bundles of connective tissue. (Klein.)

Development of Fibrous Tissues.—In the embryo the place of the fibrous tissues is at first occupied by a mass of roundish cells, derived from the "mesoblast."

These develop either into a network of branched cells, or into groups of fusiform cells (Fig. 33).

The cells are imbedded in a semi-fluid albuminous substance derived either from the cells themselves or from the neighboring blood-vessels; this afterward forms the cement substance. In it fibres are developed, either by part of the cells becoming fibrils, the others remaining as connective-tissue corpuscles, or by the fibrils being developed from the outside layers of the protoplasm of the cells, which grow up again to their original size and remain imbedded among the fibres. This process gives rise to fibres arranged in the one case in interlacing networks (areolar tissue), in the other in parallel bundles (white fibrous tissue). In the mature forms of purely fibrous tissue not only the remnants of the cell-substance, but even the nuclei may disappear. The embryonic tissue, from which *elastic* fibres are developed, is composed of fusiform cells, and a structureless intercellular substance by the gradual fibrillation of which elastic fibres are formed. The fusiform cells dwindle in size and eventually disappear so completely that in mature elastic tissue hardly a trace of them is to be found: meanwhile the elastic fibres steadily increase in size.

Another theory of the development of the connective-tissue fibrils supposes that they arise from deposits in the intercellular substance and not from the cells themselves; these deposits, in the case of elastic fibres, appearing first of all in the form of rows of granules, which, joining together, form long fibrils. It seems probable that even if this view be correct, the cells themselves have a considerable influence in the production of the deposits outside them.

Functions of Areolar and Fibrous Tissue.—The main function of connective tissue is mechanical rather than vital: it fulfils the subsidiary but important use of supporting and connecting the various tissues and organs of the body.

In glands the trabeculæ of connective tissue form an interstitial framework in which the parenchyma or secreting gland-tissue is lodged: in muscles and nerves the septa of connective tissue support the bundles of fibres, which form the essential part of the structure.

Elastic tissue, by virtue of its elasticity, has other important uses: these, again, are mechanical rather than vital. Thus the ligamentum nuchæ of the horse or ox acts very much as an India-rubber band in the same position would. It maintains the head in a proper position without any muscular exertion; and when the head has been lowered by the action of the flexor muscles of the neck, and the ligamentum nuchæ thus stretched, the head is brought up again to its normal position by the relaxation of the flexor muscles which allows the elasticity of the ligamentum nuchæ to come again into play.

(*a.*) *Adipose Tissue.*

Distribution.—In almost all regions of the human body a larger or smaller quantity of adipose or fatty tissue is present; the chief exceptions being the subcutaneous tissue of the eyelids, penis, and scrotum, the nymphæ, and the cavity of the cranium. Adipose *tissue* is also absent from the substance of many organs, as the lungs, liver, and others.

Fatty matter, not in the form of a distinct tissue, is also widely present in the body, *e.g.*, in the liver and brain, and in the blood and chyle.

Adipose tissue is almost always found seated in areolar tissue, and forms in its meshes little masses of unequal size and irregular shape, to which the term *lobules* is commonly applied.

Structure.—Under the microscope adipose tissue is found to consist essentially of little vesicles or cells which present dark, sharply-defined edges when viewed with transmitted light: they are about $\frac{1}{400}$ or $\frac{1}{500}$ of an inch in diameter, each composed of a structureless and colorless membrane or bag, filled with fatty matter, which is liquid during life, but in part solidified after death (Fig. 34). A nucleus is always present in some part or other of the cell-wall, but in the ordinary condition of the cell it is not easily or always visible.

 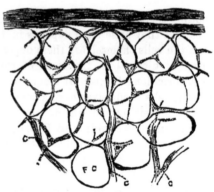

FIG. 34.   FIG. 35.

FIG. 34.—Ordinary fat-cells of a fat tract in the omentum of a rat.   (Klein.)
FIG. 35.—Group of fat-cells (FC) with capillary vessels (C).   (Noble Smith.)

This membrane and the nucleus can generally be brought into view by staining the tissue: it can be still more satisfactorily demonstrated by extracting the contents of the fat-cells with ether, when the shrunken, shriveled membranes remain behind. By mutual pressure, fat-cells come to assume a polyhedral figure (Fig. 35).

The ultimate cells are held together by capillary blood-vessels (Fig. 35); while the little clusters thus formed are grouped into small masses, and held so, in most cases, by areolar tissue.

The oily matter contained in the cells is composed chiefly of the compounds of fatty acids with glycerin, which are named *olein, stearin,* and *palmitin.*

Development of Adipose Tissue.—Fat-cells are developed from connective-tissue corpuscles: in the infra-orbital connective-tissue cells may be found exhibiting every intermediate gradation between an ordinary branched connective-tissue corpuscle and a mature fat-cell. The process of development is as follows: a few small drops of oil make their appearance in the protoplasm: by their confluence a larger drop is produced (Fig. 37): this gradually increases in size at the expense of the original protoplasm of the cell, which becomes correspondingly diminished in quantity till in the mature cell it only forms a thin crescentic film, closely pressed against the cell-wall, and with a nucleus imbedded in its substance (Figs. 34 and 37).

Under certain circumstances this process may be reversed and fat-cells may be changed back into connective-tissue corpuscles. (Kölliker, Virchow.)

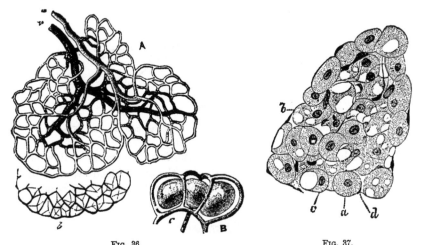

Fig. 36.   Fig. 37.

Fig. 36.—Blood-vessels of adipose tissue. A. Minute flattened fat-lobule, in which the vessels only are represented. a, the terminal artery; v, the primitive vein; b, the fat vesicles of one border of the lobule separately represented. × 100. B. Plan of the arrangement of the capillaries (c) on the exterior of the vesicles: more highly magnified. (Todd and Bowman.)

Fig. 37.—A lobule of developing adipose tissue from an eight months' fœtus. a. Spherical, or, from pressure, polyhedral cells with large central nucleus, surrounded by a finely reticulated substance staining uniformly with hæmatoxylin. b. Similar cells with spaces from which the fat has been removed by oil of cloves. c. Similar cells showing how the nucleus with enclosing protoplasm is being pressed towards periphery. d. Nucleus of endothelium of investing capillaries. (McCarthy.) Drawn by Treves.

Vessels and Nerves.—A large number of blood-vessels are found in adipose tissue, which subdivide until each lobule of fat contains a fine meshwork of capillaries ensheathing each individual fat-globule. Although nerve fibres pass through the tissue, no nerves have been demonstrated to terminate in it.

The Uses of Adipose Tissue.—Among the uses of adipose tissue, these are the chief:—

a. It serves as a store of combustible matter which may be re-absorbed into the blood when occasion requires, and, being burnt, may help to preserve the heat of the body.

b. That part of the fat which is situate beneath the skin must, by its want of conducting power, assist in preventing undue waste of the heat of the body by escape from the surface.

c. As a packing material, fat serves very admirably to fill up spaces, to form a soft and yielding yet elastic material wherewith to wrap tender and delicate structures, or form a bed with like qualities on which such structures may lie, not endangered by pressure.

As good examples of situations in which fat serves such purposes may be mentioned the palms of the hands and soles of the feet, and the orbits.

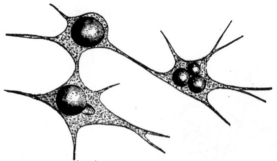

FIG. 38.—Branched connective-tissue corpuscles, developing into fat-cells. (Klein.)

d. In the long bones, fatty tissue, in the form known as yellow marrow, fills the medullary canal, and supports the small blood-vessels which are distributed from it to the inner part of the substance of the bone.

II. CARTILAGE.

Cartilage or gristle exists in three different forms in the human body, viz., 1, *Hyaline cartilage,* 2, *Yellow elastic cartilage,* and 3, *White fibro-cartilage.*

Structure of Cartilage.—All kinds of cartilage are composed of cells imbedded in a substance called the *matrix:* and the apparent differences of structure met with in the various kinds of cartilage are more due to differences in the character of the matrix than of the cells. Among the latter, however, there is also considerable diversity of form and size.

With the exception of the articular variety, cartilage is invested by a thin but tough firm fibrous membrane called the *perichondrium.* On the surface of the articular cartilage of the fœtus, the perichondrium is represented by a film of epithelium; but this is gradually worn away up to the margin of the articular surfaces, when by use the parts begin to suffer friction.

Nerves are probably not supplied to any variety of cartilage.

1. Hyaline Cartilage.

Distribution.—This variety of cartilage is met with largely in the human body—investing the articular ends of bones, and forming the costal cartilages, the nasal cartilages, and those of the larynx with the exception of the epiglottis and cornicula laryngis. The cartilages of the trachea and bronchi are also hyaline.

Structure.—Like other cartilages it is composed of cells imbedded in a matrix. The cells, which contain a nucleus with nucleoli, are irregular in shape, and generally grouped together in patches (Fig. 39). The patches are of various shapes and sizes, and placed at unequal distances apart. They generally appear flattened near the free surface of the mass of cartilage in which they are placed, and more or less perpendicular to the surface in the more-deeply seated portions.

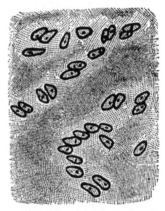

The matrix of hyaline cartilage has a dimly granular appearance like that of ground glass, and in man and the higher animals has no apparent structure. In some cartilages of the frog, however, even when examined in the fresh state,

FIG. 39.—Ordinary hyaline cartilage from trachea of a child. The cartilage cells are enclosed singly or in pairs in a capsule of hyaline substance. × 150 diams. (Klein and Noble Smith.)

it is seen to be mapped out into polygonal blocks or cell-territories, éach containing a cell in the centre, and representing what is generally called the capsule of the cartilage cells (Fig. 40). Hyaline cartilage in man has really the same structure, which can be demonstrated by the use of certain reagents. If a piece of human hyaline cartilage be macerated

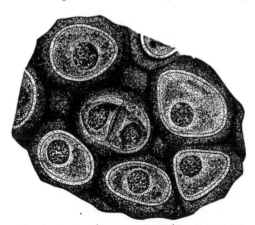

FIG. 40.—Fresh cartilage from the Triton. (A. Rollett.)

for a long time in dilute acid or in hot water 95°—113° F. (35° to 45° C.), the matrix, which previously appeared quite homogeneous, is found to be resolved into a number of concentric lamellæ, like the coats of an onion, arranged round each cell or group of cells. It is thus shown to consist of nothing but a number of large systems of capsules which have become fused with one another

The cavities in the matrix in which the cells lie are connected together by a series of branching canals, very much resembling those in the cornea: through these canals fluids may make their way into the depths of the tissue.

In the hyaline cartilage of the ribs, the cells are mostly larger than in the articular variety, and there is a tendency to the development of fibres in the matrix. The costal cartilages also frequently become calcified in old age, as also do some of those of the larynx. Fat-globules may also be seen in many cartilages.

In articular cartilage the cells are smaller, and arranged vertically in narrow lines like strings of beads.

Temporary Cartilage.—In the fœtus, cartilage is the material of which the bones are first constructed; the "model" of each bone being laid down, so to speak, in this substance. In such cases the cartilage is termed *temporary*. It closely resembles the ordinary hyaline kind; the cells, however, are not grouped together after the fashion just described, but are more uniformly distributed throughout the matrix.

A variety of temporary hyaline cartilage which has scarcely any matrix is found in the human subject only in early fœtal life, when it constitutes the *chorda dorsalis*.

Nutrition of Cartilage.—Hyaline cartilage is reckoned among the so-called *non-vascular* structures, no blood-vessels being supplied directly to its own substance; it is nourished by those of the bone beneath. When hyaline cartilage is in thicker masses, as in the case of the cartilages of the ribs, a few blood-vessels traverse its substance. The distinction, however, between all so-called *vascular* and *non-vascular* parts, is at the best a very artificial one.

2. Yellow Elastic Cartilage.

Distribution.—In the external ear, in the epiglottis and cornicula laryngis, and in the Eustachian tube.

Structure.—The cells are rounded or oval, with well-marked nuclei and nucleoli (Fig. 41). The matrix in which they are seated is composed almost entirely of fine elastic fibres, which form an intricate interlacement about the cells, and in their general characters are allied to the yellow variety of fibrous tissue: a small and variable quantity of hyaline intercellular substance is also usually present.

A variety of elastic cartilage, sometimes called *cellular*, may be obtained from the external ear of rats, mice, or other small mammals. It is composed almost entirely of cells (hence the name), which are packed very closely, with little or no matrix. When present the matrix consists of very fine fibres, which twine about the cells in various directions and enclose them in a kind of network.

3. White Fibro-Cartilage.

Distribution.—The different situations in which white fibro-cartilage is found have given rise to the following classification:—

1. *Inter-articular* fibro-cartilage, *e.g.*, the semilunar cartilages of the knee-joint.

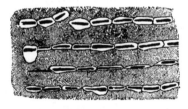

Fig. 41.   Fig. 42.

Fig. 41.—Section of the epiglottis. (Baly.)
Fig. 42.—Tranverse section through the intervertebral cartilage of the tail of mouse, showing lamellæ of fibrous tissue with cartilage cells arranged in rows between them. The cells are seen in profile, and being flattened, appear staff-shaped. Each cell lies in a capsule. × 350. (Klein and Noble Smith.)

2. *Circumferential* or marginal, as on the edges of the acetabulum and glenoid cavity.

3. *Connecting, e.g.,* the inter-vertebral fibro-cartilages.

4. In the *sheaths of tendons,* and sometimes in their substance. In the latter situation, the nodule of fibro-cartilage is called a *sesamoid* fibro-cartilage, of which a specimen may be found in the tendon of the tibialis posticus, in the sole of the foot, and usually in the neighboring tendon of the peroneus longus.

Structure.—White fibro-cartilage (Fig. 43), which is much more widely distributed throughout the body than the foregoing kind, is composed, like it, of cells and a matrix; the latter, however, being made up almost entirely of fibres closely resembling those of white fibrous tissue.

In this kind of fibro-cartilage it is not unusual to find a great part of its mass

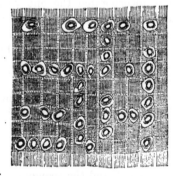

Fig. 43.—White fibro-cartilage from an intervertebral ligament. (Klein and Noble Smith.)

composed almost exclusively of fibres, and deriving the name of cartilage only from the fact that in another portion, continuous with it, cartilage cells may be pretty freely distributed.

Functions of Cartilage.—Cartilage not only represents in the fœtus the bones which are to be formed (temporary cartilage), but also offers a firm, but more or less yielding, framework for certain parts in the developed body, possessing at the same time strength and elasticity. It maintains the shape of tubes as in the larynx and tráchea. It affords attachment to muscles and ligaments; it binds bones together, yet allows a certain degree of movement, as between the vertebræ; it forms a firm framework and protection, yet without undue stiffness or weight, as in the pinna, larynx, and chest walls; it deepens joint cavities, as in the acetabulum, without unduly restricting the movements of the bones.

Development of Cartilage.—Cartilage is developed out of an embryonal tissue, consisting of cells with a very small quantity of intercellular substance: the cells multiply by fission within the cell-capsules (Fig. 6); while the capsule of the parent cell becomes gradually fused with the surrounding intercellular substance. A repetition of this process in the young cells causes a rapid growth of the cartilage by the multiplication of its cellular elements and corresponding increase in its matrix.

III. Bone.

Chemical Composition.—Bone is composed of *earthy* and *animal* matter in the proportion of about 67 per cent. of the former to 33 per cent. of the latter. The earthy matter is composed chiefly of calcium phosphate, but besides there is a small quantity (about 11 of the 67 per cent.) of calcium carbonate and fluoride, and magnesium phosphate.

The animal matter is resolved into *gelatin* by boiling.

The earthy and animal constituents of bone are so intimately blended and incorporated the one with the other, that it is only by chemical action, as, for instance, by heat in one case and by the action of acids in another, that they can be separated. Their close union, too, is further shown by the fact that when by acids the earthy matter is dissolved out, or, on the other hand, when the animal part is burnt out, the shape of the bone is alike preserved.

The proportion between these two constituents of bone varies in different bones in the same individual, and in the same bone at different ages.

Structure.—To the naked eye there appear two kinds of structure in different bones, and in different parts of the same bone, namely, the *dense* or *compact*, and the *spongy or cancellous* tissue.

Thus, in making a longitudinal section of a long bone, as the humerus or femur, the articular extremities are found capped on their surface by a thin shell of compact bone, while their interior is made up of the spongy or cancellous tissue. The *shaft*, on the other hand, is formed almost entirely of a thick layer of the compact bone, and this surrounds a central canal, the *medullary* cavity—so called from its containing the *medulla* or marrow.

In the flat bones, as the parietal bone or the scapula, one layer of the cancellous structure lies between two layers of the compact tissue, and in the short and irregular bones, as those of the *carpus* and *tarsus*, the cancellous tissue alone fills the interior, while a thin shell of compact bone forms the outside.

Marrow.—There are two distinct varieties of marrow—the *red* and *yellow.*

Red marrow is that variety which occupies the spaces in the cancellous tissue; it is highly vascular, and thus maintains the nutrition of the spongy bone, the interstices of which it fills. It contains a few fat-cells and a large number of marrow-cells, many of which are undistinguishable

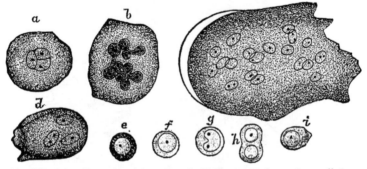

Fig. 44.—Cells of the red marrow of the guinea pig, highly magnified. *a,* a large cell, the nucleus of which appears to be partly divided into three by constrictions; *b,* a cell, the nucleus of which shows an appearance of being constricted into a number of smaller nuclei; *c,* a so-called giant cell, or myeloplaxe, with many nuclei; *d,* a smaller myeloplaxe, with three nuclei; *e—i,* proper cells of the marrow. (E. A. Schäfer.)

from lymphoid corpuscles, and has for a basis a small amount of fibrous tissue. Among the cells are some nucleated cells of very much the same tint as colored blood-corpuscles. There are also a few large cells with many nuclei, termed "giant-cells" (myeloplaxes) which are derived from over-growth of the ordinary marrow-cells (Fig. 44).

Yellow marrow fills the medullary cavity of long bones, and consists chiefly of fat-cells with numerous blood-vessels; many of its cells also are in every respect similar to lymphoid corpuscles.

From these marrow-cells, especially those of the red marrow, are derived, as we shall presently show, large quantities of red blood-corpuscles.

Periosteum and Nutrient Blood-vessels.—The surfaces of bones, except the part covered with articular cartilage, are clothed by a tough, fibrous membrane, the *periosteum;* and it is from the blood-vessels which are distributed in this membrane, that the bones, especially their more compact tissue, are in great part supplied with nourishment,—minute branches from the periosteal vessels entering the little foramina on the surface of the bone, and finding their way to the Haversian canals, to be immediately described. The long bones are supplied also by a proper nutrient artery which, entering at some part of the shaft so as to reach the medullary canal, breaks up into branches for the supply of the marrow, from which again small vessels are distributed to the interior of the bone. Other small blood-vessels pierce the articular extremities for the supply of the cancellous tissue.

Microscopic Structure of Bone.—Notwithstanding the differences of arrangement just mentioned, the structure of all bone is found under the microscope to be essentially the same.

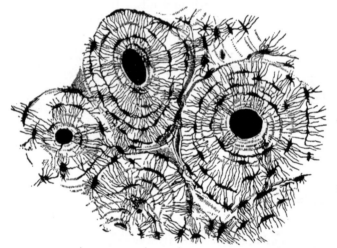

Fig. 45.—Transverse section of compact bony tissue (of humerus). Three of the Haversian canals are seen, with their concentric rings; also the corpuscles or lacunæ, with the canaliculi extending from them across the direction of the lamellæ. The Haversian apertures had got filled with debris in grinding down the section, and therefore appear black in the figure, which represents the object as viewed with transmitted light. The Haversian systems are so closely packed in this section, that scarcely any *interstitial* lamellæ are visible. × 150. (Sharpey.)

Examined with a rather high power its substance is found to contain a multitude of little irregular spaces, approximately fusiform in shape, called *lacunæ*, with very minute canals or *canaliculi*, as they are termed, leading from them, and anastomosing with similar little prolongations from other lacunæ (Fig. 45). In very thin layers of bone, no other canals than these may be visible; but on making a transverse section of the compact tissue as of a long bone, *e.g.*, the humerus or ulna, the arrangement shown in Fig. 45 can be seen.

The bone seems mapped out into small circular districts, at or about the centre of each of which is a hole, and around this an appearance as of concentric layers—the *lacunæ* and *canaliculi* following the same concentric plan of distribution around the small hole in the centre, with which, indeed, they communicate.

On making a longitudinal section, the central holes are found to be simply the cut extremities of small canals which run lengthwise through the bone, anastomosing with each other by lateral branches (Fig. 46), and are called·Haversian canals, after the name of the physician, Clopton Havers, who first accurately described them. The Haversian canals, the average diameter of which is $\frac{1}{500}$ of an inch, contain blood-vessels, and by means of them blood is conveyed to all, even the densest parts of the bone; the minute canaliculi and lacunæ absorbing nutrient matter from the Haversian blood-vessels, and conveying it still more intimately to the very substance of the bone which they traverse.

The blood-vessels enter the Haversian canals both from without, by

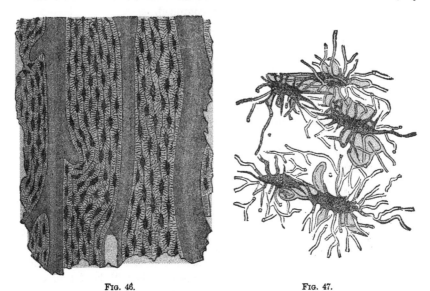

FIG. 46.   FIG. 47.

FIG. 46.—Longitudinal section of human ulna, showing Haversian canal, lacunæ, and canaliculi. (Rollett.)
FIG. 47.—Bone corpuscles with their processes as seen in a thin section of human bone. (Rollett.)

traversing the small holes which exist on the surface of all bones beneath the periosteum, and from within by means of small channels which extend from the medullary cavity, or from the cancellous tissue. The arteries and veins usually occupy separate canals, and the veins, which are the larger, often present, at irregular intervals, small pouch-like dilatations.

The *lacunæ* are occupied by branched cells (bone-cells, or bone-corpuscles) (Fig. 47), which very closely resemble the ordinary branched connective-tissue corpuscles; each of these little masses of protoplasm ministering to the nutrition of the bone immediately surrounding it, and one lacunar corpuscle communicating with another, and with its surrounding district, and with the blood-vessels of the Haversian canals, by means of the minute streams of fluid nutrient matter which occupy the canaliculi.

It will be seen from the above description that bone is essentially connective-tissue impregnated with lime salts: it bears a very close resemblance to what may be termed typical connective-tissue such as the substance of the cornea. The bone-corpuscles with their processes, occupying the lacunæ and canaliculi, correspond exactly to the cornea-corpuscles lying in branched spaces; while the finely fibrillated structure of the bone-lamellæ, to be presently described, resembles the fibrillated substance of the cornea in which the branching spaces lie.

Lamellæ of Compact Bone.—In the shaft of a long bone three distinct sets of lamellæ can be clearly recognized.

(1.) *General* or fundamental lamellæ; which are most easily traceable just beneath the periosteum, and around the medullary cavity, forming around the latter a series of concentric rings. At a little distance from the medullary and periosteal surfaces (in the deeper portions of the bone) they are more or less interrupted by

(2.) *Special* or Haversian lamellæ, which are concentrically arranged around the Haversian canals to the number of six to eighteen around each.

(3.) *Interstitial* lamellæ, which connect the systems of Haversian lamellæ, filling the spaces between them, and consequently attaining

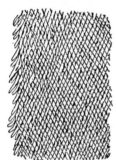

FIG. 48.—Thin layer peeled off from a softened bone. This figure, which is intended to represent the reticular structure of a lamella, gives a better idea of the object when held rather farther off than usual from the eye. × 400. (Sharpey.)

their greatest development where the Haversian systems are few, and *vice versâ*.

The ultimate structure of the *lamellæ* appears to be reticular. If a thin film be peeled off the surface of a bone, from which the earthy matter has been removed by acid, and examined with a high power of the microscope, it will be found composed of a finely reticular structure, formed apparently of very slender fibres decussating obliquely, but coalescing at the points of intersection, as if here the fibres were fused rather than woven together (Fig. 48). (Sharpey.)

In many places these reticular lamellæ are perforated by tapering fibres (*Claviculi* of Gagliardi), resembling in character the ordinary white or rarely the elastic fibrous tissue, which bolt the neighboring lamellæ together, and may be drawn out when the latter are torn asunder (Fig. 49). These perforating fibres originate from ingrowing processes of the periosteum, and in the adult still retain their connection with it.

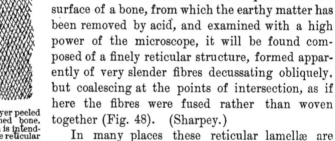

Development of Bone.—From the point of view of their development, all bones may be subdivided into two classes.

(*a.*) Those which are ossified directly in *membrane, e.g.*, the bones forming the vault of the skull, parietal, frontal.

(*b.*) Those whose form, previous to ossification, is laid down in *hyaline cartilage, e.g.*, humerus, femur.

The process of development, pure and simple, may be best studied in bones which are not preceded by cartilage—"membrane-bones" (*e.g.*, parietal); and without a knowledge of this process (ossification in *membrane*), it is impossible to understand the much more complex series of

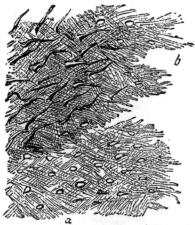

Fig. 49.—Lamellæ torn off from a decalcified human parietal bone at some depth from the surface. *a*, a lamella, showing reticular fibres; *b, b*, darker part, where several lamellæ are superposed; *c*, perforating fibres. Apertures through which perforating fibres had passed, are seen especially in the lower part, *a, a*, of the figure. (Allen Thomson.)

changes through which such a structure as the cartilaginous femur of the fœtus passes in its transformation into the body femur of the adult (ossification in *cartilage*).

Ossification in Membrane.—The membrane or periosteum from which such a bone as the parietal is developed consists of two layers—an external *fibrous*, and an internal *cellular* or *osteogenetic*.

The external one consists of ordinary connective-tissue, being composed of layers of fibrous tissue with branched connective-tissue corpuscles here and there between the bundles of fibres. The internal layer consists of a network of fine fibrils with a large number of nucleated cells, some of which are oval, others drawn out into a long branched process, and others branched: it is more richly supplied with capillaries than the outer layer. The relatively large number of its cellular elements, their variability in size and shape, together with the abundance of its blood-vessels, clearly mark it out as the portion of the periosteum which is immediately concerned in the formation of bone.

In such a bone as the parietal, the deposition of bony matter, which is preceded by increased vascularity, takes place in radiating spiculæ, starting from a "centre of ossification," and shooting out in all directions toward the periphery; while the bone increases in thickness by the deposition of successive layers beneath the periosteum. The finely fibrillar network of the deeper or *osteogenetic* layer of the periosteum becomes transformed into bone-matrix (the minute structure of which has been already (p. 46) described as reticular), and its cells into bone-corpuscles. On the young bone trabeculæ thus formed, fresh layers of cells (osteoblasts) from the osteogenetic layer are developed side by side, lining the irregular spaces like an epithelium (Fig. 50, *b*). Lime-salts are deposited in the circumferential part of each osteoblast, and thus a ring of osteoblasts gives rise to a ring of bone with the remaining uncalcified portions of the osteoblasts imbedded in it as bone-corpuscles (Fig. 50).

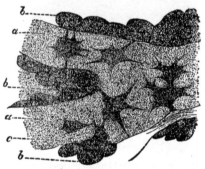

FIG. 50.—Osteoblasts from the parietal bone of a human embryo, thirteen weeks old. *a*, bony septa with the cells of the lacunæ; *b*, layers of osteoblasts; *c*, the latter in transition to bone corpuscles. Highly magnified. (Gegenbaur.)

Thus, the primitive spongy bone is formed, whose irregular branching spaces are occupied by processes from the osteogenetic layer of the periosteum with numerous blood-vessels and osteoblasts. Portions of this primitive spongy bone are re-absorbed; the osteoblasts being arranged in concentric successive layers and thus giving rise to concentric Haversian lamellæ of bone, until the irregular space in the centre is reduced to a well-formed Haversian canal, the portions of the primitive spongy bone between the Haversian systems remaining as *interstitial* or ground-lamellæ (p. 46). The bulk of the primitive spongy bone is thus gradually converted into compact bony-tissue with Haversian canals. Those portions of the in-growths from the deeper layer of the periosteum which are not converted into bone remain in the spaces of the cancellous tissue as the red marrow.

Ossification in Cartilage.—Under this heading, taking the femur as a typical example, we may consider the process by which the solid cartilaginous rod which represents it in the fœtus is converted into the hollow cylinder of compact bone with expanded ends of cancellous tissue which forms the adult femur; bearing in mind the fact that this fœtal cartilaginous femur is many times smaller than the medullary cavity even of the shaft of the mature bone, and, therefore, that not a trace of the original cartilage can be present in the femur of the adult. Its purpose is indeed purely temporary; and, after its calcification, it is gradually and entirely re-absorbed as will be presently explained.

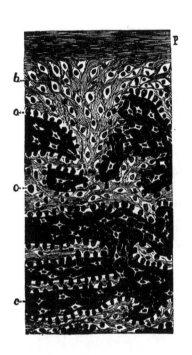

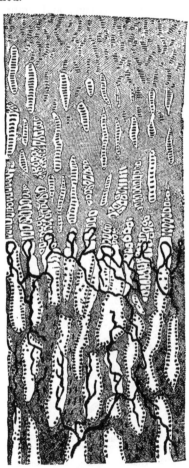

FIG. 51.   FIG. 52.

FIG. 51.—From a transverse section through part of fœtal jaw near the extreme periosteum, in the state of spongy bone. *p*, fibrous layer of periosteum; *b*, osteogenetic layer of periosteum; *o*, osteoblasts; *c*, osseous substance, containing many bone corpuscles. × 300. (Schofield.)
FIG. 52.—Ossifying cartilage showing loops of blood-vessels.

The cartilaginous rod which forms the fœtal femur is sheathed in a membrane termed the *perichondrium*, which so far resembles the periosteum described above, that it consists of two layers, in the deeper one of which spheroidal cells predominate and blood-vessels abound, while the outer layer consists mainly of fusiform cells which are in the mature tissue gradually transformed into fibres. Thus, the differences between

VOL. I.—4.

the fœtal perichondrium and the periosteum of the adult are such as usually exist between the embryonic and mature forms of connective-tissue.

Between the hyaline cartilage of which the fœtal femur consists and the bony tissue forming the adult femur, two intermediate stages exist—viz., calcified cartilage, and embryonic spongy bone. These tissues, which successively occupy the place of the fœtal cartilage, are in succession entirely re-absorbed, and their place taken by true bone.

The process by which the cartilaginous is transformed into the bony

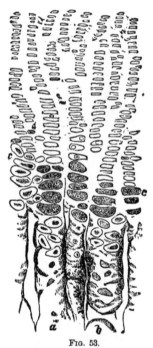 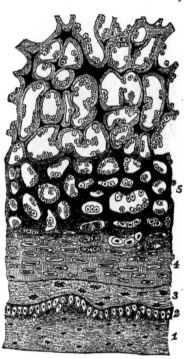

FIG. 53.   FIG. 54.

FIG. 53.—Longitudinal section of ossifying cartilage from the humerus of a fœtal sheep. Calcified trabeculæ are seen extending between the columns of cartilage cells. c, cartilage cells. × 140. (Sharpey.)

FIG. 54.—Transverse section of a portion of a metacarpal bone of a fœtus, showing—1, fibrous layer of periosteum; 2, osteogenetic layer of ditto; 3, periosteal bone; 4, cartilage with matrix gradually becoming calcified, as at 5, with cells in primary areolæ; beyond 5 the calcified matrix is being entirely replaced by spongy bone. × 200. (V. D. Harris.)

femur may be divided for the sake of clearness into the following six stages:

Stage i.—Vascularization of the Cartilage.—Processes from the osteogenetic or cellular layer of the perichondrium containing blood-vessels grow into the substance of the cartilage much as ivy insinuates itself into the cracks and crevices of a wall. Thus the substance of the cartilage, which previously contained no vessels, is traversed by a number of

branched anastomosing channels formed by the enlargement and coales-
cence of the spaces in which the cartilage-cells lie, and containing loops
of blood-vessels (Fig. 52) and spheroidal-cells which will become osteoblasts.

Stage 2.—Calcification of Cartilaginous Matrix.—Lime-salts
are next deposited in the form of fine granules in the hyaline matrix of
the cartilage, which thus becomes gradually transformed into a number
of calcified trabeculæ (Fig. 54, ᵇ), forming alveolar spaces (*primary areolæ*)
containing cartilage cells. By the absorption of some of the trabeculæ
larger spaces arise, which contain cartilage-cells for a very short time
only, their places being taken by the so-called osteogenetic layer of the
perichondrium (before referred to in Stage 1) which constitutes the primary marrow. The cartilage-cells, gradually enlarging, become more
transparent and finally undergo disintegration.

**Stage 3.—Substitution of Embryonic Spongy Bone for Car-
tilage.**—The cells of the primary marrow arrange themselves as a continuous layer like epithelium on the
calcified trabeculæ and deposit a layer
of bone, which ensheathes the calcified
trabeculæ: these calcified trabeculæ,
encased in their sheaths of young bone,
become gradually absorbed, so that
finally we have trabeculæ composed entirely of spongy bone, all trace of the
original calcified cartilage having disappeared. It is probable that the large
multinucleated giant-cells termed "os-
teoclasts" by Kölliker, which are derived from the osteoblasts by the multiplication of their nuclei, are the
agents by which the absorption of calcified cartilage, and subsequently of
embryonic spongy bone, is carried on
(Fig. 55, G). At any rate they are
almost always found wherever absorption is in progress.

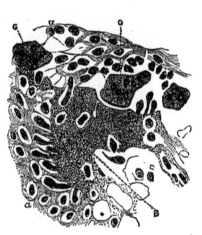

FIG. 55.—A small isolated mass of bone next
the periosteum of the lower jaw of human
fœtus. *a*, osteogenetic layer of periosteum.
G, multinuclear giant cells, the one on the left
acting here probably like an osteoclast. Above
c, the osteoblasts are seen to become surrounded by an osseous matrix. (Klein and
Noble Smith.)

Stages 2 and 3 are precisely similar to what goes on in the growing
shaft of a bone which is increasing in length by the advance of the process of ossification into the intermediary cartilage between the diaphysis
and epiphysis. In this case the cartilage-cells become flattened and,
multiplying by division, are grouped into regular columns at right angles
to the plane of calcification, while the process of calcification extends
into the hyaline matrix between them (Figs. 52 and 53).

Stage 4.—Substitution of Periosteal Bone for the Primary

Embryonic Spongy Bone.—The embryonic spongy bone, formed as above described, is simply a temporary tissue occupying the place of the fœtal rod of cartilage, once representing the femur; and the stages 1, 2, and 3 show the successive changes which occur *at the centre* of the shaft. Periosteal bone is now deposited in successive layers beneath the perios-teum, *i.e., at the circumference* of the shaft, exactly as described in the

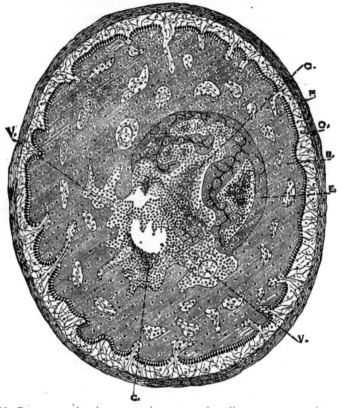

Fig. 56.—Transverse section through the tibia of a fœtal kitten semi-diagrammatic. × 60. P, Periosteum. O, osteogenetic layer of the periosteum, showing the osteoblasts arranged side by side, represented as pear-shaped black dots on the surface of the newly-formed bone. B, the periosteal bone deposited in successive layers beneath the periosteum and ensheathing E, the spongy endochondral bone; represented as more deeply shaded. Within the trabeculæ of endochondral spongy bone are seen the remains of the calcified cartilage trabeculæ represented as dark wavy lines. C, the medulla, with V, V, veins. In the lower half of the figure the endochondral spongy bone has been completely absorbed. (Klein and Noble Smith.)

section on "ossification in membrane," and thus a casing of periosteal bone is formed around the embryonic endochondral spongy bone: this casing is thickest at the centre, where it is first formed, and thins out toward each end of the shaft. The embryonic spongy bone is absorbed, its trabeculæ becoming gradually thinned and its meshes enlarging, and finally coalescing into one great cavity—the medullary cavity of the shaft.

Stage 5.—Absorption of the Inner Layers of the Periosteal

Bone.—The absorption of the endochondral spongy bone is now complete, and the medullary cavity is bounded by periosteal bone: the inner layers of this periosteal bone are next absorbed, and the medullary cavity is thereby enlarged, while the deposition of bone beneath the periosteum continues as before. The first-formed periosteal bone is spongy in character.

Stage 6.—Formation of Compact Bone.—The transformation of spongy periosteal bone into compact bone is effected in a manner exactly similar to that which has been described in connection with ossification in membrane (p. 47). The irregularities in the walls of the areolæ in the spongy bone are absorbed, while the osteoblasts which line them are developed in concentric layers, each layer in turn becoming ossified till the comparatively large space in the centre is reduced to a well-formed Haversian canal (Fig. 57). When once formed, bony tissue grows to some extent interstitially, as is evidenced by the fact that the lacunæ are rather further apart in fully-formed than in young bone.

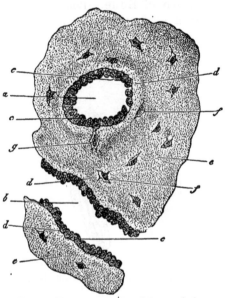

From the foregoing description of the development of bone, it will be seen that the common terms "ossification in cartilage" and "ossification in membrane" are apt to mislead, since they seem to imply two processes radically distinct. The process of ossification, however, is in all

FIG. 57.—Transverse section of femur of a human embryo about eleven weeks old. *a*, rudimentary, Haversian canal in cross section; *b*, in longitudinal section; *c*, osteoblasts; *d*, newly formed osseous substance of a lighter color; *e*, that of greater age; *f*, lacunæ with their cells; *g*, a cell still united to an osteoblast. (Frey.)

cases one and the same, all true bony tissue being formed from membrane (perichondrium or periosteum); but in the development of such a bone as the femur, which may be taken as the type of so-called "ossification in cartilage," lime-salts are deposited in the cartilage, and this calcified cartilage is gradually and entirely re-absorbed, being ultimately replaced by bone formed from the periosteum, till in the adult structure nothing but true bone is left. Thus, in the process of "ossification in cartilage," calcification of the cartilaginous matrix precedes the real formation of bone. We must, therefore, clearly distinguish between *calcification* and *ossification*. The former is simply the infiltration of an animal tissue with lime-salts, and is, therefore, a change of chemical composition rather

than of structure; while ossification is the formation of true bone—a
tissue more complex and more highly organized than that from which it
is derived.

Centres of Ossification.—In all bones ossification commences at
one or more points, termed "centres of ossification." The long bones,
e.g., femur, humerus, etc., have at least three such points—one for the
ossification of the *shaft* or *diaphysis,* and one for each articular extremity
or *epiphysis.* Besides these three primary centres which are always present in long bones, various secondary centres may be superadded for the
ossification of different *processes.*

Growth of Bone.—Bones increase *in length* by the advance of the
process of ossification into the cartilage intermediate between the diaphysis and epiphysis. The increase in length indeed is due entirely to

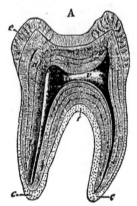
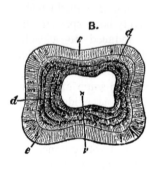

FIG. 58.—A. Longitudinal section of a human molar tooth; *c*, cement; *d*, dentine; *e*, enamel; *v*,
pulp cavity. (Owen.)
B. Transverse section. The letters indicate the same as in A.

growth at the two ends of the *shaft.* This is proved by inserting two
pins into the shaft of a growing bone: after some time their distance
apart will be found to be unaltered though the bone has gradually increased in length, the growth having taken place beyond and not between them. If now one pin be placed in the shaft, and the other in the
epiphysis, of a growing bone, their distance apart will increase as the bone
grows in length.

Thus it is that if the epiphyses with the intermediate cartilage be removed from a young bone, growth in length is no longer possible; while the
natural termination of growth of a bone in length takes place when the
epiphyses become united in bony continuity with the shaft.

Increase *in thickness* in the shaft of a long bone, occurs by the deposition of successive layers beneath the periosteum.

If a thin metal plate be inserted beneath the periosteum of a growing
bone, it will soon be covered by osseous deposit, but if it be put between the

fibrous and osteogenetic layers, it will never become enveloped in bone, for all the bone is formed beneath the latter.

. Other varieties of connective tissue may become ossified, *e.g.*, the tendons in some birds. ·

Functions of Bones.—Bones form the framework of the body; for this they are fitted by their hardness and solidity together with their comparative lightness; they serve both to protect internal organs in the trunk and skull, and as levers worked by muscles in the limbs; notwithstanding their hardness they possess a considerable degree of elasticity, which often saves them from fractures.

TEETH.

The principal part of a tooth, viz., *dentine,* is called by some a connective tissue, and on this account the structure of the teeth is considered here.

A tooth is generally described as possessing a *crown, neck,* and *fang* or *fangs.*

The *crown* is the portion which projects beyond the level of the gum. The *neck* is that constricted portion just below the crown which is embraced by the free edges of the gum, and the *fang* includes all below this.

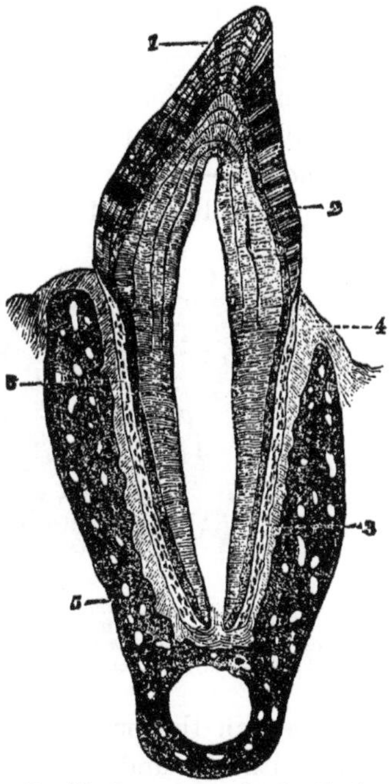

On making a longitudinal section through the centre of a tooth (Figs. 58, 59), it is found to be principally composed of a hard matter, *dentine* or ivory; while in the centre this dentine is hollowed out into a cavity resembling in general shape the outline of the tooth, and called the

FIG. 59.—Premolar tooth of cat in situ. Vertical section. 1, Enamel with decussating and parallel striæ. 2. Dentine with Schreger's lines. 3. Cement. 4. Periosteum of alveolus. 5, Inferior maxillary bone showing canal for the inferior dental nerve and vessels which appears nearly circular in transverse section. (Waldeyer.)

pulp cavity, from its containing a very vascular and sensitive little mass, composed of connective-tissue, blood-vessels, and nerves, which is called the *tooth-pulp.*

The blood-vessels and nerves enter the pulp through a small opening at the extremity of the fang.

Capping that part of the dentine which projects beyond the level of the gum, is a layer of very hard calcareous matter, the *enamel;* while sheathing the portion of dentine which is beneath the level of the gum, is a layer of true bone, called the *cement* or *crusta petrosa.*

At the neck of the tooth, where the enamel and cement come into contact, each is reduced to an exceedingly thin layer. The covering of enamel becomes thicker as we approach the crown, and the cement as we approach the lower end or apex of the fang.

I.—Dentine.

Chemical composition.—Dentine or ivory in chemical composition closely resembles bone. It contains, however, rather less animal matter; the proportion in a hundred parts being about twenty-eight *animal* to seventy-two of *earthy*. The former, like the animal matter of bone, may be resolved into gelatin by boiling. The earthy matter is made up chiefly of calcium phosphate, with a small portion of the carbonate, and traces of calcium fluoride and magnesium phosphate.

Structure.—Under the microscope dentine is seen to be finely channeled by a multitude of delicate tubes, which, by their inner ends, com-

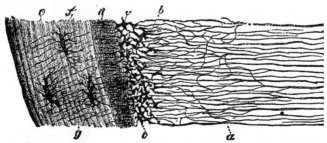

FIG. 60.—Section of a portion of the dentine and cement from the middle of the root of an incisor tooth. *a*, dental tubuli ramifying and terminating, some of them in the interglobular spaces *b* and *c*, which somewhat resemble bone lacunae; *d*, inner layer of the cement with numerous closely set canaliculi; *e*, outer layer of cement; *f*, lacunæ; *g*, canaliculi. ✕ 350. (Kölliker.)

municate with the pulp-cavity, and by their outer extremities come into contact with the under part of the enamel and cement and sometimes even penetrate them for a greater or less distance (Fig. 60).

In their course from the pulp-cavity to the surface of the dentine, the minute tubes form gentle and nearly parallel curves and divide and subdivide dichotomously, but without much lessening of their calibre until they are approaching their peripheral termination.

From their sides proceed other exceedingly minute secondary canals, which extend into the dentine between the tubules, and anastomose with each other. The tubules of the dentine, the average diameter of which at their inner and larger extremity is $\frac{1}{4500}$ of an inch, contain fine prolongations from the tooth-pulp, which give the dentine a certain faint sensitiveness under ordinary circumstances, and, without doubt, have to do also with its nutrition. These prolongations from the tooth-pulp are really processes of the dentine-cells or *odontoblasts* which are branched cells lining the pulp-cavity; the relation of these processes to the tubules in which they lie being precisely similar to that of the processes of the bone-corpuscles to the canaliculi of bone. The outer portion of the dentine, underlying both the cement and enamel, forms a more or less distinct layer termed the *granular* or *interglobular* layer. It is characterized by the presence of a number of minute cell-like cavities, much more closely packed than the lacunæ in the cement, and communicating with one another and with the ends of the dentine-tubes (Fig. 60), and containing cells like bone-corpuscles.

II.—*Enamel.*

Chemical composition.—The *enamel*, which is by far the hardest portion of a tooth, is composed, chemically, of the same elements that enter

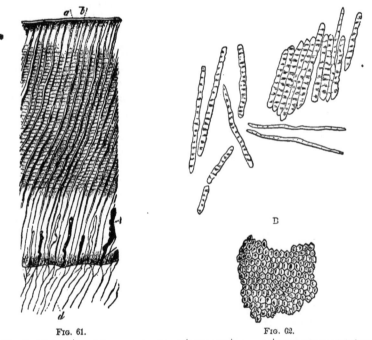

FIG. 61. FIG. 62.

FIG. 61.—Thin section of the enamel and a part of the dentine. *a*, cuticular pellicle of the enamel; *b*, enamel fibres, or columns with fissures between them and cross striæ; *c*, larger cavities in the enamel, communicating with the extremities of some of the tubuli (*d*). × 350. (Kölliker.)

FIG. 62.—Enamel fibres. A, fragments and single fibres of the enamel, isolated by the action of hydrochloric acid. B, surface of a small fragment of enamel, showing the hexagonal ends of the fibres. × 350. (Kölliker.)

into the composition of dentine and bone. Its animal matter, however, amounts only to about 2 or 3 per cent. It contains a larger proportion of inorganic matter and is harder than any other tissue in the body.

Structure.—Examined under the microscope, enamel is found composed of fine hexagonal fibres (Figs. 61, 62) $\frac{1}{5000}$ of an inch in diameter, which are set on end on the surface of the dentine, and fit into corresponding depressions in the same.

They radiate in such a manner from the dentine that at the top of the tooth they are more or less vertical, while toward the sides they tend to the horizontal direction. Like the dentine tubules, they are not straight, but disposed in wavy and parallel curves. The fibres are marked by transverse lines, and are mostly solid, but some of them contain a very minute canal.

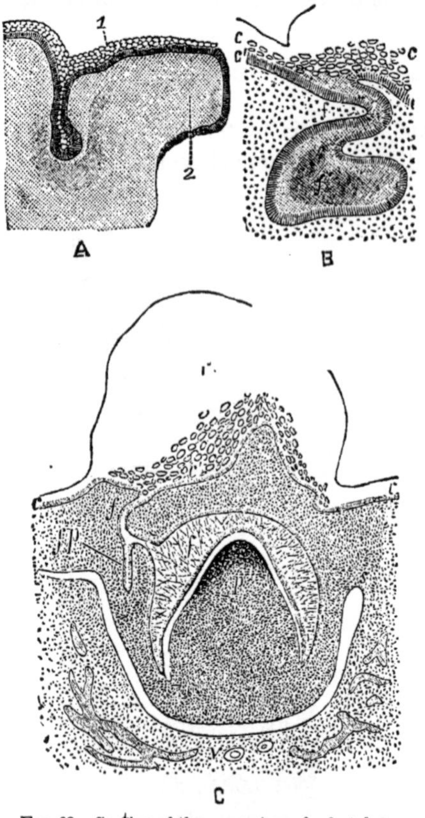

The enamel-prisms are connected together by a very minute quantity of hyaline cement-substance. In the deeper part of the enamel, between the prisms, are small *lacunæ*, which communicate with the "interglobular spaces" on the surface of the dentine.

The enamel itself is coated on the outside by a very thin calcified membrane, sometimes termed the *cuticle* of the enamel.

III.—*Crusta Petrosa.*

The *crusta petrosa*, or *cement* (Fig. 60, *c*, *d*), is composed of true bone, and in it are lacunæ (*f*) and canaliculi (*g*) which sometimes communicate with the outer finely branched ends of the dentine tubules. Its laminæ are as it were bolted together by perforating fibres like those of ordinary bone, but it differs in possessing Haversian canals only in the thickest part.

FIG. 63.—Section of the upper jaw of a fœtal sheep. A.—1, common enamel-germ dipping down into the mucous membrane; 2, palatine process of jaw. B.— Section similar to A, but passing through one of the special enamel-germs here becoming flask-shaped; *c*, *c′*, epithelium of mouth; *f*, neck; *f′*, body of special enamel-germ. C.—A later stage; *c*, outline of epithelium of gum; *f*, neck of enamel-germ; *f′*, enamel organ; *p*, papilla; *s*, dental sac forming; *f p*, the enamel-germ of permanent tooth. (Waldeyer and Kölliker.) Copied from Quain's Anatomy.

DEVELOPMENT OF TEETH.

Development of the Teeth.—The first step in the development of the teeth consists in a downward growth (Fig. 63, A, 1) from the stratified epithelium of the mucous membrane of the mouth, now thickened in the neighborhood of the maxillæ which are in the course of formation. This process passes downward into a recess (enamel groove) of the imperfectly developed tissue of which the chief part of the jaw consists. The downward epithelial growth forms the *primary enamel organ* or *enamel germ,* and its position is indicated by a slight groove in the mucous membrane of the jaw. The next step in the process consists in the elongation downward of the enamel groove and ·of the enamel germ and the inclination outward of the deeper part (Fig. 63, B, f'), which is now inclined at an angle with the upper portion or neck (f), and has become bulbous. After this, there is an increased development at certain points corresponding to the situations of the future milk teeth, and the enamel germ, or common enamel germ, as it may be called, becomes divided at its deeper portion, or extended by further growth, into a number of special enamel germs corresponding to each of the above-mentioned milk teeth, and connected to the common germ by a narrow neck, each tooth being placed in its own special recess in the embryonic jaw (Fig. 63, B, $f f'$).

As these changes proceed, there grows up from the underlying tissue into each enamel germ (Fig. 63, C, p), a distinct vascular *papilla* (dental papilla), and upon it the enamel germ becomes moulded and presents the appearance of a cap of two layers of epithelium separated by an interval (Fig. 63, C, f'). Whilst part of the sub-epithelial tissue is elevated to form the dental papillæ, the part which bounds the embryonic teeth forms the dental sacs (Fig. 63, C, s); and the rudiment of the jaw, at first a bony gutter in which the teeth germs lie, sends up processes forming partitions between the teeth. In this way small chambers are produced in which the dental sacs are contained, and thus the sockets of the

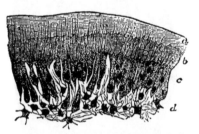

FIG. 64.—Part of section of developing tooth of a young rat, showing the mode of deposition of the dentine. Highly magnified. *a*, outer layer of fully formed dentine; *b*, uncalcified matrix with one or two nodules of calcareous matter near the calcified parts; *c*, odontoblasts sending processes into the dentine; *d*, pulp. The section is stained in carmine, which colors the uncalcified matrix but not the calcified part. (E. A. Schäfer.)

teeth are formed. The papilla, which is really part of the dental sac, if one thinks of this as the whole of the sub-epithelial tissue surrounding the enamel organ and interposed between the enamel germ and the developing bony jaw, is composed of nucleated cells arranged in a meshwork, the outer or peripheral part being covered with a layer of columnar nucleated cells called *odontoblasts.* The odontoblasts form the dentine, while the remainder of the papilla forms the tooth-pulp. The method of the formation of the dentine from the odontoblasts is as follows:—The cells elongate at their outer part, and these processes are directly converted into the tubules of dentine (Fig. 64). The continued formation of dentine proceeds by the·elongation of the odontoblasts, and their subsequent conversion by a process of calcification into dentine tubules. The most recently formed tubules are not immediately calcified. The dentine fibres contained in the tubules are said to be formed from processes of the deeper layer of odontoblasts, which are wedged in between the cells of the superficial layer (Fig. 64) which form the tubules only.

Since the papillæ are to form the main portion of each tooth, *i.e.*, the dentine, each of them early takes the shape of the crown of the tooth it is to form. As the dentine increases in thickness, the papillæ diminish, and at last when the tooth is cut, only a small amount of the papilla remains as the dental pulp, and is supplied by vessels and nerves which enter at the end of the fang. The shape of the crown of the tooth is taken by the corresponding papilla, and that of the single or double fang by the subsequent constriction below the crown, or by division of the lower part of the papilla.

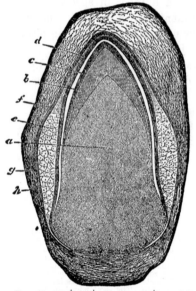

FIG. 65.—Vertical transverse section of the dental sac, pulp, etc., of a kitten, *a*, dental papilla or pulp; *b*, the cap of dentine formed upon the summit; *c*, its covering of enamel; *d*, inner layer of epithelium of the enamel organ; *e*, gelatinous tissue; *f*, outer epithelial layer of the enamel organ; *g*, inner layer, and *h*, outer layer of dental sac. × 14. (Thiersch.)

The enamel cap is found later on to consist (Fig. 65) of three parts: (*a*) an inner membrane, composed of a layer of columnar epithelium in contact with the dentine, called *enamel* cells, and outside of these one or more layers of small polyhedral nucleated cells (*stratum intermedium* of Hannover); (*b*) an outer membrane of several layers of epithelium; (*c*) a middle membrane formed of a matrix of non-vascular, gelatinous tissue, containing a hyaline interstitial substance. The enamel is formed by the enamel cells of the inner membrane, by the elongation of their distal extremities, and the direct conversion of these processes into enamel. The calcification of the enamel processes or prisms takes place first at the periphery, the centre remaining for a time transparent. The cells of the stratum intermedium are used for the regeneration of the enamel cells, but these and the middle membrane after a time disappear. The cells of the outer membrane give origin to the cuticle of the enamel.

The *cement* or *crusta petrosa* is formed from the tissue of the tooth sac, the structure and function of which are identical with those of the osteogenetic layer of the periosteum.

In this manner the first set of teeth, or the milk-teeth, are formed; and each tooth, by degrees developing, presses at length on the wall of the sac enclosing it, and, causing its absorption, is *cut*, to use a familiar phrase.

The *temporary* or *milk-teeth* have only a very limited term of existence.

This is due to the growth of the permanent teeth, which push their way up from beneath, absorbing in their progress the whole of the fang of each milk-tooth and leaving at length only the crown as a mere shell, which is shed to make way for the eruption of the permanent teeth (Fig. 66).

The temporary teeth are ten in each jaw, namely, four *incisors*, two *canines*, and four *molars*, and are replaced by ten permanent teeth, each of which is developed in a way almost exactly similar to the manner of development already described, from a small process or sac set by, so to speak, from the enamel germ of the temporary tooth which precedes it, and called the *cavity of reserve*.

The number of permanent teeth in each jaw is, however, increased to sixteen, by the development of three others on each side of the jaw after much the same fashion as that by which the milk-teeth were themselves formed.

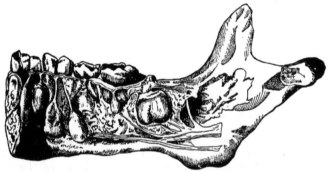

FIG. 66.—Part of the lower jaw of a child of three or four years old, showing the relations of the temporary and permanent teeth, The specimen contains all the milk teeth of the right side, together with the incisors of the left; the inner plate of the jaw has been removed, so as to expose the sacs of all the permanent teeth of the right side, except the eighth or wisdom tooth, which is not yet formed. The large sac near the ascending ramus of the jaw is that of the first permanent molar, and above and behind it is the commencing rudiment of the second molar. (Quain.)

The beginning of the development of the permanent teeth of course takes place long before the *cutting* of those which they are to succeed. One of the first steps in the development of a milk-tooth is the outgrowth of a lateral process of epithelial cells from its primitive enamel organ (Fig. 63, c, *f p*). This epithelial outgrowth ultimately becomes the enamel organ of the permanent tooth, and is indented from below by a primitive dental papilla, precisely as described above.

The following formula shows, at a glance, the comparative arrangement and number of the temporary and permanent teeth:—

	Mo.	Ca.	In.	Ca.	Mo.	
Upper	2	1	4	1	2	=10
Lower	2	1	4	1	2	=10

Temporary Teeth . . . . =20

	Mo.	Bi.	Ca.	In.	Ca.	Bi.	Mo.	
Upper	3	2	1	4	1	2	3	=16
Lower	3	2	1	4	1	2	3	=16

Permanent Teeth . . . . =32

From this formula it will be seen that the two bicuspid teeth in the adult are the successors of the two molars in the child. They differ from them, however, in some respects, the *temporary* molars having a stronger likeness to the *permanent* than to their immediate descendants, the so-called bicuspids.

The temporary incisors and canines differ from their successors but little except in their smaller size.

The following tables show the average times of eruption of the Temporary and Permanent teeth. In both cases, the eruption of any given tooth of the lower jaw precedes, as a rule, that of the corresponding tooth of the upper.

Temporary or Milk Teeth.

The figures indicate in *months* the age at which each tooth appears.

Molars.	Canines.	Incisors.	Canines.	Molars.
24 12	18	9 7 7 9	18	12 24

Permanent Teeth.

The age at which each tooth is cut is indicated in this table in *years*.

Molars.	Bicuspid.	Canines.	Incisors.	Canines.	Bicuspid.	Molars.
17 12 to ₊to 6 25 13	10 9	11 to 12	8 7 7 8	11 to 12	9 10	12 17 6 to to 13 25

The times of eruption put down in the above tables are only approximate: the limits of variation being tolerably wide. Some children may cut their first teeth before the age of six months, and others not till nearly the twelfth month. In nearly all cases the two central incisors of the lower jaw are cut first; these being succeeded after a short interval by the four incisors of the upper jaw, next follow the lateral incisors of the lower jaw, and so on as indicated in the table till the completion of the milk dentition at about the age of two years.

The milk-teeth usually come through in batches, each period of eruption being succeeded by one of quiescence lasting sometimes several months. The milk-teeth are in use from the age of two up to five and a half years: at about this age the first permanent molars (four in number) make their appearance *behind* the milk-molars, and for a short time the child has four permanent and twenty temporary teeth in position at once.

It is worthy of note that from the age of five years to the shedding of the first milk-tooth the child has no fewer than forty-eight teeth, twenty milk-teeth and twenty-eight calcified germs of permanent teeth (all in fact except the four wisdom teeth).

CHAPTER IV.

THE BLOOD.

THE blood of man, as indeed of the great majority of vertebrate animals, is a more or less viscid fluid, of a red color. The exact shade of red is variable, for whereas that taken from the arteries, from the left side of the heart or from the pulmonary veins, is of a bright scarlet hue, that obtained from the systemic veins, from the right side of the heart, or from the pulmonary artery, is of a much darker color, and varies from bluish-red to reddish-black. To the naked eye, the red color appears to belong to the whole mass of blood, but on examination with the microscope it is found that this is not the case. By the aid of this instrument the blood is shown to consist in reality of an almost colorless fluid, called *Liquor Sanguinis* or *Plasma*, in which are suspended numerous minute rounded masses of protoplasm, called *Blood Corpuscles*. The corpuscles are, for the most part, colored, and it is to their presence that the red color of the blood is due.

Even when examined in very thin layers blood is *opaque*, on account of the different refractive powers possessed by its two constituents, viz., the plasma and the corpuscles. On treatment with chloroform and other reagents, however, it becomes transparent, and assumes a lake color, in consequence of the coloring matter of the corpuscles having been, by these means, discharged into the plasma. The average *specific gravity* of blood at 60° F. (15° C.) is 1055, the extremes consistent with health being 1045–1062. The *reaction* of blood is faintly alkaline. Its *temperature* varies within narrow limits, the average being 100° F. (37·8° C.). The blood stream is slightly warmed by passing through the muscles, nerve centres, and glands, but is somewhat cooled on traversing the capillaries of the skin. Recently drawn blood has a distinct *odor*, which in many cases is characteristic of the animal from which it has been taken; the odor may be further developed by adding to blood a mixture of equal parts of sulphuric acid and water.

Quantity of the Blood.—The quantity of blood in any animal under normal conditions bears a pretty constant relation to the body weight. The methods employed for estimating it are not so simple as might at first sight be thought. For example, it would not be possible to get any accurate information on the point from the amount obtained by rapidly bleeding an animal to death, for then an indefinite quantity would remain in the vessels, as well as in the tissues; nor, on the other hand, would it be possible to obtain a correct estimate by less rapid bleeding, as, since life would be more prolonged, time would be allowed for the passage into the blood of lymph from the lymphatic vessels and from the tissues. In the former case, therefore, we should under-estimate, and in the latter over-estimate the total amount of the blood.

Of the several methods which have been employed, the most accurate appears to be the following. A small quantity of blood is taken from an animal by venesection; it is defibrinated and measured, and used to make standard solutions of blood. The animal is then rapidly bled to death, and the blood which escapes is collected. The blood-vessels are next washed out with water or saline solution until the washings are no longer colored, and these are added to the previously withdrawn blood; lastly the whole animal is finely minced with water or saline solution. The fluid obtained from the mincings is carefully filtered, and added to the diluted blood previously obtained, and the whole is measured. The next step in the process is the comparison of the color of the diluted blood with that of standard solutions of blood and water of a known strength, until it is discovered to what standard solution the diluted blood corresponds. As the amount of blood in the corresponding standard solution is known, as well as the total quantity of diluted blood obtained from the animal, it is easy to calculate the absolute amount of blood which the latter contained, and to this is added the small amount which was withdrawn to make the standard solutions. This gives the total amount of blood which the animal contained. It is contrasted with the weight of the animal, previously known. The result of many experiments shows that the quantity of blood in various animals averages $\frac{1}{12}$ to $\frac{1}{14}$ of the total body weight.

An estimate of the quantity in man which corresponded nearly with the above, was made some years ago from the following data. A criminal was weighed before and after decapitation; the difference in the weight representing, of course, the quantity of blood which escaped. The blood-vessels of the head and trunk were then washed out by the injection of water, until the fluid which escaped had only a pale red or straw color. This fluid was then also weighed; and the amount of blood which it represented was calculated by comparing the proportion of solid matter contained in it with that of the first blood which escaped on decapitation. Two experiments of this kind gave precisely similar results. (Weber and Lehmann.)

It should be remembered, however, in connection with these estimations, that the quantity of the blood must vary, even in the same animal, very considerably with the amount of both the ingesta and egesta of the period immediately preceding the experiment; and it has been found, indeed, that the quantity of blood obtainable from a fasting animal barely exceeds a half of that which is present soon after a full meal.

Coagulation of the Blood.—One of the most characteristic properties which the blood possesses is that of *clotting* or *coagulating*, when removed from the body. This phenomenon may be observed under the most favorable conditions in blood which has been drawn into an open vessel. In about two or three minutes, at the ordinary temperature of the air, the surface of the fluid is seen to become semi-solid or jelly-like; this change next taking place, in a minute or two, at the sides of the vessel in which it is contained, and then extending throughout the entire mass.

The time which is required for the blood to become solid is about eight or nine minutes. The solid mass occupies exactly the same volume as the previously liquid blood, and adheres so closely to the sides of the contain-

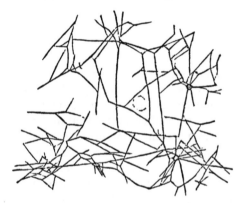

Fig. 67.—Reticulum of fibrin, from a drop of human blood, after treatment with rosanilin.
(Ranvier.)

ing vessel that if it be inverted none of its contents escape. The solid mass is the *crassamentum* or *clot*. If the clot be watched for a few minutes, drops of a light straw-colored fluid, the *serum*, may be seen to make their appearance on the surface, and, as they become more and more numerous, run together, forming a complete superficial stratum above the solid clot. At the same time the fluid begins to transude at the sides and at the under surface of the clot, which in the course of an hour or two floats in the liquid. The first drops of serum appear on the surface about eleven or twelve minutes after the blood has been drawn; and the fluid continues to transude for from thirty-six to forty-eight hours.

The clotting of blood is due to the development in it of a substance called *fibrin*, which appears as a meshwork (Fig. 67) of fine fibrils. This meshwork entangles and encloses within it the blood corpuscles, as clotting takes place too quickly to allow them to sink to the bottom of the plasma. The first clot formed, therefore, includes the whole of the constituents of the blood in an apparently solid mass, but soon the fibrinous meshwork begins to contract, and the serum which does not belong to the clot is squeezed out. When the whole of the serum has transuded, the clot is found to be smaller, but firmer and harder, as it is now made up of fibrin and blood corpuscles only. It will be noticed that coagulation rearranges the constituents of the blood according to the following scheme, liquid blood being made up of plasma and blood-corpuscles, and clotted blood of serum and clot.

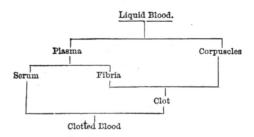

Buffy Coat.—Under ordinary circumstances coagulation occurs, as we have mentioned above, before the red corpuscles have had time to sub-side; and thus from their being entangled in the meshes of the fibrin, the clot is of a deep red color throughout, somewhat darker, it may be, at the most dependent part, from accumulation of red corpuscles, but not to any very marked degree. When, however, coagulation is delayed from any cause, as when blood is kept at a temperature of 32° F. (0° C.), or when clotting is normally a slow process, as in the case of horse's blood, or, lastly, in certain diseased conditions of the blood in which clotting is naturally delayed, time is allowed for the colored corpuscles to sink to the bottom of the fluid. When clotting does occur, the upper layers of the blood, being free of colored corpuscles and consisting chiefly of fibrin, form a superficial stratum differing in appearance from the rest of the clot, in that it is of a grayish yellow color. This is known as the "*buffy coat*."

Cupped appearance of the Clot.—When the buffy coat has been produced in the manner just described, it commonly contracts more than the rest of the clot, on account of the absence of colored corpuscles from its meshes, and because contraction is less interfered with by adhesion to the interior of the containing vessel in the vertical than the horizontal direction. This produces a cup-like appearance of the buffy coat, and the clot is not only buffed but cupped on the surface. The buffed and cupped appearance of the clot is well marked in certain states of the system, especially in inflammation, where the fibrin-forming constituents are in excess, and it is also well marked in chlorosis where the corpuscles are deficient in quantity.

Formation of Fibrin.—In describing the coagulation of the blood in the preceding paragraphs, it was stated that this phenomenon was due to the development in the clotting blood of a meshwork of fibrin. This may be demonstrated by taking recently-drawn blood, and whipping it with a bundle of twigs; the fibrin is found to adhere to the twigs as a reddish-white, stringy mass, having been thus obtained from the fluid nearly free from colored corpuscles. The defibrinated blood no longer retains the power of spontaneous coagulability.

The fibrin which makes its appearance in the blood when it is undergoing coagulation is derived chiefly, if not entirely, from the plasma or liquor sanguinis; for although the colorless corpuscles are intimately connected with the process in a way which will be presently explained, the colored corpuscles appear to take no active part in it whatever. This may be shown by experimenting with plasma free from colored corpuscles. Such plasma may be procured by delaying coagulation in blood, by keeping it at a low temperature, 32° F. (0° C.), until the colored corpuscles which are of higher specific gravity than the other constituents of blood, have had time to sink to the bottom of the containing vessel, and to leave an upper stratum of colorless plasma, in the lower layers of which are many colorless corpuscles. The blood of the horse is specially suited for the purposes of this experiment; and the upper stratum of colorless plasma derived from it, if decanted into another vessel and exposed to the ordinary temperature of the air, will coagulate just as though it were the entire blood, producing a clot similar in all respects to blood clot, except that it is almost colorless from the absence of red corpuscles. If some of the plasma be diluted with neutral saline solution,[1] coagulation is delayed, and the stages of the gradual formation of fibrin may be more conveniently watched. The viscidity which precedes the complete coagulation may be seen to be due to fibrin fibrils developing in the fluid—first of all at the circumference of the containing vessel, and gradually extending throughout the mass. Again, if plasma be whipped with a bundle of twigs, the fibrin may be obtained as a solid, stringy mass, just in the same way as from the entire blood, and the resulting fluid no longer retains its power of spontaneous coagulability. Evidently, therefore, fibrin is derived from the plasma and not from the colored corpuscles. In these experiments, it is not necessary that the plasma shall have been obtained by the process of cooling above described, as plasma obtained in any other way, *e.g.*, by allowing blood to flow direct from the vessels of an animal into a vessel containing a third or a fourth of the bulk of the blood of a saturated solution of a neutral salt (preferably of magnesium sulphate) and mixing carefully, will answer the purpose, and, just as in the other case, the colored corpuscles will subside, leaving the clear super-

---

[1] Neutral saline solution commonly consists of a ·75 solution of common salt (sodium chloride) in water.

stratum of (salted) plasma. In order to cause this plasma to coagulate, it is necessary to get rid of the salts by dialysis, or to dilute it with several times its bulk of water.

The antecedent of Fibrin.—If plasma be saturated with *solid* magnesium sulphate or sodium chloride, a white, sticky precipitate, called *plasmine*, is thrown down, after the removal of which, by filtration, the plasma will not spontaneously coagulate. This *plasmine* is soluble in dilute neutral saline solutions, and the solution of it speedily coagulates, producing a clot composed of fibrin. From this we see that blood plasma contains a substance without which it cannot coagulate, and a solution of which is spontaneously coagulable. This substance is very soluble in dilute saline solutions, and is not, therefore, fibrin, which is insoluble in these fluids. We are, therefore, led to the belief that plasmine produces or is converted into fibrin, when clotting of fluids containing it takes place.

Nature of Plasmine.—There seems distinct evidence that plasmine is a compound body made up of two or more substances, and that it is not mere soluble fibrin. This view is based upon the following observations:—There exists in all the serous cavities of the body in health, *e.g.*, the pericardium, the peritoneum, and the pleura, a certain small amount of transparent fluid, generally of a pale straw color, which in diseased conditions may be greatly increased. It somewhat resembles serum in appearance, but in reality differs from it, and is probably identical with plasma. This serous fluid is not, as a rule, spontaneously coagulable, but may be made to clot on the addition of serum, which is also a fluid which has no tendency of itself to coagulate. The clot produced consists of fibrin, and the clotting is identical with the clotting of plasma. From the serous fluid (that from the inflamed *tunica vaginalis testis* or *hydrocele* fluid is mostly used) we may obtain, by saturating it with *solid* magnesium sulphate or sodium chloride, a white viscid substance as a precipitate which is called *fibrinogen*, which may be separated by filtration, and is then capable of being dissolved in water, as a certain amount of the neutral salt is entangled with the precipitate sufficient to produce a dilute saline solution in which it is soluble. This body belongs to the *globulin* class of proteid substances. Its solution has no tendency to clot of itself. Fibrinogen may also be obtained as a viscid precipitate from hydrocele fluid by diluting it with water, and passing a brisk stream of carbon dioxide gas through the solution. Now if serum be added to a solution of fibrinogen, the mixture clots.

From serum may be obtained another globulin very similar in properties to fibrinogen, if it be subjected to treatment similar to either of the two methods by which fibrinogen is obtained from hydrocele fluid; this substance is called *paraglobulin,* and it may be separated by filtration and dissolved in a dilute saline solution in a manner similar to fibrinogen.

If the solutions of fibrinogen and paraglobulin be mixed, the mixture cannot be distinguished from a solution of plasmine, and like that solution (in a great majority of cases) firmly clots; whereas a mixture of the hydrocele fluid and serum, from which they have been respectively taken, no longer does so. In addition to this evidence of the compound nature of plasmine, it may be further shown that, if sufficient care be taken, both fibrinogen and paraglobulin may be obtained from plasma: fibrinogen, as a flaky precipitate, by adding carefully 13 per cent. of crystalline sodium chloride; and after the removal of fibrinogen from the plasma by filtration, paraglobulin may be afterward precipitated, on the further addition of the same salt or of magnesium sulphate to the filtrate. It is evident, therefore, that both these substances must be thrown down together when plasma is saturated with sodium chloride or magnesium sulphate, and that the mixture of the two corresponds with plasmine.

Presence of a Fibrin Ferment.—So far it has been shown that *plasmine*, the antecedent of fibrin in blood, to the possession of which blood owes its power of coagulating, is not a simple body, but is composed of at least two factors—viz., fibrinogen and paraglobulin; there is reason for believing that yet another body is associated with them in plasmine to produce coagulation; this is what is known under the name of *fibrin ferment* (Schmidt). It was at one time thought that the reason why hydrocele fluid coagulated when serum was added to it was that the latter fluid supplied the paraglobulin which the former lacked; this, however, is not the case, as hydrocele does not lack this body, and if paraglobulin, obtained from serum by the carbonic acid method, be added to it, it will not coagulate, neither will a mixture of solutions of fibrinogen and paraglobulin obtained in the same way. But if paraglobulin, obtained by the saturation method, be added to hydrocele fluid, it will clot, as will also, as we have seen above, a mixed solution of fibrinogen and paraglobulin, when obtained by the saturation method. From this it is evident that in plasmine there is something more than the two bodies above mentioned, and that this something is precipitated with the paraglobulin by the saturation method, and is not precipitated by the carbonic acid method. The following experiments show that it is of the nature of a ferment. If defibrinated blood or serum be kept in a stoppered bottle with its own bulk of alcohol for some weeks, all the proteid matter is precipitated in a coagulated form; if the precipitate be then removed by filtration, dried over sulphuric acid, finely powdered, and then suspended in water, a watery extract may be obtained by further filtration, containing extremely little, if any, proteid matter. Yet a little of this watery extract will determine coagulation in fluids, *e.g.*, hydrocele fluid or diluted plasma, which are not spontaneously coagulable, or which coagulate slowly and with difficulty. It will also cause a mixture of fibrinogen and paraglobulin, obtained by the carbonic acid method, to clot. This watery extract appears to contain the body which is precipitated with the paraglobulin by the saturation method. Its active properties are entirely destroyed by boiling. The amount of the extract added does not influence the amount of the clot formed, but only the rapidity of clotting, and moreover the active substance contained in the extract evidently does not form part of the clot, as it may be obtained from the serum after blood has clotted. So that the third factor, which is contained in the aqueous extract of blood, belongs to that class of bodies which promote the union of other bodies, or cause changes in other bodies, without themselves entering into union or undergoing change, *i.e. ferments*. The third substance has, therefore, received the name *fibrin ferment*. This ferment is developed in blood soon after it has been shed, and its amount appears to increase for a certain time afterward (p. 74).

The part played by Paraglobulin.—So far we have seen that plasmine is a body composed of three substances, viz., fibrinogen, paraglobulin, and fibrin ferment. The question presents itself, are these three bodies actively concerned in the formation of fibrin? Here we come to a point about which two distinct opinions prevail, and which it will be necessary to mention. Schmidt holds that fibrin is produced by the interaction of the two proteid bodies, viz., fibrinogen and paraglobulin, brought about by the presence of a special fibrin ferment. Also, that when coagulation does not occur in serum, which contains paraglobulin and the fibrin ferment, the non-coagulation is accounted for by lack of fibrinogen, and when it does not occur in fluids which contain fibrinogen, it is due to the absence of paraglobulin, or of the ferment, or of both. It will be seen that, according to this view, paraglobulin has a very important fibrino-plastic property. The other opinion, held by Hammersten, is that paraglobulin is not an essential in coagulation, or at any rate does not take an active part in the process. He believes that paraglobulin possesses the property in common with many other bodies of combining with—or decomposing, and so rendering inert—certain substances which have the power of preventing the formation or precipitation of fibrin, this power of preventing coagulation being well known to belong to the free alkalies, to the alkaline carbonates, and to certain salts; and he looks upon fibrin as formed from fibrinogen, which is either (1) decomposed into that substance with the production of some other substances; or (2) bodily converted into it under the action of a ferment, which is frequently precipitated with paraglobulin.

Influence of Salts on Coagulation.—It is believed that the presence of a certain but small amount of salts, especially of sodium chloride, is necessary for coagulation, and that without it, clotting cannot take place.

Sources of the Fibrin Generators.—It has been previously remarked that the colorless corpuscles which are always present in smaller or greater numbers in the plasma, even when this has been freed from colored corpuscles, have an important share in the production of the clot. The proofs of this may be briefly summarized as follows:—(1) That all strongly coagulable fluids contain colorless corpuscles almost in direct proportion to their coagulability; (2) That clots formed on foreign bodies, such as needles inserted into the interior of living blood-vessels, are preceded by an aggregation of colorless corpuscles; (3) That plasma in which the colorless corpuscles happen to be scanty, clots feebly; (4) That if horse's blood be kept in the cold, so that the corpuscles subside, it will be found that the lowest stratum, containing chiefly colored corpuscles, will, if removed, clot feebly, as it contains little of the fibrin factors; whereas the colorless plasma, especially the lower layers of it in which the colorless corpuscles are most numerous, will clot well, but if filtered in the cold will not clot so well, indicating that when filtered nearly free from colorless corpuscles even the plasma does not contain sufficient of all the fibrin factors to produce thorough coagulation; (5) In a drop of coagulating blood, observed under the miscroscope, the fibrin fibrils are seen to start from the colorless corpuscles.

Although the intimate connection of the colorless corpuscles with the process of coagulation seems indubitable, for the reasons just given, the exact share which they have in contributing the various fibrin factors remains still uncertain. It is generally believed that the fibrin-ferment at any rate is contributed by them, inasmuch as the quantity of this substance obtainable from plasma bears a direct relation to the numbers of colorless corpuscles which the plasma contains. Many believe that the fibrinogen also is wholly or in part derived from them.

Conditions affecting Coagulation.—The coagulation of the blood is **hastened** by the following means:—

1. **Moderate warmth,**—from about 100° to 120° F. (37·8—49° C.).

2. **Rest** is favorable to the coagulation of blood. Blood, of which the whole mass is kept in uniform motion, as when a closed vessel completely filled with it is constantly moved, coagulates very slowly and imperfectly.

3. **Contact with foreign matter,** and especially multiplication of the points of contact. Thus, coagulated fibrin may be quickly obtained from liquid blood by stirring it with a bundle of small twigs; and even in the living body the blood will coagulate upon rough bodies projecting into the vessels; as, for example, upon threads passed through them, or upon the heart's valves roughened by inflammatory deposits or calcareous accumulations.

4. **The free access of air.**—Coagulation is quicker in shallow than in tall and narrow vessels.

5. **The addition of less than twice the bulk of water.**

The blood last drawn is said to coagulate more quickly than the first.

The coagulation of the blood is **retarded, suspended, or prevented** by the following means:—

1. **Cold** retards coagulation; and so long as blood is kept at a temperature, 32° F. (0° C.), it will not coagulate at all. Freezing the blood, of course, prevents its coagulation; yet it will coagulate, though not firmly, if thawed after being frozen; and it will do so, even after it has been frozen for several months. A higher temperature than 120° F. (49° C.) retards coagulation, or, by coagulating the albumen of the serum, prevents it altogether.

2. The addition of **water in greater proportion than twice the bulk** of the blood.   •

3. **Contact with living tissues,** and especially with the interior of a living blood-vessel.

4. **The addition of neutral salts** in the proportion of 2 or 3 per cent. and upward. When added in large proportion most of these saline substances prevent coagulation altogether. Coagulation, however, ensues on dilution with water. The time during which blood can be thus preserved in a liquid state and coagulated by the addition of water, is quite indefinite.

5. **Imperfect Aeration,**—as in the blood of those who die by asphyxia.

6. In **inflammatory states of the system** the blood coagulates more slowly although more firmly.

7. Coagulation is retarded **by exclusion of the blood from the air,** as by pouring oil on the surface, etc. In vacuo, the blood coagulates quickly; but Lister thinks that the rapidity of the process is due to the bubbling which ensues from the escape of gas, and to the blood being thus brought more freely into contact with the containing vessel.

8. The coagulation of the blood is prevented altogether **by the addition of strong acids and caustic alkalies.**

9. It has been believed, and chiefly on the authority of Hunter, **that after certain modes of death the blood does not coagulate;** he enumerates the death by lightning, over-exertion (as in animals hunted to death), blows on the stomach, fits of anger. He says, "I have seen instances of them all." Doubtless he had done so; but the results of such events are not constant. The blood has been often observed coagulated in the bodies of animals killed by lightning or an electric shock; and Gulliver has published instances in which he found clots in the hearts of hares and stags hunted to death, and of cocks killed in fighting.

Cause of the fluidity of the blood within the living body.— Very closely connected with the problem of the coagulation of the blood arises the question,—why does the blood remain liquid within the living body? We have certain pathological and experimental facts, apparently opposed to one another, which bear upon it, and these may be, for the sake of clearness, classed under two heads:—

(1) *Blood will coagulate within the living body under certain conditions*,—for example, on ligaturing an artery, whereby the inner and middle coats are generally ruptured, a clot will form within it, or by passing a needle through the coats of the vessel into the blood stream a clot will gradually form upon it. Other foreign bodies, *e.g.* wire, thread, etc., produce the same effect. It is a well-known fact that small clots are apt to form upon the roughened edges of the valves of the heart when the roughness has been produced by inflammation, as in endocarditis, and it is also equally true that aneurisms of arteries are sometimes spontaneously cured by the deposition within them, layer by layer, of fibrin from the blood stream, which natural cure it is the aim of the physician or surgeon to imitate.

(2) *Blood will remain liquid under certain conditions outside the body,* without the addition of any re-agent, even if exposed to the air at the ordinary temperature. It is well known that blood remains fluid in the body for some time after death, and it is only after rigor mortis has occurred that the blood is found clotted. It has been demonstrated by Hewson, and also by Lister, that if a large vein in the horse or similar animal be ligatured in two places some inches apart, and after some time be opened, the blood contained within it will be found fluid, and that coagulation will occur only after a considerable time. But this is not due to occlusion from the air simply. Lister further showed that if the vein with the blood contained within it be removed from the body and then be carefully opened, the blood might be poured from the vein into another similarly prepared, as from one test-tube into another, thereby suffering free exposure to the air, without coagulation occurring as long as the vessels retain their vitality. If the endothelial lining of the vein, however, be injured, the blood will not remain liquid. Again, blood will remain liquid for days in the heart of a turtle, which continues to beat for a very long time after removal from the body.

Any theory which aims at explaining the fluidity under the usual conditions of the blood within the living body must reconcile the above apparently contradictory facts, and must at the same time be made to include all the other known facts concerning the coagulation of the blood. We may therefore dismiss as insufficient the following;—that coagulation is due to exposure to the air or oxygen; that it is due to the cessation of the circulatory movement; that it is due to evolution of various gases, or to the loss of heat.

Two theories, those of Lister and Brücke, remain. The former supposes that the blood has no natural tendency to clot, but that its coagulation out of the body is due to the action of foreign matter with which it happens to be brought into contact, and in the body to conditions of the tissues which cause them to act toward it like foreign matter. The latter, on the other hand, supposes that there is a natural tendency on the part of the blood to clot, but that this is restrained in the living body by some inhibitory power resident in the walls of the containing vessels..

Support was once thought to be given to Brücke's and like theories by cases of injury, in which blood extravasated in the living body has seemed to remain uncoagulated for weeks, or even months, on account of its contact with living tissues. But the supposed facts have been shown to be without foundation. The blood-like fluid in such cases is not uncoagulated blood, but a mixture of serum and blood-corpuscles, with a certain proportion of clot in various stages of disintegration. (Morrant Baker.)

As the blood must contain the substances from which fibrin is formed, and as the re-arrangement of these substances occurs very quickly whenever the blood is shed, so that it is somewhat difficult to prevent coagulation, it seems more reasonable to hold with Brücke, that the blood has a strong tendency to clot, rather than with Lister, that it has no special tendency thereto.

It has been recently suggested that the reason why blood does not coagulate in the living vessels, is that the factors which we have seen are necessary for the formation of fibrin are not in the exact state required for its production, and that the fibrin ferment is not formed or is not, at any rate, free in the living blood, but that it is produced (or set free) at the moment of coagulation by the disintegration of the colorless corpuscles. This supposition is certainly plausible, but if it be a true one, it must be assumed either that the living blood-vessels exert a restraining influence upon the disintegration of the corpuscles in sufficient numbers to form a clot, or that they render inert any small amount of fibrin ferment which may have been set free by the disintegration of a few corpuscles; as it is certain that corpuscles of all kinds must from time to time disintegrate in the blood without causing it to clot; and, secondly, that shed and defibrinated blood which contains blood corpuscles, broken down and disintegrated, will not, when injected into the vessels of an animal, produce clotting. There must be a distinct difference, therefore, if only in amount, between the normal disintegration of a few colorless corpuscles in the living uninjured blood-vessels and the abnormal disintegration of a large number which occurs whenever the blood is shed without suitable precaution, or when coagulation is unrestrained by the neighborhood of the living uninjured blood-vessels.

The Blood Corpuscles or Blood-Cells.

There are two principal forms of corpuscles, the **red** and the **white**, or, as they are now frequently named, the **colored** and the **colorless**. In the moist state, the red corpuscles form about 45 per cent. by weight, of the whole mass of the blood. The proportion of colorless corpuscles is only as 1 to 500 or 600 of the colored.

Red or Colored Corpuscles.—Human red blood-corpuscles are circular, biconcave disks with rounded edges, from $\frac{1}{3000}$ to $\frac{1}{4000}$ inch in diameter, and $\frac{1}{12000}$ inch in thickness, becoming flat or convex on addition of water. When viewed singly, they appear of a pale yellowish tinge; the deep red color which they give to the blood being observable in them only when they are seen *en masse*. They are composed of a colorless, structureless, and transparent filmy framework or *stroma*, infiltrated in all parts by a red coloring matter termed *hæmoglobin*. The *stroma* is tough and elastic, so that, as the cells circulate, they admit of elongation and other changes of form, in adaptation to the vessels, yet recover their natural shape as soon as they escape from compression. The term cell, in the sense of a bag or sac, is inapplicable to the red blood corpuscle; and it must be considered, if not solid throughout, yet as having no such variety of consistence in different parts as to justify the notion of its being a membranous sac with fluid contents. The stroma exists in all parts of its substance, and the coloring-matter uniformly pervades this, and is not merely surrounded by and mechanically enclosed within the outer wall of the corpuscle. The red corpuscles have no nuclei, although, in their usual state, the unequal refraction of transmitted light gives the appearance of a central spot, brighter or darker than the border, according as it is viewed in or out of focus. Their specific gravity is about 1088.

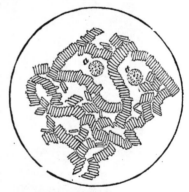

Fig. 68.—Red corpuscles in rouleaux. At *a, a,* are two white corpuscles.

Varieties.—The red corpuscles are not all alike, some being rather larger, paler, and less regular than the majority, and sometimes flat or slightly convex, with a shining particle apparent like a nucleolus. In almost every specimen of blood may be also observed a certain number of corpuscles smaller than the rest. They are termed *microcytes,* and are probably immature corpuscles.

A peculiar property of the red corpuscles, exaggerated in inflammatory blood, may be here again noticed, *i.e.,* their great tendency to adhere together in rolls or columns, like piles of coins. These rolls quickly fasten together by their ends, and cluster; so that, when the blood is spread out thinly on a glass, they form a kind of irregular network, with crowds of corpuscles at the several points corresponding with the knots of the net (Fig. 68). Hence, the clot formed in such a thin layer of blood looks mottled with blotches of pink upon a white ground, and in a larger quan-

tity of such blood help, by the consequent rapid subsidence of the corpuscles, in the formation of the buffy coat already referred to.

This tendency on the part of the red corpuscles, to form rouleaux, is
probably only a physical phenomenon, comparable to the collection into
somewhat similar rouleaux of discs of corks when they are partially immersed in water. (Norris.)

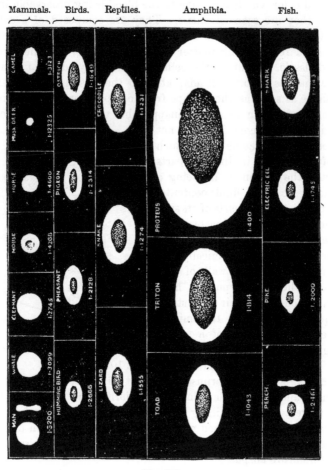

FIG. 69.[1]

_____

[1] The above illustration is somewhat altered from a drawing by Gulliver, in the
Proceed. Zool. Society, and exhibits the typical characters of the red blood-cells in the
main divisions of the Vertebrata. The fractions are those of an inch, and represent
the average diameter. In the case of the oval cells, only the long diameter is here
given. It is remarkable, that although the size of the red blood-cells varies so much
in the different classes of the vertebrate kingdom, that of the white corpuscles remains comparatively uniform, and thus they are, in some animals, much greater, in
others much less than the red corpuscles existing side by side with them.

Action of Reagents.—Considerable light has been thrown on the physical and chemical constitution of red blood-cells by studying the effects produced by mechanical means and by various reagents: the following is a brief summary of these reactions:—

Pressure.—If the red blood-cells of a frog or man are gently squeezed, they exhibit a wrinkling of the surface, which clearly indicates that there is a superficial pellicle partly differentiated from the softer mass within; again, if a needle be rapidly drawn across a drop of blood, several corpuscles will be found cut in two, but this is not accompanied by any escape of cell contents; the two halves, on the contrary, assume a rounded form, proving clearly that the corpuscles are not mere membranous sacs with fluid contents like fat-cells.

Fluids.— Water.—When water is added gradually to frog's blood, the oval disc-shaped corpuscles become spherical, and gradually discharge their hæmoglobin, a pale, transparent stroma being left behind; human red blood-cells change from a discoidal to a spheroidal form, and discharge their cell-contents, becoming quite transparent and all but invisible.

Saline solution (dilute) produces no appreciable effect on the red

FIG. 70.   FIG. 71.   FIG. 72.

blood-cells of the frog. In the red blood-cells of man the discoid shape is exchanged for a spherical one, with spinous projections, like a horse-chestnut (Fig. 70). Their original forms can be at once restored by the use of carbonic acid.

Acetic acid (dilute) causes the nucleus of the red blood cells in the frog to become more clearly defined; if the action is prolonged, the nucleus becomes strongly granulated, and all the coloring matter seems to be concentrated in it, the surrounding cell-substance and outline of the cell becoming almost invisible; after a time the cells lose their color altogether. The cells in the figure (Fig. 71) represent the successive stages of the change. A similar loss of color occurs in the red cells of human blood, which, however, from the absence of nuclei, seem to disappear entirely.

Alkalies cause the red blood-cells to swell and finally disappear.

Chloroform added to the red blood-cells of the frog causes them to part with their hæmoglobin; the stroma of the cells becomes gradually broken up. A similar effect is produced on the human red blood-cell.

Tannin.—When a 2 per cent. solution of tannic acid is applied to frog's blood it causes the appearance of a sharply-defined little knob, projecting from the free surface: the coloring matter becomes at the same time concentrated in the nucleus, which grows more distinct (Fig. 72).

A somewhat similar effect is produced on the human red blood-cell. (Roberts.) *Magenta*, when applied to the red blood-cells of the [frog, produces a similar little knob or knobs, at the same time staining the nucleus and causing the discharge of the hæmoglobin. (Roberts.) The first effect of the magenta is to cause the discharge of the hæmoglobin, then the nucleus becomes suddenly stained, and lastly a finely granular matter issues through the wall of the corpuscle, becoming stained by the magenta, and a *macula* is formed at the point of escape. A similar macula is produced in the human red blood-cell.

Boracic acid.—A 2 per cent. solution applied to nucleated red blood-cells (frog) will cause the concentration of all the coloring matter in the nucleus; the colored body thus formed gradually quits its central position, and comes to be partly, sometimes entirely, protruded from the surface of the now colorless cell (Fig. 73). The result of this experiment led Brücke to distinguish the colored contents of the cell (zooid) from its colorless stroma (œcoid). When applied to the non-nucleated mammalian corpuscle its effect merely resembles that of other dilute acids.

Gases—Carbonic acid.—If the red blood-cells of a frog be first exposed

FIG. 73.　　　　FIG. 74.　　　　FIG. 75.　　　　FIG. 76.

to the action of water-vapor (which renders their outer pellicle more readily permeable to gases), and then acted on by carbonic acid, the nuclei immediately become clearly defined and strongly granulated; when air or oxygen is admitted the original appearance is at once restored. The upper and lower cell in Fig. 74 show the effect of carbonic acid; the middle one the effect of the re-admission of air. These effects can be reproduced five or six times in succession. If, however, the action of the carbonic acid be much prolonged, the granulation of the nucleus becomes permanent; it appears to depend on a coagulation of the paraglobulin. (Stricker.)

Ammonia.—Its effects seem to vary according to the degree of concentration. Sometimes the outline of the corpuscles becomes distinctly crenated; at other times the effect resembles that of boracic acid, while in other cases the edges of the corpuscles begin to break up. (Lankester.)

Heat.—The effect of heat up to 120°—140° F. (50°—60° C.) is to cause the formation of a number of bud-like processes (Fig. 75).

Electricity causes the red blood-corpuscles to become crenated, and at length mulberry-like. Finally they recover their round form and become quite pale.

The **general conclusions** to be drawn from these observations have been summed up as follows by Prof. Ray Lankester:—

"The red blood-corpuscle of the vertebrata is a viscid, and at the same time elastic disc, oval or round in outline, its surface being differentiated somewhat from the underlying material, and forming a pellicle or membrane of great tenuity, not distinguishable with the highest powers (whilst the corpuscle is normal and living), and having no pronounced inner limitation. The viscid mass consists of (or rather *yields*, since the state of combination of the components is not known) a variety of albuminoid and other bodies, the most easily separable of which is hæmoglobin; *secondly*, the matter which segregates to form Roberts's macula; and *thirdly*, a residuary stroma, apparently homogeneous in the mammalia (excepting as far as the outer surface or pellicle may be of a different chemical nature), but containing in the other vertebrata a sharply definable nucleus, this nucleus being already differentiated, but not sharply delineated during life, and consisting of, or separable into, at least two components, one (paraglobulin) precipitable by carbon dioxide, and removable by the action of weak ammonia; the other pellucid, and not granulated by acids."

The White or Colorless Corpuscles.—In human blood the white or colorless corpuscles or *leucocytes* are nearly spherical masses of granular protoplasm without cell wall. The granular appearance, more marked in some than in others (*vide infra*), is due to the presence of particles probably of a fatty nature. In all cases one or more nuclei exist in each corpuscle. The size of the corpuscle averages $\frac{1}{2500}$ of an inch in diameter.

In health, the proportion of white to red corpuscles, which, taking an average, is about 1 to 500 or 600, varies considerably even in the course of the same day. The variations appear to depend chiefly on the amount and probably also on the kind of food taken; the number of leucocytes being very considerably increased by a meal, and diminished again on fasting. Also in young persons, during pregnancy, and after great loss of blood, there is a larger proportion of colorless blood-corpuscles, which

FIG. 77.—A. Three colored blood-corpuscles. B. Three colorless blood-corpuscles acted on by acetic acid; the nuclei are very clearly visible. × 900.

probably shows that they are more rapidly formed under these circumstances. In old age, on the other hand, their proportion is diminished.

Varieties.—The colorless corpuscles present greater diversities of form than the red ones do. Two chief varieties are to be seen in human blood; one which contains a considerable number of granules, and the other which is paler and less granular. In size the variations are great, for in most specimens of blood it is possible to make out, in addition to the full-sized varieties, a number of smaller corpuscles, consisting of a large spherical nucleus surrounded by a variable amount of more or less granular protoplasm. The small corpuscles are, in all probability, the undeveloped forms of the others, and are derived from the cells of the lymph. Besides the above-mentioned varieties, Schmidt describes another form which he looks upon as intermediate between the colored and the colorless forms, viz., certain corpuscles which contain red granules of hæmoglobin in their protoplasm. The different varieties of colorless corpuscles are especially well seen in the blood of frogs, newts, and other cold-blooded animals.

Amœboid movement.—A remarkable property of the colorless corpuscles consists in their capability of spontaneously changing their shape. This was first demonstrated by Wharton Jones in the blood of the skate. If a drop of blood be examined with a high power of the microscope on a warm stage, or, in other words, under conditions by which loss of moisture is prevented, and at the same time the temperature is maintained at about that of the blood in its natural state within the walls of the living vessels, 100° F. (37·8° C.), the colorless corpuscles will be observed slowly altering their shapes, and sending out processes at various parts of their circumference. This alteration of shape, which can be most conveniently

Fig. 78.—Human colorless blood-corpuscle, showing its successive changes of outline within ten minutes when kept moist on a warm stage. (Schofield.)

studied in the newt's blood, is called amœboid, inasmuch as it strongly resembles the movement of the lowly organized *amœba*. The processes which are sent out are either lengthened or withdrawn. If lengthened, the protoplasm of the whole corpuscle flows as it were into its process, and the corpuscle changes its position; if withdrawn, protrusion of another process at a different point of the circumference speedily follows. The change of position of the corpuscle can also take place by a flowing movement of the whole mass, and in this case the locomotion is comparatively rapid. The activity both in the processes of change of shape and also of change in position, is much more marked in some corpuscles, viz., in the granular variety, than in others. Klein states that in the newt's blood the changes are especially likely to occur in a variety of the colorless corpuscle, which consists of masses of finely granular protoplasm with jagged outline, containing three or four nuclei, or of large irregular masses of protoplasm containing from five to twenty nuclei. Another phenomenon may be observed in such a specimen of blood, viz., the division of the corpuscles, which occurs in the following way. A cleft takes place in the protoplasm at one point, which becomes deeper and deeper, and then by the lengthening out and attenuation of the connection, and finally by its rupture, two corpuscles result. The nuclei have previously undergone division. The cells so formed are said to be remarkably active in their movements. Thus we see that the rounded form which the colorless corpuscles present in ordinary microscopic specimens must be looked upon as the shape natural to a dead corpuscle or to one whose vitality is dormant rather than as the shape proper to one living and active.

Action of re-agents upon the colorless corpuscles.—*Feeding the corpuscles.*— If some fine pigment granules, *e.g.*, powdered vermilion, be added to a fluid containing colorless blood-corpuscles, on a glass slide, these will be observed, under the microscope, to take up the pigment. In some cases colorless corpuscles have been seen with fragments of colored ones thus embedded in their substance. This property of the colorless corpuscles is especially interesting as helping still further to connect them with the lowest forms of animal life, and to connect both with the organized cells of which the higher animals are composed.

The property which the colorless corpuscles possess of passing through the walls of the blood-vessels will be described later on.

Enumeration of the Red and White Corpuscles.—Several methods are employed for counting the blood-corpuscles, most of them depending upon the same principle, *i.e.*, the dilution of a minute volume of blood with a given volume of a colorless solution similar in specific gravity to blood serum, so that the size and shape of the corpuscles is altered as little as possible. A minute quantity of the well-mixed solution is then taken, examined under the microscope, either in a flattened capillary tube (Malassez) or in a cell (Hayem & Nachet, Gowers) of known capacity, and the number of corpuscles in a measured length of the tube, or in a given area of the cell is counted. The length of the tube and the area of the cell are ascertained by means of a micrometer scale in the microscope ocular; or in the case of Gowers' modification, by the division of the cell area into squares of known size. Having ascertained the number of corpuscles in the diluted blood, it is easy to find out the number in a given volume of normal blood. Gowers' modification of Hayem & Nachet's instrument, called by him "*Hæmacytometer*," appears to be the most convenient form of instrument for counting the corpuscles, and as such will alone be described (Fig. 79). It consists of a small pipette (A), which, when filled up to a mark on its stem, holds 995 cubic millimetres. It is furnished with an india-rubber tube and glass mouth-piece to facilitate filling and emptying; a capillary tube (B) marked to hold 5 cubic millimetres, and also furnished with an india-rubber tube and mouthpiece; a small glass jar (D) in which the dilution of the blood is performed; a glass stirrer (E) for mixing the blood thoroughly, (F) a needle, the length of which can be regulated by a screw; a brass stage plate (c) carrying a glass slide, on which is a cell one-fifth of a millimetre deep, and the bottom of which is divided into one-tenth millimetre squares. On the top of the cell rests the cover glass, which is kept in its place by the pressure of two springs proceeding from the stage plate. A standard saline solution of sodium sulphate, or similar salt, of specific gravity 1025, is made, and 995 cubic millimetres are measured by means of the pipette into the glass jar, and with this five cubic millimetres of blood, obtained by pricking the finger with a needle, and measured in the capillary pipette (B), are thoroughly mixed by the

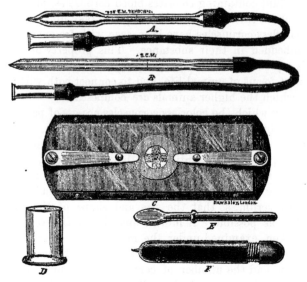

FIG. 79.—Hæmacytometer.

glass stirring-rod. A drop of this diluted blood is then placed in the cell and covered with a cover-glass, which is fixed in position by means of the two lateral springs. The preparation is then examined under a microscope with a power of about 400 diameters, and focussed until the lines dividing the cell into squares are visible.

After a short delay, the red corpuscles which have sunk to the bottom of the cell, and are resting on the squares, are counted in ten squares, and the number of white corpuscles noted. By adding together the numbers counted in ten (one-tenth millimetre) squares the number of corpuscles in one-cubic millimetre of blood is obtained. The average number of corpuscles per each cubic millimetre of healthy blood, according to Vierordt and Welcker, is 5,000,000 in adult men, and rather fewer in women.

Chemical Composition of the Blood in Bulk.—

Water	784
Solids—	
Corpuscles	130
Proteids (of serum)	70
Fibrin (of clot)	2·2
Fatty matters (of serum) . . .	1·4
Inorganic salts (of serum) . . .	6
Gases, kreatin, urea and other extractive matter, glucose and accidental substances .'	6·4—
	216
	1,000

Chemical Composition of the Red Corpuscles.—Analysis of a thousand parts of moist blood corpuscles shows the following as the result:—

Water	688
Solids { Organic	303.88
{ Mineral	8.12—312
	1,000

Of the solids the most important is *Hæmoglobin,* the substance to which the blood owes its color. It constitutes, as will be seen from the appended Table, more than 90 per cent. of the organic matter of the corpuscles. Besides hæmoglobin there are proteid [1] and fatty matters, the former chiefly consisting of *globulins,* and the latter of *cholesterin* and *lecithin.*

In 1000 parts **organic** matter are found:—

Hæmoglobin	905·4
Proteids	86·7
Fats	7·9
	1,000·

Of the **inorganic salts of the corpuscles**, with the iron omitted—

In 1000 parts corpuscles (Schmidt) are found :—

Potassium Chloride		3·679
" **Phosphate**		2·343
" sulphate		·132
Sodium "		·633
Calcium "		·094
Magnesium "		·060
Soda		·341
		7·282

---

[1] An account of the proteid bodies, etc., will be found in the Appendix, and should be referred to for explanation of the terms employed in the text.

The properties of hæmoglobin will be considered in relation to the Gases of the blood.

Chemical Composition of the Colorless Corpuscles.—In consequence of the difficulty of obtaining colorless corpuscles in sufficient number to make an analysis, little is accurately known of their chemical composition; in all probability, however, the stroma of the corpuscles is made up of proteid matter, and the nucleus of *nuclein*, a nitrogenous phosphorus-containing body akin to *mucin*, capable of resisting the action of the gastric juice. The proteid matter (globulin) is soluble in a ten per cent. solution of sodium chloride, and the solution is precipitated on the addition of water, by heat and by the mineral acids. The stroma contains *fatty granules*, and in it also the presence of *glycogen* has been demonstrated. The salts of the corpuscles are chiefly *potassium*, and of these the phosphate is in greatest amount.

Chemical Composition of the Plasma or Liquor Sanguinis.—The liquid part of the blood, the plasma or liquor sanguinis in which the corpuscles float, may be obtained in the ways mentioned under the head of the Coagulation of the Blood. In it are the fibrin factors, inasmuch as when exposed to the ordinary temperature of the air it undergoes coagulation and splits up into fibrin and serum. It differs from the serum in containing fibrinogen, but in appearance and in reaction it closely resembles that fluid; its alkalinity, however, is less than that of the serum obtained from it. It may be freed from white corpuscles by filtration at a temperature below 41°F. (5°C.)

Fibrin.—The part played by fibrin in the formation of a clot has been already described (p. 66), and it is only necessary to consider here its general properties. It is a stringy elastic substance belonging to the proteid class of bodies. It is insoluble in water and in weak saline solutions, it swells up into a transparent jelly when placed in dilute-hydrochloric acid, but does not dissolve, but in strong acid it dissolves, producing acid-albumin;[1] it is also soluble on boiling in strong saline solutions. Blood contains only ·2 per cent. of fibrin. It can be converted by the gastric or pancreatic juice into peptone. It possesses the power of liberating the oxygen from solutions of hydric peroxide H_2O_2. This may be shown by dipping a few shreds of fibrin in tincture of guaiacum and then immersing them in a solution of hydric peroxide. The fibrin becomes of a bluish color, from its having liberated from the solution oxygen, which oxidizes the resin of guaiacum contained in the tincture and thus produces the coloration.

---

[1] The use of the two words *albumen* and *albumin* may need explanation. The former is the *generic* word which may include several albuminous or proteid bodies, *e.g.*, albumen of blood; the latter, which requires to be qualified by another word, is the specific form, and is applied to varieties, *e.g.*, egg-albumin, serum-albumin.

Salts of the Plasma.—In 1000 parts plasma there are:—

Sodium Chloride	5·546
Soda	1·532
Sodium Phosphate	·271
Potassium chloride	·359
" sulphate	·281
Calcium phosphate	·298
Magnesium phosphate	·218
	8.505

Serum.—The serum is the liquid part of the blood or of the plasma remaining after the separation of the clot. It is an alkaline, yellowish, transparent fluid, with a specific gravity of from 1025 to 1032. In the usual mode of coagulation, part of the serum remains in the clot, and the rest, squeezed from the clot by its contraction, lies around it. Since the contraction of the clot may continue for thirty-six or more hours, the quantity of serum in the blood cannot be even roughly estimated till this period has elapsed. There is nearly as much, by weight, of serum as there is clot in coagulated blood.

Chemical Composition of the Serum.—

Water . . . . about	900
Proteids:	
α. Serum-albumin	} 80
β. Paraglobulin	}
Salts.	
Fats—including fatty acids, cholesterin, lecithin; and some soaps	
Grape sugar in small amount	
Extractives—kreatin, kreatinin, urea, etc.	} 20
Yellow pigment, which is independent of hæmoglobin	
Gases—small amounts of oxygen, nitrogen, and carbonic acid	
	1000

Water.—The water of the serum varies in amount according to the amount of food, drink, and exercise, and with many other circumstances.

Proteids.—α. Serum-albumin is the chief proteid found in serum.

It is precipitated on heating the serum to 140° F. (60° C.), and entirely coagulates at (167° F. 75° C.), and also by the addition of strong acids, such as nitric and hydrochloric; by long contact with alcohol it is precipitated. It is not precipitated on addition of ether, and so differs from the other native albumin, viz., *egg*-albumin. When dried at 104°F. (40° C.) serum-albumin is a brittle, yellowish substance, soluble in water, possessing a lævo-rotary power of —56°. It is with great difficulty freed from its salts, and is precipitated by solutions of metallic salts, *e.g.*, of mercuric chloride, copper sulphate, lead acetate, sodium tungstate, etc. If dried at a temperature over 167° F. (75° C.) the residue is insoluble' in water, having been changed into *coagulated proteid*.

β. Paraglobulin can be obtained as a white precipitate from cold serum by adding a considerable excess of water and passing through it a current of carbonic acid gas or by the cautious addition of dilute acetic acid. It can also be obtained by saturating serum with crystallized sulphate magnesium or chloride sodium. When obtained in the latter way precipitation seems to be much more complete than by means of the former method. Paraglobulin belongs to the class of proteids called *globulins*.

The proportion of serum-albumin to paraglobulin in human blood serum is as 1·511 to 1.

The salts of sodium predominate in serum as in plasma, and of these the chloride generally forms by far the largest proportion.

Fats are present partly as fatty acids and partly emulsified. The fats are *triolein, tristearin,* and *tripalmitin.* The amount of fatty matter varies according to the time after, and the ingredients of, a meal. Of *cholesterin* and *lecithin* there are mere traces.

Grape sugar is found principally in the blood of the hepatic vein, about one part in a thousand.

The **extractives** vary from time to time; sometimes uric and hippuric acids are found in addition to urea, kreatin and kreatinin. Urea exists in proportion from ·02 to ·04 per cent.

The yellow *pigment* of the serum and the *odorous* matter which gives the blood of each particular animal a peculiar smell, have not yet been properly isolated.

VARIATIONS IN HEALTHY BLOOD UNDER DIFFERENT CIRCUMSTANCES.

The conditions which appear most to influence the composition of the blood in health are these: Sex, Pregnancy, Age, and Temperament. The composition of the blood is also, of course, much influenced by diet.

1. *Sex.*—The blood of men differs from that of women, chiefly in being of somewhat higher specific gravity, from its containing a relatively larger quantity of red corpuscles.

2. *Pregnancy.*—The blood of pregnant women has a rather lower specific gravity than the average, from deficiency of red corpuscles. The quantity of white corpuscles, on the other hand, and of fibrin, is increased.

3. *Age.*—It appears that the blood of the fœtus is very rich in solid matter, and especially in red corpuscles; and this condition, gradually diminishing, continues for some weeks after birth. The quantity of solid matter then falls during childhood below the average, again rises during adult life, and in old age falls again.

4. *Temperament.*—But little more is known concerning the connection of this with the condition of the blood, than that there appears to be a relatively larger quantity of solid matter, and particularly of red corpuscles, in those of a plethoric or sanguineous temperament.

5. *Diet.*—Such differences in the composition of the blood as are due to the temporary presence of various matters absorbed with the food and drink, as well as the more lasting changes which must result from generous or poor diet respectively, need be here only referred to.

Effects of Bleeding.—The result of bleeding is to diminish the specific gravity of the blood; and so quickly, that in a single venesection, the portion of blood last drawn has often a less specific gravity than that of the blood that flowed first.   This is, of course, due to absorption of fluid from the tissues of the body.   The physiological import of this fact, namely, the instant absorption of liquid from the tissues, is the same as that of the intense thirst which is so common after either loss of blood, or the abstraction from it of watery fluid, as in cholera, diabetes, and the like.

For some little time after bleeding, the want of red corpuscles is well marked; but with this exception, no considerable alteration seems to be produced in the composition of the blood for more than a very short time: the loss of the other constituents, including the pale corpuscles, being very quickly repaired.

VARIATIONS IN THE COMPOSITION OF THE BLOOD, IN DIFFERENT PARTS OF THE BODY.

The composition of the blood, as might be expected, is found to vary in different parts of the body.   Thus arterial blood differs from venous; and although its composition and general characters are uniform throughout the whole course of the systemic arteries, they are not so throughout the venous system,—the blood contained in some veins differing remarkably from that in others.

Differences between Arterial and Venous Blood.—The differences between arterial and venous blood are these:—

(*a.*) Arterial blood is bright red, from the fact that almost all its hæmoglobin is combined with oxygen (Oxyhæmoglobin, or scarlet hæmoglobin), while the purple tint of venous blood is due to the deoxidation of a certain quantity of its oxyhæmoglobin, and its consequent reduction to the purple variety (Deoxidized, or purple hæmoglobin).

(*b.*) Arterial blood coagulates somewhat more quickly.

(*c.*) Arterial blood contains more oxygen than venous, and less carbonic acid.

Some of the veins contain blood which differs from the ordinary standard considerably.   These are the Portal, the Hepatic, and the Splenic veins.

Portal vein.—The blood which the portal vein conveys to the liver is supplied from two chief sources; namely, that in the gastric and mesenterio veins, which contains the soluble elements of food absorbed from the stomach and intestines during digestion, and that in the splenic vein; it must, therefore, combine the qualities of the blood from each of these sources.

The blood in the gastric and mesenteric veins will vary much according to the stage of digestion and the nature of the food taken, and can therefore be seldom exactly the same. Speaking generally, and without considering the sugar, dextrin, and other soluble matters which may have been absorbed from the alimentary canal, this blood appears to be deficient in solid matters, especially in red corpuscles, owing to dilution by the quantity of water absorbed, to contain an excess of albumin, and to yield a less tenacious kind of fibrin than that of blood generally.

The blood from the *splenic vein* is generally deficient in red corpuscles, and contains an unusually large proportion of proteids. The fibrin obtainable from the blood seems to vary in relative amount, but to be almost always above the average. The proportion of colorless corpuscles is also unusually large. The whole quantity of solid matter is decreased, the diminution appearing to be chiefly in the proportion of red corpuscles.

The blood of the *portal vein,* combining the peculiarities of its two factors, the splenic and mesenteric venous blood, is usually of lower specific gravity than blood generally, is more watery, contains fewer red corpuscles, more proteids, and yields a less firm clot than that yielded by other blood, owing to the deficient tenacity of its fibrin.

Guarding (by ligature of the portal vein) against the possibility of an error in the analysis from regurgitation of hepatic blood into the portal vein, recent observers have determined that *hepatic venous blood* contains less water, albumen, and salts, than the blood of the portal vein; but that it yields a much larger amount of extractive matter, in which is one constant element, namely, grape-sugar, which is found, whether saccharine or farinaceous matter have been present in the food or not.

The Gases of the Blood.

The gases contained in the blood are Carbonic acid, Oxygen, and Nitrogen, 100 volumes of blood containing from 50 to 60 volumes of these gases collectively.

Arterial blood contains relatively more oxygen and less carbonic acid than venous. But the absolute quantity of carbonic acid is in both kinds of blood greater than that of the oxygen.

	Oxygen.	Carbonic Acid.	Nitrogen.
Arterial Blood . .	20 vol. per cent.	39 vol. per cent.	1 to 2 vols.
Venous " (from muscles at rest)	8 to 12 " " "	46 " " "	1 to 2 vols.

The Extraction of the Gases from the Blood.—As the ordinary air-pumps are not sufficiently powerful for the purpose, the extraction of the gases from the blood is accomplished by means of a mercurial air-pump, of which there are many varieties, those of Ludwig, Alvergnidt, Geissler, and Sprengel being the chief. The principle of action in all is much the same. Ludwig's pump, which may be taken as a type, is represented in the figure. It consists of two fixed globes, C and F, the upper one communicating by means of the stopcock D, and a stout india-rubber tube with another-glass globe, L, which can be raised or lowered by means of a pulley; it also communicates by means of a stop-cock, B, and a bent glass tube, A, with a gas receiver (not represented in the figure), A dipping into a bowl of mercury, so that the gas may be received over mercury. The lower globe, F, communicates with C by means of the stopcock, E, with I in which the blood is contained by the stopcock G, and with a movable glass globe, M, similar to L, by means of the stopcock, H, and the stout india-rubber tube, K.

In order to work the pump, L and M are filled with mercury, the blood from which the gases are to be extracted is placed in the bulb I, the stopcocks, H, E, D, and B, being open, and G closed. M is raised by means of the pulley until F is full of mercury, and the air is driven out. E is then closed, and L is raised so that C becomes full of mercury, and the air driven off. B is then closed. On lowering L the mercury runs into it from C, and a vacuum is established in C. On opening E and lowering M, a vacuum is similarly established in F; if G be now opened, the blood in I will enter into ebullition, and the gases will pass off into F and C, and on raising M and then L, the stopcock B being opened, the gas is driven through A, and is received into the receiver over mercury. By repeating the experiment several times the whole of the gases of the specimen of blood is obtained, and may be estimated.

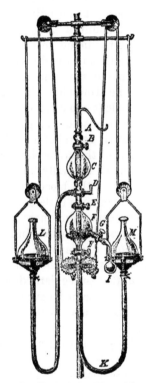

FIG. 80.—Ludwig's Mercurial Pump.

The Oxygen of the Blood.—It has been found that a very small proportion of the oxygen which can be obtained, by the aid of the mercurial pump, from the blood, exists in a state of simple solution in the plasma. If the gas were in simple solution, the amount of oxygen in any given quantity of blood exposed to any given atmosphere ought to vary with the amount of oxygen contained in the atmosphere. Since, speaking generally, the amount of any gas absorbed by a liquid such as plasma would depend upon the proportion of the gas in the atmosphere to which the liquid was exposed — if the proportion were great, the absorption would be great; if small, the absorption would be similarly small. The absorption would continue until the proportion of the gas in the liquid and in the atmosphere became equal. Other things would, of course, influence the absorption, such as the *kind of gas* employed, *nature of the liquid,* and the *temperature* of both, but *cæteris paribus,* the amount of a gas which a liquid absorbs depends upon the proportion of the gas—the so-called partial pressure—of the gas in the atmosphere to which the liquid is subjected. And conversely, if a liquid containing a gas in solution be exposed to an atmosphere containing none of the gas, the gas will be given up to the atmosphere until its amount in the liquid and in the atmosphere becomes equal. This condition is called a condition of equal tensions. The condition may be understood by a simple illustration. A large amount of carbonic acid gas is dissolved in a bottle of water by exposing the liquid to extreme pressure of the gas, and a cork is placed in the bottle and wired down. The gas exists in the water in a condition of extreme tension, and therefore there is a tendency of the gas to escape into the atmosphere, in order that the tension may be relieved; this causes the violent expulsion of the cork when the wire is removed, and if the water be placed in a glass the gas will continue to be evolved until it is almost all got rid of, and the tension of the gas in the water approximates to that of the atmosphere in which, it should be remembered, the carbon dioxide is, naturally, in very small amount, viz., ·04 per cent. Now the oxygen of the blood does not obey this law of pressure. For if blood which contains little or no oxygen be exposed to a succession of atmospheres containing more and more of that gas, we find that the absorption is at first very great, but soon becomes relatively very small, not being therefore regularly in proportion to the increased amount (or tension) of the oxygen of the atmospheres, and that conversely, if arterial blood be submitted to regularly diminishing pressures of oxygen, at first very little of the contained oxygen is given off to the atmosphere, then suddenly the gas escapes with great rapidity, again disobeying the law of pressures.

Very little oxygen can be obtained from serum freed from blood corpuscles, even by the strongest mercurial air-pump, neither can serum be made to absorb a large quantity of that gas; but the small quantity which is so given up or so absorbed follows the laws of absorption according to pressure.

It must be, therefore, evident that the chief part of the oxygen is contained in the corpuscles, and not in a state of simple solution. The chief solid constituent of the colored corpuscles is *hæmoglobin,* which constitutes more than 90 per cent. of their bulk. This body has a very remarkable affinity for oxygen, absorbing it to a very definite extent under favorable circumstances, and giving it up when subjected to the action of reducing agents, or to a sufficiently low oxygen pressure. From these facts it is inferred that the oxygen of the blood is combined with hæmoglobin, and not simply dissolved; but inasmuch as it is comparatively easy to cause the hæmoglobin to give up its oxygen, it is believed that the oxygen is but loosely combined with the substance.

Hæmoglobin.—Hæmoglobin is a crystallizable body which constitutes by far the largest portion of the colored corpuscles. It is intimately distributed throughout their stroma, and must be dissolved out of it before it will undergo crystallization. Its percentage composition is C. 53·85; H. 7·32; N. 16·17; O. 21·84; S. ·63; Fe. ·42; and if the molecule be supposed to contain one atom of iron the formula would be C_{600}, H_{960}, N_{154}, $Fe\ S_3$, O_{179}. The most interesting of the properties of hæmoglobin are its powers of crystallizing and its attraction for oxygen and other gases.

Crystals.—The hæmoglobin of the blood of various animals possesses the power of crystallizing to very different extents (blood-crystals). In some animals the formation of crystals is almost spontaneous, whereas in others crystals are formed either with great difficulty or not at all. Among the animals whose blood coloring-matter crystallizes most readily are the guinea-pig, rat, squirrel, and dog; and in these cases to obtain crystals it is generally sufficient to dilute a drop of recently-drawn blood with water and expose it for a few minutes to the air. Light seems to favor the formation of the crystals. In many instances other means must be adopted, *e.g.*, the addition of alcohol, ether, or chloroform, rapid freezing, and then thawing, an electric current, a temperature of 140° F. (60° C.), or the addition of sodium sulphate.

Human blood crystallizes with difficulty, as does also that of the ox, the pig, the sheep, and the rabbit.

FIG. 81.—Crystals of oxy-hæmoglobin—prismatic from human blood.

The forms of hæmoglobin crystals, as will be seen from the appended figures, differ greatly.

Hæmogloblin crystals are soluble in water. Both the crystals themselves and also their solutions have the characteristic color of arterial blood.

A dilute solution of hæmoglobin gives a characteristic appearance with the spectroscope. Two absorption bands are seen between the solar lines D and E (see Plate), one toward the red, with its middle line some little way to the blue side of D, is very intense, but narrower than the other, which lies near to the red side of E. Each band is darkest in the middle and fades away at the sides. As the strength of the solution increases the bands become broader and deeper, and both the red and the blue ends of the spectrum become encroached upon until the bands coalesce to form one very broad band, and only a slight amount of the green remains unabsolved, and part of the red, and on further increase of strength the former disappears.

If the crystals of oxy-hæmoglobin be subjected to a mercurial air-pump they give off a definite amount of oxygen (1 gramme giving off 1·59

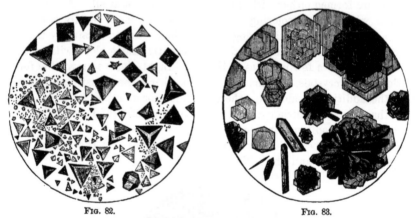

FIG. 82.   FIG. 83.

FIG. 82.—Oxy-hæmoglobin crystals—tetrahedral, from blood of the guinea-pig.
FIG. 83.—Hexagonal oxy-hæmoglobin crystals, from blood of squirrel. On these hexagonal plates, prismatic crystals, grouped in a stellate manner, not unfrequently occur (after Funke).

c. cm. of oxygen), and they become of a purple color; and a solution of oxy-hæmoglobin may be made to give up oxygen and to become purple in a similar manner.

This change may be also effected by passing through it hydrogen or nitrogen gas, or by the action of reducing agents, of which Stokes's fluid[1] is the most convenient.

With the spectroscope a solution of deoxidized hæmoglobin is found to give an entirely different appearance from that of oxidized hæmoglobin. Instead of the two bands at D and E we find a single broader but fainter band occupying a position midway between the two, and at the

[1] *Stokes's Fluid* consists of a solution of *ferrous sulphate*, to which ammonia has been added and sufficient tartaric acid to prevent precipitation. Another reducing agent is a solution of *stannous chloride*, treated in a way similar to the ferrous sulphate, and a third reagent of like nature is an aqueous solution of *ammonium sulphide*.

same time less of the blue end of the spectrum is absorbed. Even in strong solutions this latter appearance is found, thereby differing from the strong solution of oxidized hæmoglobin which lets through only the red and orange rays; accordingly to the naked eye the one (reduced hæmoglobin solution) appears purple, the other (oxy-hæmoglobin solution) red. The deoxidized crystals or their solutions quickly absorb oxygen on exposure to the air, becoming scarlet. If solutions of blood be taken instead of solutions of hæmoglobin, results similar to the whole of the foregoing can be obtained.

Venous blood never, except in the last stages of asphyxia, fails to show the oxy-hæmoglobin bands, inasmuch as the greater part of the hæmoglobin even in venous blood exists in the more highly oxidized condition.

Action of Gases on Hæmoglobin.—*Carbonic oxide*, passed through a solution of hæmoglobin, causes it to assume a bluish color, and the spectrum is slightly altered; two bands are still visible, but are somewhat nearer the blue end than those of oxy-hæmoglobin (see Plate). The amount of carbonic oxide is equal to the amount of the oxygen displaced. Although the carbonic oxide gas readily displaces oxygen, the reverse is not the case, and upon this property depends the dangerous effect of coal gas poisoning. Coal gas contains much carbonic oxide, and this at once, when breathed, combines with the hæmoglobin of the blood, producing a compound which cannot easily be reduced, and since it is by no means an oxygen carrier, death may result from suffocation from want of oxygen notwithstanding the free entry into the lungs of pure air. Crystals of carbonic-oxide hæmoglobin closely resemble those of oxyhæmoglobin.

Nitric oxide produces a similar compound to the carbonic-oxide hæmoglobin, which is even less easily reduced.

Nitrous oxide reduces oxyhæmoglobin, and therefore leaves the reduced hæmoglobin in a condition to actively take up oxygen.

Sulphuretted Hydrogen.—If this gas be passed through a solution of oxyhæmoglobin, the hæmoglobin is reduced and an additional band appears in the red. If the solution be then shaken with air, the two bands of oxyhæmoglobin replace that of reduced hæmoglobin, but the band in the red persists.

Products of the Decomposition of Hæmoglobin.

Methæmoglobin.—If an aqueous solution of oxyhæmoglobin be exposed to the air for some time, its spectrum undergoes a change; the two D and E bands become faint, and a new line in the red at c is developed. The solution, too, has become brown and acid in reaction, and is precipitable by basic lead acetate. This change is due to the decomposition of hæmoglobin, and to the production of *methæmoglobin*. On adding ammonium sulphide, reduced hæmoglobin is produced, and on shaking this up with air, oxyhæmoglobin is reproduced.

Hæmatin.—By the action of heat, or of acids or alkalies in the presence of oxygen, hæmoglobin can be split up into a substance called *Hæmatin*, which contains all the iron of the hæmoglobin from which it was derived, and a proteid residue. Of the latter it is impossible to say more than that it is probably made up of one or more bodies of the globulin class. If there be no oxygen present, instead of hæmatin a body called *hæmochromogen* is produced, which, however, will speedily undergo oxidation into hæmatin.

Hæmatin is a dark brownish or black non-crystallizable substance of metallic lustre. Its percentage composition is C. 64·30; H. 5·50; N. 9·06; Fe, 8·82; O. 12·32; which gives the formula C_{68}, H_{70}, N_8, Fe_2, O_{10} (Hoppe-Seyler). It is insoluble in water, alcohol, and ether; soluble in the caustic alkalies; soluble with difficulty in hot alcohol to which is added sulphuric acid. The iron may be removed from hæmatin by heating it with fuming hydrochloric acid to 320° F. (160° C.), and a new body, *hæmatoporphyrin*, is produced.

In acid solution.—If to blood an excess of acetic acid be added, the color alters to brown from decomposition of hæmoglobin, and the setting free of hæmatin; by shaking this solution with ether, solution of the hæmatin is obtained. The spectrum of the etherial solution shows no less than four absorption bands, viz., one in the red between c and D, one faint and narrow close to D, and then two broader bands, one between D and E, and another nearly midway between b and F. The first band is by far the most distinct, and the acid solution of hæmatin without ether shows it plainly.

In alkaline solution.—The absorption band is still in the red, but nearer to D, and the blue end of the spectrum is partially absorbed to a considerable extent. If a reducing agent be added, two bands resembling those of oxyhæmoglobin, but nearer to the blue, appear; this is the spectrum of *reduced hæmatin*. On shaking the reduced hæmatin with air or oxygen the two bands are replaced by the single band of alkaline hæmatin.

Hæmatoidin.—This substance is found in the form of yellowish crystals in old blood extravasations, and is derived from the hæmoglobin. Their crystalline form and the reaction they give with nitric acid seem to show them to be identical with *Bilirubin*, the chief coloring matter of the Bile.

Hæmin.—One of the most important derivatives of næmatin is Hæmin. It is usually called *Hydrochlorate of Hæmatin* (or hydrochloride), but its exact chemical composition is uncertain. Its formula is C_{68}, H_{70}, N_8, Fe_2, O_{10}, 2 Hcl, and it contains 5·18 per cent. of chlorine, but by some it is looked upon as simply crystallized hæmatin. Although difficult to obtain in bulk, a specimen may be easily made for the microscope in the following way:—A small drop of dried blood is finely powdered with a few crystals of common salt on a glass slide, and spread out; a cover glass is then plàced upon it, and glacial acetic acid added by means of a capillary pipette. The blood at once turns of a brownish color. The slide is then heated, and the acid mixture evaporated to dryness at a high temperature. The excess of salt is washed away with water from the dried residue, and the specimen may then be mounted. A large number of small, dark, reddish black crystals of a rhombic shape, sometimes arranged in bundles, will be seen if the slide be subjected to microscopic examination.

The formation of these hæmin crystals is of great interest and importance from a medico-legal point of view, as it constitutes the most cer-

Fig. 84.—Hæmatoidin crystals. (Frey.)   Fig. 85.—Hæmin crystals. (Frey.)

tain and delicate test we have for the presence of blood (not of necessity the blood of man) in a stain on clothes, etc. It exceeds in delicacy even the spectroscopic test.

Estimation of Hæmoglobin.—The most exact method is by the estimation of the amount of iron in a given specimen of blood, but as this is a somewhat complicated process, a method has been proposed which, though not so exact, has the advantage of simplicity. This consists in comparing the color of a given small amount of diluted blood with glycerine jelly tinted with carmine and picrocarmine to represent a standard solution of blood diluted one hundred times. The amount of dilution which the given blood requires will thus approximately represent the quantity of hæmoglobin it contains. (Gowers.)

Distribution of Hæmoglobin.—In connection with the ascertained function of hæmoglobin as the great oxygen-carrier, the following facts with regard to its distribution are of importance.

It occurs not only in the red blood-cells of all Vertebrata (except one fish (leptocephalus) whose blood-cells are all colorless), but also in similar cells in many Worms: moreover, it is found diffused in the vascular fluid of some other worms and certain Crustacea; it also occurs in all the striated muscles of Mammals and Birds. It is generally absent from unstriated muscle except that of the rectum. It has also been found in Mollusca in certain muscles which are specially active, viz., those which work the rasp-like tongue.

In the muscles of Fish it has hitherto only been met with in the very active muscle which moves the dorsal fin of the Hippocampus (Ray Lan-kester).

The Carbon Dioxide Gas in the Blood.—Of this gas in the blood, part exists in a state of simple solution in the serum, and the rest in a state of weak chemical combination. It is believed that the latter is combined with the sodium carbonate in a condition of bicarbonate. Some observers consider that part of the gas is associated with the cor-puscles.

The Nitrogen in the Blood.—It is believed that the whole of the small quantity of the nitrogen contained in the blood is simply dissolved in the fluid plasma.

DEVELOPMENT OF THE BLOOD.

The first formed blood-corpuscles of the human embryo differ much in their general characters from those which belong to the later periods

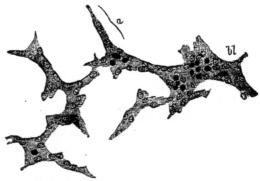

FIG. 86.—Part of the network of developing blood-vessels in the vascular area of a guinea-pig. *bl*, blood corpuscles becoming free in an enlarged and hollowed out part of the network; *a*, process of protoplasm. (E. A. Schäfer.)

of intra-uterine, and to all periods of extra-uterine life. Their manner of origin is at first very simple.

Surrounding the early embryo is a circular area, called the vascular area, in which the first rudiments of the blood-vessels and blood-corpuscles are developed. Here the nucleated embryonal cells of the mesoblast, from which the blood-vessels and corpuscles are to be formed, send out processes in various directions, and these joining together, form an irregular meshwork. The nuclei increase in number, and collect chiefly in the larger masses of protoplasm, but partly also in the processes. These nuclei gather around them a certain amount of the protoplasm, and becoming colored, form the red blood corpuscles. The protoplasm of the cells and their branched network in which these corpuscles lie then becomes hollowed out into a system of canals enclosing fluid, in which the red nucleated corpuscles float. The corpuscles at first are from about $\frac{1}{2500}$ to $\frac{1}{1500}$ of an inch in diameter, mostly spherical, and with granular contents, and a well-marked nucleus. Their nuclei, which are about $\frac{1}{5000}$ of an inch in diameter, are central, circular, very little prominent on the surfaces of the corpuscle, and apparently slightly granular or tuberculated.

The corpuscles then strongly resemble the colorless corpuscles of the fully developed blood, but are colored. They are capable of amœboid movement and multiply by division.

When, in the progress of embryonic development, the liver begins to be formed, the multiplication of blood-cells in the whole mass of blood ceases, and new blood-cells are produced by this organ, and also by the lymphatic glands, thymus and spleen. These are at first colorless and nucleated, but afterward acquire the ordinary blood-tinge, and resemble very much those of the first set. They also multiply by division. In whichever way produced, however, whether from the original formative cells of the embryo, or by the liver and the other organs mentioned above, these colored nucleated cells begin very early in fœtal life to be mingled with colored *non*-nucleated corpuscles resembling those of the adult, and at about the fourth or fifth month of embryonic existence are completely replaced by them.

Origin of the Mature Red Corpuscles.—The non-nucleated red corpuscles may possibly be derived from the nucleated, but in all probability are an entirely new formation, and the methods of their origin are

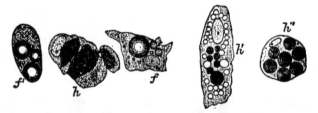

Fig. 87.—Development of red corpuscles in connective-tissue cells. From the subcutaneous tissue of a new-born rat. *h*, a cell containing hæmoglobin in a diffused form in the protoplasm; *h'*, one containing colored globules of varying size and vacuoles; *h"*, a cell filled with colored globules of nearly uniform size; *f, f'*, developing fat cells. (E. A. Schäfer.)

the following:—(1.) During fœtal life and possibly in some animals, *e.g.*, the rat, which are born in an immature condition, for some little time after birth, the blood discs arise in the connective tissue cells in the following way. Small globules, of varying size, of coloring matter arise in the protoplasm of the cells, and the cells themselves become branched, their branches joining the branches of similar cells. The cells next become vacuolated, and the red globules are free in a cavity filled with fluid (Fig. 88); by the extension of the cavity of the cells into their processes anastomosing vessels are produced, which ultimately join with the previously existing vessels, and the globules, now having the size and appearance of the ordinary red corpuscles, are passed into the general circulation. This method of formation is called *intracellular* (Schäfer).

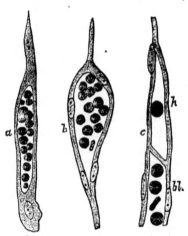

FIG. 88.—Further development of blood-corpuscles in connective-tissue cells and transformation of the latter into capillary blood-vessels. *a,* an elongated cell with a cavity in the protoplasm occupied by fluid and by blood-corpuscles which are still globular; *b,* a hollow cell, the nucleus of which has multiplied. The new nuclei are arranged around the wall of the cavity, the corpuscles in which have now become discord; *c,* shows the mode of union of a "hæmapoietic" cell, which, in this instance, contains only one corpuscle, with the prolongation (*bl*) of a previously existing vessel; *a* and. *c,* from the new-born rat; *b,* from the fœtal sheep. (E. A. Schäfer.)

(2.) *From the white corpuscles.*—The belief that the red corpuscles are derived from the white is still very general, although no new evidence has been recently advanced in favor of this view. It is, however, uncertain whether the nucleus of the white corpuscle becomes the red corpuscle, or whether the whole white corpuscle is bodily converted into the red by the gradual clearing up of its contents with a disappearance of the nucleus. Probably the latter view is the correct one.

FIG. 89.—Colored nucleated corpuscles, from the red marrow of the guinea-pig. (E. A. Schäfer.)

(3.) *From the medulla of bones.*—Red corpuscles are to a very large extent derived during adult life from the large pale cells in the red marrow of bones, especially of the ribs (Figs. 44, 89). These cells become colored from the formation of hæmoglobin chiefly in one part of their protoplasm. This colored part becomes separated from the rest of the cell and forms a red corpuscle, being at first cup-shaped, but soon taking on the normal appearance of the mature corpuscle. It is supposed that the protoplasm may grow up again and form a number of red corpuscles in a similar way.

(4.) *From the tissue of the spleen.*—It is probable that red as well as white corpuscles may be produced in the spleen.

(5.) *From Microcytes.*—Hayem describes the small particles (microcytes), previously mentioned as contained in the blood (p. 75), and which he calls hæmatoblasts, as the precursors of the red corpuscles. They acquire color, and enlarge to the normal size of red corpuscles.

Without doubt, the red corpuscles have, like all other parts of the organism, a tolerably definite term of existence, and in a like manner die and waste away when the portion of work allotted to them has been performed. Neither the length of their life, however, nor the fashion of their decay has been yet clearly made out. It is generally believed that a certain number of the red corpuscles undergo disintegration in the spleen; and indeed corpuscles in various degrees of degeneration have been observed in this organ.

Origin of the Colorless Corpuscles.—The colorless corpuscles of the blood are derived from the lymph corpuscles, being, indeed, indistinguishable from them; and these come chiefly from the lymphatic glands. Their number is increased by division.

Colorless corpuscles are also in all probability derived from the spleen and thymus, and also from the germinating endothelium of serous membranes, and from connective tissue. The corpuscles are carried into the blood either with the lymph and chyle, or pass directly from the lymphatic tissue in which they have been formed into the neighboring blood-vessels.

USES OF THE BLOOD.

1. To be a medium for the reception and storing of matter (ordinary food, drink, and oxygen) from the outer world, and for its conveyance to all parts of the body.

2. To be a source whence the various tissues of the body may take the materials necessary for their nutrition and maintenance; and whence the secreting organs may take the constituents of their various secretions.

3. To be a medium for the absorption of refuse matters from all the tissues, and for their conveyance to those organs whose function it is to separate them and cast them out of the body.

4. To warm and moisten all parts of the body.

USES OF THE VARIOUS CONSTITUENTS OF THE BLOOD.

Albumen.—Albumen, which exists in so large a proportion among the chief constituents of the blood, is without doubt mainly for the nourishment of those textures which contain it or other compounds nearly allied to it.

Fibrin.—In considering the functions of fibrin, we may exclude the notion of its existence, as such, in the blood in a fluid state, and of its use in the nutrition of certain special textures, and look for the explanation of its functions to those circumstances, whether of health or disease, under which it is produced. In hæmorrhage, for example, the formation of fibrin in the clotting of blood, is the means by which, at least for a time, the bleeding is restrained or stopped; and the material or *blastema* which is produced for the permanent healing of the injured part, contains a coagulable material identical, or very nearly so, with the fibrin of clotted blood.

Fatty matters.—The fatty matters of the blood subserve more than one purpose. For while they are the means, in part, by which the fat of the body, so widely distributed in the proper adipose and other textures, is replenished, they also, by their union with oxygen, assist in maintaining the temperature of the body. To certain secretions also, notably the milk and bile, fat is contributed.

Saline Matter.—The uses of the saline constituents of the blood are, first, to enter into the composition of such textures and secretions as naturally contain them, and, secondly, to assist in preserving the due specific gravity and alkalinity of the blood, and in preventing its decomposition. The phosphate and carbonate of sodium, to which the blood owes its alkaline reaction, increase the absorptive power of the serum for gases.

Corpuscles.—The important use of the red corpuscles is in relation to the absorption of oxygen in the lungs, and its conveyance to the tissues. How far the red corpuscles are actually concerned in the nutrition of the tissues is quite unknown.

The relation of the colorless corpuscles to the coagulation of the blood has been already considered; of their functions, other than are concerned in this phenomenon, and in the regeneration of the red corpuscles, nothing is positively known.

CHAPTER V.

THE CIRCULATION OF THE BLOOD.

THE Heart is a hollow muscular organ containing four chambers, two auricles and two ventricles, arranged in pairs. On each side (right and left) of the heart is an auricle joined to and communicating with a ventricle, but the chambers on the right side do not directly communicate with those on the left side. The circulation of the blood is chiefly

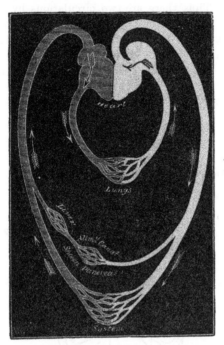

FIG. 90.—Diagram of the Circulation.

carried on by the contraction of the muscular walls of these chambers of the heart, the auricles contracting simultaneously, and their contraction being followed by the simultaneous contraction of the ventricles. The blood is conveyed away from the left side of the heart by the *arteries*, and returned to the right side of the heart by the *veins*, the arteries and veins being continuous with each other at one end by means of the heart, and at the other by a fine network of vessels called the *capillaries*. The

blood, therefore, in its passage from the heart passes first into the arteries, then into the capillaries, and lastly into the veins, by which it is conveyed back again to the heart, thus completing a *revolution* or *circulation.*

The right side of the heart does not directly communicate with the left to complete the entire circulation, but the blood has to pass from the right side to the lungs, through the pulmonary artery, then through the pulmonary capillary-vessels and through the pulmonary veins to the left side of the heart. Thus there are two circulations by which the blood *must* pass; the one, a shorter circuit from the right side of the heart to the lungs and back again to the left side of the heart; the other and larger circuit, from the left side of the heart to all parts of the body and back again to the right side; but more strictly speaking, there is only *one* complete circulation, which may be diagrammatically represented by a double loop, as in the accompanying figure (Fig. 90).

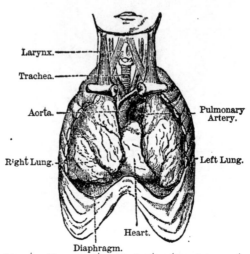

Fig. 91.—View of heart and lungs in situ. The front portion of the chest-wall, and the outer or *parietal* layers of the pleuræ and pericardium have been removed. The lungs are partly collapsed.

On reference to this figure, and noticing the direction of the arrows, which represent the course of the stream of blood, it will be observed that while there is a smaller and a larger circle, both of which pass through the heart, yet that these are not distinct, one from the other, but are formed really by one continuous stream, the whole of which must, at one part of its course, pass through the lungs. Subordinate to the two principal circulations, the *Pulmonary* and *Systemic,* as they are named, it will be noticed also in the same figure that there is another, by which a portion of the stream of blood having been diverted once into the capillaries of the intestinal canal, and some other organs, and gathered up again into a single stream, is a second time divided in its passage through the liver, before it finally reaches the heart and completes a revolution. This subordinate stream through the liver is called the *Portal* circulation.

The Forces concerned in the Circulation of the Blood.—(1) The principal force provided for constantly moving the blood through the course of the circulation is that of the muscular substance of the heart; other assistant forces are (2) those of the elastic walls of the arteries, (3) the pressure of the muscles among which some of the veins run, (4) the movements of the walls of the chest in respiration, and probably, to some extent, (5) the interchange of relations between the blood and the tissues which occurs in the capillary system during the nutritive processes.

THE HEART.

The Pericardium.—The heart is invested by a membranous sac—the *pericardium*, which is made up of two distinct parts, an *external* fibrous membrane, composed of closely interlacing fibres, which has its base attached to the diaphragm—both to the central tendon and to the adjoining muscular fibres, while the smaller and upper end is lost on the large blood-vessels by mingling its fibres with that of their external coats; and an *internal* serous layer, which not only lines the fibrous sac, but also is reflected on to the heart, which it completely invests. The part which lines the fibrous membrane is called the parietal layer, and that enclosing the heart, the visceral layer, and these being continuous for a short distance along the great vessels of the base of the heart, form a closed sac, the cavity of which in health contains just enough fluid to lubricate the two surfaces, and thus enable them to glide smoothly over each other during the movements of the heart. Most of the vessels passing in and out of the heart receive more or less investment from this sac.

The heart is situated in the chest behind the sternum and costal cartilages, being placed obliquely from right to left, quite two-thirds to the left of the mid-sternal line. It is of pyramidal shape, with the apex pointing downward, outward, and toward the left, and the base backward, inward, and toward the right. It rests upon the diaphragm, and its pointed apex, formed exclusively of the left side of the heart, is in contact with the chest wall, and during life beats against it at a point called the *apex beat*, situated in the fifth intercostal space, about two inches below the left nipple, and an inch and a half to the sternal side. The heart is suspended in the chest by the large vessels which proceed from its base, but, excepting the base, the organ itself lies free in the sac of the pericardium. The part which rests upon the diaphragm is flattened, and is known as the *posterior* surface, whilst the free upper part is called the *anterior* surface. The margin toward the left is thick and obtuse, whilst the lower margin toward the right is thin and acute.

On examination of the external surface the division of the heart into parts which correspond to the chambers inside of it may be traced, for a deep transverse groove called the auriculo-ventricular groove divides the auricles which form the base of the heart from the ventricles which form the remainder, including the apex, the ventricular portion being by far the greater; and, again, the inter-ventricular groove runs between the

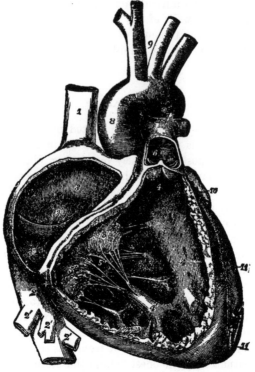

Fig. 92.—The right auricle and ventricle opened, and a part of their right and anterior walls removed, so as to show their interior. ⅙.—1, superior vena cava; 2, inferior vena cava; 2', hepatic veins cut short; 3, right auricle; 3', placed in the fossa ovalis, below which is the Eustachian valve; 3', is placed close to the aperture of the coronary vein; +, +, placed in the auriculo-ventricular groove, where a narrow portion of the adjacent walls of the auricle and ventricle has been preserved; 4, 4, cavity of the right ventricle, the upper figure is immediately below the semilunar valves; 4', large columna carnea or musculus papillaris; 5, 5', 5', tricuspid valve; 6, placed in the interior of the pulmonary artery, a part of the anterior wall of that vessel having been removed, and a narrow portion of it preserved at its commencement, where the semilunar valves are attached; 7, concavity of the aortic arch close to the cord of the ductus arteriosus; 8, ascending part or sinus of the arch covered at its commencement by the auricular appendix and pulmonary artery; 9, placed between the innominate and left carotid arteries; 10, appendix of the left auricle; 11, 11, the outside of the left ventricle, the lower figure near the apex. (Allen Thomson.)

ventricles both front and back, and separates the one from the other. The anterior groove is nearer the left margin and the posterior nearer the right, as the front surface of the heart is made up chiefly of the right ventricle and the posterior surface of the left ventricle. In the furrows run the coronary vessels, which supply the tissue of the heart itself with blood, as well as nerves and lymphatics imbedded in more or less fatty tissue.

The Chambers of the Heart.—The interior of the heart is divided by a partition in such a manner as to form two chief chambers or cavities —right and left. Each of these chambers is again subdivided into an upper and a lower portion, called respectively, as already incidentally mentioned, auricle and ventricle, which freely communicate one with the other; the aperture of communication, however, being guarded by valves, so disposed as to allow blood to pass freely from the auricle into the ventriole, but not in the opposite direction. There are thus four cavities altogether in the heart—two auricles and two ventricles; the auricle and ventricle of one side being quite separate from those of the other (Fig. 90).

Right Auricle.—The right auricle is situated at the right part of the base of the heart as viewed from the front. It is a thin walled cavity of more or less quadrilateral shape prolonged at one corner into a tongue-shaped portion, the right auricular *appendix*, which slightly overlaps the exit of the great artery, the aorta, from the heart.

The interior is smooth, being lined with the general lining of the heart, the *endocardium*, and into it open the superior and inferior venæ cavæ, or great veins, which convey the blood from all parts of the body to the heart. The former is directed downward and forward, the latter upward and inward; between the entrances of these vessels is a slight tubercle called *tubercle of Lower*. The opening of the inferior cava is protected and partly covered by a membrane called the *Eustachian valve*. In the posterior wall of the auricle is a slight depression called the *fossa ovalis*, which corresponds to an opening between the right and left auricles which exists in fœtal life. The right auricular appendix is of oval form, and admits three fingers. Various veins, including the coronary *sinus*, or the dilated portion of the right coronary vein, open into this chamber. In the appendix are closely set elevations of the muscular tissue covered with endocardium, and on the anterior wall of the auricle are similar elevations arranged parallel to one another, called *musculi pectinati*.

Right Ventricle.—The right ventricle occupies the chief part of the anterior surface of the heart, as well as a small part of the posterior surface: it forms the right margin of the heart. It takes no part in the formation of the apex. On section its cavity, in consequence of the encroachment upon it of the septum ventriculorum, is semilunar or crescentic (Fig. 94); into it are two openings, the auriculo-ventricular at the base, and the opening of the pulmonary artery also at the base, but more to the left; the part of the ventricle leading to it is called the *conus arteriosus* or *infundibulum;* both orifices are guarded by valves, the former called *tricuspid* and the latter *semilunar* or *sigmoid*. In this ventricle are also the projections of the muscular tissue called *columnæ carneæ* (described at length p. 110).

Left Auricle.—The left auricle is situated at the left and posterior part of the base of the heart, and is best seen from behind. It is quadrilateral, and receives on either side two pulmonary veins. The auricular appendix is the only part of the auricle seen from the front, and corre-

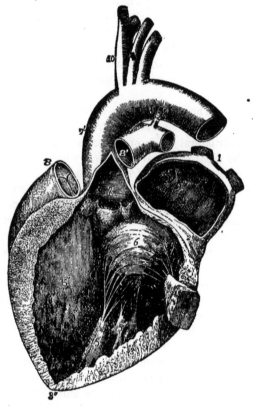

Fig. 93.—The left auricle and ventricle opened, and a part of their anterior and left walls removed. ½.—The pulmonary artery has been divided at its commencement; the opening into the left ventricle carried a short distance into the aorta between two of the segments of the semilunar valves, and the left part of the auricle with its appendix has been removed. The right auricle is out of view. 1, the two right pulmonary veins cut short; their openings are seen within the auricle; 1′, placed within the cavity of the auricle on the left side of the septum and on the part which forms the remains of the valve of the foramen ovale, of which the crescentic fold is seen toward the left hand of 1′; 2, a narrow portion of the wall of the auricle and ventricle preserved round the auriculo-ventricular orifice; 3, 3′, the cut surface of the walls of the ventricle, seen to become very much thinner toward 3″, at the apex; 4, a small part of the anterior wall of the left ventricle which has been preserved with the principal anterior columna carnea or musculus papillaris attached to it; 5, 5, musculi papillares; 5′, the left side of the septum, between the two ventricles, within the cavity of the left ventricle; 6, 6′, the mitral valve; 7, placed in the interior of the aorta near its commencement and above the three segments of its semilunar valve which are hanging loosely together; 7′, the exterior of the great aortic sinus; 8, the root of the pulmonary artery and its semilunar valves; 8′. the separated portion of the pulmonary artery remaining attached to the aorta by 9,the cord of the ductus arteriosus; 10, the arteries rising from the summit of the aortic arch. (Allen Thomson.)

sponds with that on the right side, but is thicker, and the interior is more smooth. The left auricle is only slightly thicker than the right, the difference being as 1¼ lines to 1 line. The left auriculo-ventricular orifice is oval, and a little smaller than that on the right side of the heart.

There is a slight vestige of the foramen between the auricles, which exists in fœtal life, on the septum between them.

Left Ventricle.—Though taking part to a comparatively slight extent in the anterior surface, the left ventricle occupies the chief part of the posterior surface. In it are two openings very close together, viz. the auriculo-ventricular and the aortic, guarded by the valves corresponding to those of the right side of the heart, viz. the *bicuspid* or *mitral* and the *semilunar* or *sigmoid.* The first opening is at the left and back part of the base of the ventricle, and the aortic in front and toward the right. In this ventricle, as in the right, are the columnæ carneæ, which are smaller but more closely reticulated. They are chiefly found near the apex and along the posterior wall. They will be again

F'g. 94.—Transverse section of bullock's heart in a state of cadaveric rigidity. *a*, cavity of left ventricle. *b*, cavity of right ventricle. (Dalton.)

referred to in the description of the valves. The walls of the left ventriole, which are nearly half an inch in thickness, are, with the exception of the apex, twice or three times as thick as those of the right.

Capacity of the Chambers.—The *capacity* of the two ventricles is about four to six ounces of blood, the whole of which is impelled into their respective arteries at each contraction. The capacity of the auricles is rather less than that of the ventricles: the thickness of their walls is considerably less. The latter condition is adapted to the small amount of force which the auricles require in order to empty themselves into their adjoining ventricles; the former to the circumstance of the ventricles being partly filled with blood before the auricles contract.

Size and Weight of the Heart.—The heart is about 5 inches long, 3½ inches greatest width, and 2½ inches in its extreme thickness. The average weight of the heart in the adult is from 9 to 10 ounces; its weight gradually increasing throughout life till middle age; it diminishes in old age.

Structure.—The walls of the heart are constructed almost entirely of layers of muscular fibres; but a ring of connective tissue, to which some of the muscular fibres are attached, is inserted between each auricle and ventricle, and forms the boundary of the *auriculo-ventricular* opening. Fibrous tissue also exists at the origins of the pulmonary artery and aorta.

The muscular fibres of each auricle are in part continuous with those of the other, and partly separate; and the same remark holds true for the ventricles. The fibres of the auricles are, however, quite separate from those of the ventricles, the bond of connection between them being only the fibrous tissue of the auriculo-ventricular openings.

The muscular fibres of the heart, unlike those of most of the involuntary muscles, are striated; but although, in this respect, they resemble the skeletal muscles, they have distinguishing characteristics of their own. The fibres which lie side by side are united at frequent intervals by short branches (Fig. 95). The fibres are smaller than those of the ordinary striated muscles, and their striation is less marked. No sarcolemma can be discerned. The muscle-corpuscles are situate in the middle of the substance of the fibre; and in correspondence with these the fibres appear under certain conditions subdivided into oblong portions or "cells," the off-sets from which are the means by which the fibres anastomose one with another (Fig. 96).

Endocardium.—As the heart is clothed on the outside by a thin transparent layer of pericardium, so its cavities are lined by a smooth and

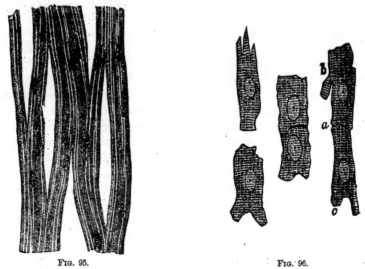

FIG. 95.   FIG. 96.

FIG. 95.—Network of muscular fibres (striated) from the heart of a pig. The nuclei of the muscle-corpuscles are well shown. × 450. (Klein and Noble Smith.)
FIG. 96.—Muscular fibre cells from the heart. (E. A. Shäfer.)

shining membrane, or *endocardium*, which is directly continuous with the internal lining of the arteries and veins. The endocardium is composed of connective tissue with a large admixture of elastic fibres; and on its inner surface is laid down a single tessellated layer of flattened endothelial cells. Here and there unstriped muscular fibres are sometimes found in the tissue of the endocardium.

Course of the Blood through the Heart.—The arrangement of the heart's valves is such that the blood can pass only in one direction, and this is as follows (Fig. 97):—From the right auricle the blood passes into the right ventricle, and thence into the pulmonary artery, by which it is conveyed to the capillaries of the lungs. From the lungs the blood, which is now purified and altered in color, is gathered by the pulmonary veins and taken to the left auricle. From the left auricle it passes into the left ventricle, and thence into the aorta, by which it is distributed to the capillaries of every portion of the body. The branches of the aorta, from being distributed to the general system, are called *systemic* arteries; and from these the blood passes into the *systemic* capillaries, where it again becomes dark and impure, and thence into the branches of the *systemic* veins, which, forming by their union two large trunks, called the superior and inferior vena cava, discharge their contents into the right auricle, whence we supposed the blood to start.

The Valves of the Heart.—The valve between the right auricle and ventricle is named *tricuspid* (5, Fig. 99), because it presents *three* principal cusps or subdivisions, and that between the left auricle and ven-

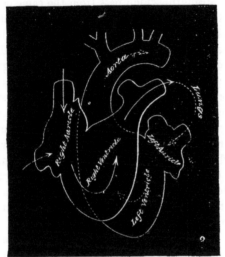

Fig. 97.—Diagram of the circulation through the heart. (Dalton.)

tricle *bicuspid* (or mitral), because it has *two* such portions (6, Fig. 93). But in both valves there is between each two principal portions a smaller one; so that more properly, the tricuspid may be described as consisting of six, and the mitral of four, portions. Each portion is of triangular form, its apex and sides lying free in the cavity of the ventricle, and its base, which is continuous with the bases of the neighboring portions, so as to form an annular membrane around the auriculo-ventricular opening, being fixed to a tendinous ring which encircles the orifice between the auricle and ventricle and receives the insertions of the muscular fibres of both. In each principal cusp may be distinguished a middle-piece, extending from its base to its apex, and including about half its width, which is thicker, and much tougher and tighter than the border-pieces or edges.

While the bases of the several portions of the valves are fixed to the tendinous rings, their ventricular surfaces and borders are fastened by slender tendinous fibres, the *chordæ tendineæ*, to the walls of the ventricles, the muscular fibres of which project into the ventricular cavity in the form of bundles or columns—the *columnæ carneæ*. These columns are not all of them alike, for while some of them are attached along their whole length on one side and by their extremities, others are attached only by their extremities; and a third set, to which the name *musculi papillares* has been given, are attached to the wall of the ventricle by one extremity only, the other projecting, papilla-like, into the cavity of the ventricle (5, Fig. 93), and having attached to it *chordæ tendineæ*. Of the tendinous cords, besides those which pass from the walls of the ventricle and the musculi papillares to the margins of the valves, there are some of especial strength, which pass from the same parts to the edges of the middle and thicker portions of the cusps before referred to. The ends of these cords. are spread out in the substance of the valve, giving its middle piece its peculiar strength and toughness; and from the sides numerous other more slender and branching cords are given off, which are attached all over the ventricular surface of the adjacent border-pieces of the principal portions of the valves, as well as to those smaller portions which have been mentioned as lying between each two principal ones. Moreover, the musculi papillares are so placed that, from the summit of each, tendinous cords proceed to the adjacent halves of two of the principal divisions, and to one intermediate or smaller division, of the valve.

The preceding description applies equally to the mitral and tricuspid valve; but it should be added that the mitral is considerably thicker and stronger than the tricuspid, in accordance with the greater force which it is called upon to resist.

It has been already said that while the ventricles communicate, on the one hand, with the auricles, they communicate, on the other, with the large arteries which convey the blood away from the heart; the right ventricle with the pulmonary artery (6, Fig. 93), which conveys blood to the lungs, and the left ventricle with the aorta, which distributes it to the general system (7, Fig. 93). And as the auriculo-ventricular orifice is guarded by valves, so are also the mouths of the pulmonary artery, and aorta (Figs. 93, 99).

The semilunar valves, three in number, guard the orifice of each of these two arteries. They are nearly alike on both sides of the heart; but those of the aorta are altogether thicker and more strongly constructed than those of the pulmonary artery, in accordance with the greater pressure which they have to withstand. Each valve is of semilunar shape, its convex margin being attached to a fibrous ring at the place of junction of the artery to the ventricle, and the concave or nearly straight border being free, so that each valve forms a little pouch like a watch-pocket (7, Fig. 93). In the centre of the free edge of the valve, which contains a fine cord of fibrous tissue, is a small fibrous nodule, the *corpus Arantii*, and from this and from the attached border fine fibres extend into every part of the mid substance of the valve, except a small lunated space just within the free edge, on each side of the *corpus Arantii.* Here the valve is thinnest, and composed of little more than the endocardium. Thus constructed and attached, the three semilunar valves are placed side by side around the arterial orifice of each ventricle, so as to form three little pouches, which can be separated by the blood passing out of the ventricle, but which immediately afterward are pressed together so as to prevent any return (7, Fig. 93, and 7, Fig. 99). This will be again referred to. Opposite each of the semilunar cusps, both in the aorta and pulmonary artery, there is a bulging outward of the wall of the vessel: these bulgings are called the *sinuses of Valsalva.*

Structure of the Valves.—The valves of the heart are formed essentially of thick layers of closely woven connective and elastic tissue, over which, on every part, is reflected the endocardium.

THE ACTION OF THE HEART.

The heart's action in propelling the blood consists in the successive alternate contraction (systole) and relaxation (diastole) of the muscular walls of its two auricles and two ventricles.

Action of the Auricles.—The description of the action of the heart may best be commenced at that period in each action which immediately precedes the beat of the heart against the side of the chest. For at this time the whole heart is in a passive state, the walls of both auricles and ventricles are relaxed, and their cavities are being dilated. The auricles are gradually filling with blood flowing into them from the veins; and a portion of this blood passes at once through them into the ventricles, the opening between the cavity of each auricle and that of its corresponding ventricle being, during all the *pause,* free and patent. The auricles, however, receiving more blood than at once passes through them to the ventricles, become, near the end of the pause, fully distended; and at the end of the pause, they contract and expel their contents into the ventricles.

The contraction of the auricles is sudden and very quick; it commences at the entrance of the great veins into them, and is thence propagated toward the auriculo-ventricular opening; but the last part which contracts is the auricular appendix. The effect of this contraction of the auricles is to quicken the flow of blood from them into the ventricles; the force of their contraction not being sufficient under ordinary circumstances to cause any back-flow into the veins. The reflux of blood into the great veins is, moreover, resisted not only by the mass of blood in the veins and the force with which it streams into the auricles, but also by the simultaneous contraction of the muscular coats with which the large veins are provided near their entrance into the auricles. Any slight regurgitation from the right auricle is limited also by the valves at the junction of the subclavian and internal jugular veins, beyond which the blood cannot move backward; and the coronary vein is preserved from it by a valve at its mouth.

In birds and reptiles regurgitation from the right auricle is prevented by valves placed at the entrance of the great veins.

During the auricular contraction the force of the blood propelled into the ventricle is transmitted in all directions, but being insufficient to separate the semilunar valves, it is expended in distending the ventricle, and, by a reflux of the current, in raising and gradually closing the auriculo-ventricular valves, which, when the ventricle is full, form a complete septum between it and the auricle.

Action of the Ventricles.—The blood which is thus driven, by the contraction of the auricles, into the corresponding ventricles, being added to that which had already flowed into them during the heart's pause, is sufficient to complete their diastole. Thus distended, they immediately contract: so immediately, indeed, that their systole looks as if it were continuous with that of the auricles. The ventricles contract much more slowly than the auricles, and in their contraction probably always thoroughly empty themselves, differing in this respect from the auricles, in which, even after their complete contraction, a small quantity of blood remains. The shape of both ventricles during systole undergoes an alteration, the left probably not altering in length but to a certain degree in breadth, the diameters in the plane of the base being diminished. The right ventricle does actually shorten to a small extent. The systole has the effect of diminishing the diameter of the base, especially in the plane of the auriculo-ventricular valves; but the length of the heart as a whole is not altered. (Ludwig.) During the systole of the ventricles, too, the aorta and pulmonary artery, being filled with blood by the force of the ventricular action against considerable resistance, elongate as well as expand, and the whole heart moves slightly toward the right and forward, twisting on its long axis, and exposing more of the left ventricle anteriorly than is usually in front. When the systole ends the heart resumes its former position, rotating to the left again as the aorta and pulmonary artery contract.

Functions of the Auriculo-Ventricular Valves.—The distension of the ventricles with blood continues throughout the whole period of their diastole. The auriculo-ventricular valves are gradually brought into play by some of the blood getting behind the cusps and floating them up; and by the time that the diastole is complete, the valves are no doubt in apposition, the completion of this being brought about by the reflex current caused by the systole of the auricles. This elevation of the auriculo-ventricular valves is, no doubt, materially aided by the action of the elastic tissue which has been shown to exist so largely in their structure, especially on the auricular surface. At any rate at the *commencement* of the ventricular systole they are completely closed. It should be recollected that the diminution in the breadth of the base of the heart in its transverse diameters during ventricular systole is especially marked in the neighborhood of the auriculo-ventricular rings, and thus aids in rendering the auriculo-ventricular valves competent to close the openings, by greatly diminishing their diameter. The margins of the cusps of the valves are still more secured in apposition with another, by the simultaneons contraction of the musculi papillares, whose chordæ tendineæ have a special mode of attachment for this object (p. 110). As in the case of the semilunar valves to be immediately described, the auriculo-ventricular valves meet not by their *edges* only, but by the opposed surfaces of their thin outer borders. The semilunar valves, on the other hand, which are closed in the intervals of the ventricle's contraction (Fig. 92, 6), are forced apart by the same pressure that tightens the auriculo-ventricular valves; and, thus, the whole force of the contracting ventricles is directed to the expulsion of blood through the aorta and pulmonary artery.

The form and position of the fleshy columns on the internal walls of the ventricle no doubt help to produce this obliteration of the cavity during their contraction; and the completeness of the closure may often be observed on making a transverse section of a heart shortly after death, in any case in which the contraction of the *rigor mortis* is very marked (Fig. 94). In such a case only a central fissure may be discernible to the eye in the place of the cavity of each ventricle.

If there were only circular fibres forming the ventricular wall, it is evident that on systole the ventricle would elongate; if there were only longitudinal fibres the ventricle would shorten on systole; but there are both. The tendency to alter in length is thus counterbalanced, and the whole force of the contraction is expended in diminishing the cavity of the ventricle; or, in other words, in expelling its contents.

On the conclusion of the systole the ventricular walls tend to expand by virtue of their elasticity, and a negative pressure is set up, which tends to suck in the blood. This negative or suctional pressure on the left side of the heart is of the highest importance in helping the pulmonary circulation. It has been found to be equal to 23 mm. of mercury, and is quite independent of the aspiration or suction power of the thorax in aiding the blood-flow to the heart, to be described in the chapter on Respiration.

Function of the Musculi Papillares.—The special function of the *musculi papillares* is to prevent the auriculo-ventricular valves from being everted into the auricle. For the chordæ tendineæ might allow the valves to be pressed back into the auricle, were it not that when the

wall of the ventricle is brought by its contraction nearer the auriculo-
ventricular orifice, the musculi papillares more than compensate for this
by their own contraction—holding the cords tight, and, by pulling down
the valves, adding slightly to the force with which the blood is expelled.

What has been said applies equally to the auriculo-ventricular valves
on both sides of the heart, and of both alike the closure is generally complete every time the ventricles contract.　But in some circumstances the
closure of the tricuspid valve is not complete, and a certain quantity of
blood is forced back into the auricle.　This has been called the *safety-
valve action* of this valve.　The circumstances in which it usually happens
are those in which the vessels of the lung are already full enough when
the right ventricle contracts, as *e.g.*, in certain pulmonary diseases, in
very active exertion, and in great efforts.　In these cases, the tricuspid
valve does not completely close, and the regurgitation of the blood may
be indicated by a pulsation in the jugular veins synchronous with that in
the carotid arteries.

Function of the Semilunar Valves.—The arterial or semilunar
valves are forced apart by the out-streaming blood, with which the contracting ventricle dilates the large arteries.　The dilation of the arteries
is, in a peculiar manner, adapted to bring the valves into action.　The
lower borders of the semilunar valves are attached to the inner surface of
a tendinous ring, which is, as it were, inlaid at the orifice of the artery,
between the muscular fibres of the ventricle and the elastic fibres of the walls
of the artery.　The tissue of this ring is tough, and does not admit of
extension under such pressure as it is commonly exposed to; the valves
are equally inextensile, being, as already mentioned, formed of tough, close-
textured, fibrous tissue, with strong interwoven cords, and covered with
endocardium.　Hence, when the ventricle propels blood through the orifice and into the canal of the artery, the lateral pressure which it exercises
is sufficient to dilate the walls of the artery, but not enough to stretch in an
equal degree, if at all, the unyielding valves and the ring to which their
lower borders are attached.　The effect, therefore, of each such propulsion of blood from the ventricle is, that the wall of the first portion of
the artery is dilated into three pouches behind the valves, while the free
margins of the valves are drawn inward toward its centre (Fig. 98, B).
Their positions may be explained by the diagrams, in which the continuous lines represent a transverse section of the arterial walls, the dotted
ones the edges of the valves, firstly, when the valves are nearest to the
walls (A), and, secondly, when, the walls being dilated, the valves are
drawn away from them (B).

This position of the valves and arterial walls is retained so long as the
ventricle continues in contraction: but, as soon as it relaxes, and the dilated arterial walls can recoil by their elasticity, the blood is forced backward toward the ventricles as onward in the course of the circulation.

Part of the blood thus forced back lies in the pouches (sinuses of Valsalva) (*a*, Fig. 98, B) between the valves and the arterial walls; and the valves are by it pressed together till their thin lunated margins meet in three

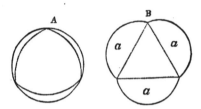

FIG. 98.—Sections of aorta, to show the action of the semilunar valves. A is intended to show the valves, represented by the dotted lines, pressed toward the arterial walls, represented by the continuous outer line. B (after Hunter) shows the arterial wall distended into three pouches (*a*), and drawn away from the valves, which are straightened into the form of an equilateral triangle. as represented by the dotted lines.

lines radiating from the centre to the circumference of the artery (7 and 8, Fig. 99).

The contact of the valves in this position, and the complete closure of the arterial orifice, are secured by the peculiar construction of their borders before mentioned. Among the cords which are interwoven in the

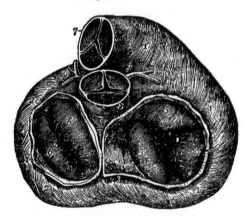

FIG. 99. —View of the base of the ventricular part of the heart, showing the relative position of the arterial and auriculo-ventricular orifices.—¾. The muscular fibres of the ventricles are exposed by the removal of the pericardium, fat, blood-vessels, etc.; the pulmonary artery and aorta have been removed by a section made immediately beyond the attachment of the semilunar valves, and the auricles have been removed immediately above the auriculo-ventricular orifices. The semilunar and auriculo-ventricular valves are in the nearly closed condition. 1, 1, the base of the right ventricle; 1', the conus arteriosus; 2, 2, the base of the left ventricle; 3, 3, the divided wall of the right auricle; 4, that of the left; 5, 5,' 5", the tricuspid valve; 6, 6', the mitral valve. In the angles between these segments are seen the smaller fringes frequently observed; 7, the anterior part of the pulmonary artery; 8, placed upon the posterior part of the root of the aorta; 9, the right, 9', the left coronary artery. (Allen Thomson.)

substance of the valves, are two of greater strength and prominence than the rest; of which one extends along the free border of each valve, and the other forms a double curve or festoon just below the free border.

Each of these cords is attached by its outer extremities to the outer end of the free margin of its valve, and in the middle to the corpus Arantii; they thus enclose a lunated space from a line to a line and a half in width, in which space the substance of the valve is much thinner and more pliant than elsewhere. When the valves are pressed down, all these parts or spaces of their surfaces come into contact, and the closure of the arterial orifice is thus secured by the apposition not of the mere edges of the valves, but of all those thin lunated parts of each which lie between the free edges and the cords next below them. These parts are firmly pressed together, and the greater the pressure that falls on them the closer and more secure is their apposition. The corpora Arantii meet at the centre of the arterial orifice when the valves are down, and they probably assist in the closure; but they are not essential to it, for, not unfrequently, they are wanting in the valves of the pulmonary artery, which are then extended in larger, thin, flapping margins. In valves of this form, also, the inlaid cords are less distinct than in those with corpora Arantii; yet the closure by contact of their surfaces is not less secure.

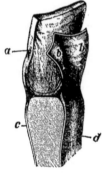

It has been clearly shown that this pressure of the blood is not entirely sustained by the valves alone, but in part by the muscular substance of the ventricle (Savory). By making vertical sections (Fig. 100) through various parts of the tendinous rings it is possible to show clearly that the aorta and pulmonary artery, expanding toward their termination, are situated upon the *outer* edge of the thick upper border of the ventricles, and that consequently the portion of each semilunar valve adjacent to the vessel passes over and rests u_{pon} the muscular substance—being thus supported, as it were, on a kind of muscular floor

FIG. 100.—Vertical section through the aorta at its junction with the left ventricle. *a*, Section of aorta. *bb*, Section of two valves. *c*, Section of wall of ventricle. *d*, Internal surface of ventricle.

formed by the upper border of the ventricle. The result of this arrangement is that the reflux of the blood is most efficiently sustained by the ventricular wall.[1]

As soon as the auricles have completed their contraction they begin again to dilate, and to be refilled with blood, which flows into them in a steady stream through the great venous trunks. They are thus filling during all the time in which the ventricles are contracting; and the contraction of the ventricles being ended, these also again dilate, and receive again the blood that flows into them from the auricles. By the time that the ventricles are thus from one-third to two-thirds full, the auricles are

---

[1] Savory's preparations, illustrating this and other points in relation to the structure and functions of the valves of the heart, are in the Museum of St. Bartholomew's Hospital.

distended; these, then suddenly contracting, fill up the ventricles, as already described (p. 111).

Cardiac Revolution.—If we suppose a cardiac revolution divided into five parts, *one* of these will be occupied by the contraction of the auricles, *two* by that of the ventricles, and two by repose of both auricles and ventricles.

Contraction of Auricles . . . $1 +$ Repose of Auricles . . . $4=5$
" Ventricles . . $2 +$ " Ventricles . . $3=5$
Repose (no contraction of either
auricles or ventricles) . . . $2 +$ Contraction (of either auri-
— cles or ventricles) . . . $3=5$
5

If the speed of the heart be quickened, the time occupied by each cardiac revolution is of course diminished, but the diminution affects only the diastole and pause. The systole of the ventricles occupies very much the same time, about $\frac{4}{10}$ sec., whatever the pulse-rate.

The periods in which the several valves of the heart are in action may be connected with the foregoing table; for the auriculo-ventricular valves are closed, and the arterial valves are open during the whole time of the ventricular contraction, while, during the dilation and distension of the ventricles the latter valves are shut, the former open. Thus whenever the auriculo-ventricular valves are open, the arterial valves are closed and *vice versâ.*

SOUNDS OF THE HEART.

When the ear is placed over the region of the heart, two *sounds* may be heard at every beat of the heart, which follow in quick succession, and are succeeded by a *pause* or period of silence. The *first* sound is dull and prolonged; its commencement coincides with the impulse of the heart, and just precedes the pulse at the wrist. The *second* is a shorter and sharper sound, with a somewhat flapping character, and follows close after the arterial pulse. The period of time occupied respectively by the two sounds taken together, and by the pause, are almost exactly equal. The relative length of time occupied by each sound, as compared with the other, is a little uncertain. The difference may be best appreciated by considering the different forces concerned in the production of the two sounds. In one case there is a strong, comparatively slow, contraction of a large mass of muscular fibres, urging forward a certain quantity of fluid against considerable resistance; while in the other it is a strong but shorter and sharper recoil of the elastic coat of the large arteries,—shorter because there is no resistance to the flapping back of the semilunar valves, as there was to their opening. The sounds may be expressed by saying the words *lŭbb—dŭp* (C. J. B. Williams).

The events which correspond, in point of time, with the *first* sound, are (1) the contraction of the ventricles, (2) the first part of the dilatation of the auricles, (3) the closure of the auriculo-ventricular valves, (4) the opening of the semilunar valves, and (5) the propulsion of blood into the arteries. The sound is succeeded, in about one-thirtieth of a second, by the pulsation of the facial arteries, and in about one-sixth of a second, by the pulsation of the arteries at the wrist. The *second* sound, in point of time, immediately follows the cessation of the ventricular contraction, and corresponds with (*a*) the closure of the semilunar valves, (*b*) the continued dilatation of the auricles, (*c*) the commencing dilatation of the ventricles, and (*d*) the opening of the auriculo-ventricular valves. The *pause* immediately follows the second sound, and corresponds *in its first part* with the completed distension of the auricles, and *in its second* with their contraction, and the completed distension of the ventricles; the auriculo-ventricular valves being, all the time of the pause, open, and the arterial valves closed.

Causes.—The chief cause of the first sound of the heart appears to be the vibration of the auriculo-ventricular valves, due to their stretching, and also, but to a less extent, of the ventricular walls, and coats of the aorta and pulmonary artery, all of which parts are suddenly put into a state of tension at the moment of ventricular contraction. The effect may be intensified by the *muscular sound* produced by contraction of the mass of muscular fibres which form the ventricle.

The cause of the *second* sound is more simple than that of the first. It is probably due entirely to the sudden closure and *consequent vibration* of the semilunar valves when they are pressed down across the orifices of the aorta and pulmonary artery. The influence of the valves in producing the sound is illustrated by the experiment performed on large animals, such as calves, in which the results could be fully appreciated. In these experiments two delicate curved needles were inserted, one into the aorta, and another into the pulmonary artery, below the line of attachment of the semilunar valves, and, after being carried upward about half an inch, were brought out again through the coats of the respective vessels, so that in each vessel one valve was included between the arterial walls and the wire. Upon applying the stethoscope to the vessels, after such an operation, the second sound had ceased to be audible. Disease of these valves, when so extensive as to interfere with their efficient action, also often demonstrates the same fact by modifying or destroying the distinctness of the second sound.

One reason for the second sound being a clearer and sharper one than the first may be, that the semilunar valves are not covered in by the thick layer of fibres composing the walls of the heart to such an extent as are the *auriculo-ventricular*. It might be expected therefore that their vibration would be more easily heard through a stethoscope applied to the walls of the chest.

. The contraction of the auricles which takes place in the end of the pause is inaudible outside the chest, but may be heard, when the heart is exposed and the stethoscope placed on it, as a slight sound preceding and continued into the louder sound of the ventricular contraction.

The Impulse of the Heart.—At the commencement of each ventricular contraction, the heart may be felt to beat with a slight shock or *impulse* against the walls of the chest. The force of the impulse, and the extent to which it may be perceived beyond this point, vary considerably in different individuals, and in the same individual under different circumstances. It is felt more distinctly, and over a larger extent of surface, in emaciated than in fat and robust persons, and more during a forced expiration than in a deep inspiration; for, in the one case, the intervention of a thick layer of fat or muscle between the heart and the surface of the chest, and in the other the inflation of the portion of lung which overlaps the heart, prevents the impulse from being fully transmitted to the surface. An excited action of the heart, and especially a hypertrophied condition of the ventricles, will increase the impulse; while a depressed condition, or an atrophied state of the ventricular walls, will diminish it.

Cause of the Impulse.—During the period which precedes the ventricular systole, the apex of the heart is situated upon the diaphragm and against the chest-wall in the fifth intercostal space. When the ventricles contract, their walls become hard and tense, since to expel their contents into the arteries is a distinctly laborious action, as it is resisted by the tension within the vessels. It is to this sudden hardening that the impulse of the heart against the chest-wall is due, and the shock of the sudden tension may be felt not only externally, but also internally, if the abdomen of an animal be opened and the finger be placed upon the under surface of the diaphragm, at a point corresponding to the under surface of the ventricle. The shock is felt, and possibly seen more distinctly, because of the partial rotation of the heart, already spoken of, along its long axis toward the right. The movement produced by the ventricular contraction may be registered by means of an instrument called the *cardiograph*, and it will be found to correspond almost exactly with a tracing obtained by the same instrument applied over the contracting ventricle itself.

The *Cardiograph* (Fig. 101) consists of a cup-shaped metal box, over the open front of which is stretched an elastic membrane, upon which is fixed a small knob of hard wood or ivory. This knob, however, may be attached instead, as in the figure, to the side of the box by means of a spring, and may be made to act upon a metal disc attached to the elastic membrane.

The knob (A) is for application to the chest-wall over the place of the greatest impulse of the heart. The box or *tympanum* communicates by

means of an air-tight elastic tube (*f*) with the interior of a second tympanum (Fig. 102, *b*), in connection with which is a long and light lever (*a*). The shock of the heart's impulse being communicated to the ivory knob, and through it to the first tympanum, the effect is, of course, at once transmitted by the column of air in the elastic tube to the interior of the second tympanum, also closed, and through the elastic and movable lid of the latter to the lever, which is placed in connection with a registering apparatus, which consists generally of a cylinder or drum covered with smoked paper, revolving according to a definite velocity by clockwork. The point of the lever writes upon the paper, and a tracing of the heart's impulse is thus obtained.

FIG. 101.
Cardiograph. (Sanderson's.)

By placing three small india-rubber air-bags in the interior respec-

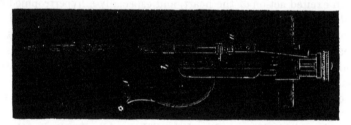

FIG. 102.—Marey's Tambour (*b*), to which the movement of the column of air in the first tympanum is conducted by the tube, *f*, and from which it is communicated by the lever, *a*, to a revolving cylinder, so that the tracing of the movement of the impulse beat is obtained.

tively of the right auricle, the right ventricle, and in an intercostal space in front of the heart of living animals (horse), and placing these bags, by means of long narrow tubes, in communication with three levers, arranged

FIG. 103.—Tracing of the impulse of the heart of man. (Marey.)

one over the other in connection with a registering apparatus (Fig. 104), MM. Chauveau and Marey have been able to measure with much accuracy the variations of the endocardial pressure and the comparative duration of the contractions of the auricles and ventricles. By means of the same apparatus, the synchronism of the impulse with the contraction of the ventricles, is also well shown; and the causes of the several vibrations of which it is really composed, have been discovered.

In the tracing (Fig 105), the intervals between the vertical lines represent periods of a tenth of a second. The parts on which any given

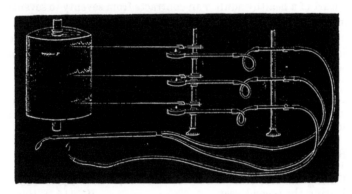

Fig. 104.—Apparatus of MM. Chauveau and Marey for estimating the variations of endocardial pressure, and production of impulse of the heart.

vertical line falls represent, of course, simultaneous events. Thus,—it will be seen that the contraction of the auricle, indicated by the upheaval of the tracing at A in first tracing, causes a slight increase of pressure in the ventricle (A′ in second tracing), and produces a tiny impulse (A″ in third tracing). So also, the closure of the semilunar valves, while it causes a momentarily increased pressure in the ventricle at D′, does not fail to affect the pressure in the auricle D″, and to leave its mark in the tracing of the impulse also, D″,

The large upheaval of the ventricular and the impulse tracings, between A′ and D′, and A″ and D″, are caused by the ventricular contraction, while the smaller undulations, between B and C, B′ and C′, B″ and C″, are caused by the vibrations consequent on the tightening and closure of the auriculo-ventricular valves.

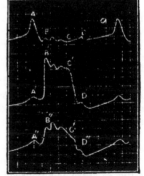

Fig. 105.—Tracings of (1), Intra-auricular, and (2), Intra-ventricular pressures, and (3), of the impulse of the heart, to be read from left to right, obtained by Chauveau and Marey's apparatus.

Although, no doubt, the method thus described may show a perfectly correct view of the endocardiac pressure variations, it should be recollected that the muscular walls may grip the air-bags, even after the complete expulsion of the contents of the chamber, and so the lever might remain for a too long time in the position of extreme tension, and would represent on the tracing not only, as it ought to do, the auricular or ventricular pressure on the blood, but, also afterward, the muscular pressure exerted upon the bags themselves. (M. Foster.)

FREQUENCY AND FORCE OF THE HEART'S ACTION.

The heart of a healthy adult man contracts from seventy to seventy-five times in a minute; but many circumstances cause this rate, which of course corresponds with that of the arterial *pulse*, to vary even in health. The chief are age, temperament, sex, food and drink, exercise, time of day, posture, atmospheric pressure, temperature.

Age.—The frequency of the heart's action gradually diminishes from the commencement to near the end of life, but is said to rise again somewhat in extreme old age, thus:—

Before birth the average number of pulses in a minute is	150
Just after birth	from 140 to 130
During the first year	" 130 " 115
During the second year	" 115 " 100
During the third year	" 100 " 90
About the seventh year	" 90 " 85
About the fourteenth year, the average number of pulses in a minute is	" 85 " 80
In adult age	" 80 " 70
In old age	" 70 " 60
In decrepitude	" 75 " 65

Temperament and Sex.—In persons of sanguine temperament, the heart acts somewhat more frequently than in those of the phlegmatic; and in the female sex more frequently than in the male.

Food and Drink. Exercise.—After a meal its action is accelerated, and still more so during bodily exertion or mental excitement; it is slower during sleep.

Diurnal Variation.—It appears that, in the state of health, the pulse is most frequent in the morning, and becomes gradually slower as the day advances. and that this diminution of frequency is both more regular and more rapid in the evening than in the morning.

Posture.—It is found that, as a general rule, the pulse, especially in the adult male, is more frequent in the standing than in the sitting posture, and in the latter than in the recumbent position; the difference being greatest between the standing and the sitting posture. The effect of change of posture is greater as the frequency of the pulse is greater, and, accordingly, is more marked in the morning than in the evening. By supporting the body in different postures, without the aid of muscular effort of the individual, it has been proved that the increased frequency of the pulse in the sitting and standing positions is dependent upon the muscular exertion engaged in maintaining them; the usual effect of these postures on the pulse being almost entirely prevented when the usually attendant muscular exertion was rendered unnecessary. (Guy.)

Atmospheric Pressure.—The frequency of the pulse increases in a corresponding ratio with the elevation above the sea.

Temperature.—The rapidity and force of the heart's contractions are largely influenced by variations of temperature. The frog's heart, when excised, ceases to beat if the temperature be reduced to 32° F. (0° C.). When heat is gradually applied to it, both the speed and force of the heart's contractions increase till they reach a maximum. If the temperature is still further raised, the beats become irregular and feeble, and the heart at length stands still in a condition of "heat-rigor."

Similar effects are produced in warm-blooded animals. In the rabbit, the number of heart-beats is more than doubled when the temperature of the air was maintained at 105° F. (40°.5 C.). At 113°—114° F. (45° C.), the rabbit's heart ceases to beat.

Relative Frequency of the Pulse to that of Respiration.—

In health there is observed a nearly uniform relation between the frequency of the pulse and of the respirations; the proportion being, on an average, one respiration to three or four beats of the heart. The same relation is generally maintained in the cases in which the pulse is naturally accelerated, as after food or exercise; but in disease this relation usually ceases. In many affections accompanied with increased frequency of the pulse, the respiration is, indeed, also accelerated, yet the degree of its acceleration may bear no definite proportion to the increased number of the heart's actions: and in many other cases, the pulse becomes more frequent without any accompanying increase in the number of respirations; or, the respiration alone may be accelerated, the number of pulsations remaining stationary, or even falling below the ordinary standard.

The Force of the Ventricular Systole and Diastole.—The

force of the left ventricular systole is more than double that exerted by the contraction of the right: this difference in the amount of force exerted by the contraction of the two ventricles, results from the walls of the left ventricle being about twice or three times as thick as those of the right. And the difference is adapted to the greater degree of resistance which the left ventricle has to overcome, compared with that to be overcome by the right: the former having to propel blood through every part of the body, the latter only through the lungs.

The actual amount of the intra-ventricular pressures during systole in the dog has been found to be 2·4 inches (60 mm.) of mercury in the right ventricle, and 6 inches (150 mm.) in the left. During diastole there is in the right ventricle a negative or suction pressure of about ⅔ of an inch (−17 to −16 mm.), and in the left ventricle from 2 inches to ⅘ of an inch (−52 to −20 mm.). Part of this fall in pressure, and possibly the greater part, is to be referred to the influence of respiration; but without this the negative pressure of the left ventricle caused by its active dilatation is about ⅘ of an inch (23 mm.) of mercury.

The right ventricle is undoubtedly aided by this suction power of the left, so that the whole of the work of conducting the pulmonary circulation does not fall upon the right side of the heart, but is assisted by the left side.

The Force of the Auricular Systole and Diastole.—The maximum pressure within the right auricle is about $\frac{4}{5}$ of an inch (20 mm.) of mercury, and is probably somewhat less in the left. It has been found that during diastole the pressure within both auricles sinks considerably below that of the atmosphere; and as some fall in pressure takes place, even when the thorax of the animal operated upon has been opened, a certain proportion of the fall must be due to active auricular dilatation independent of respiration. In the right auricle, this negative pressure is about −10 mm.

Work Done by the Heart.—In estimating the work done by any machine it is usual to express it in terms of the "unit of work." The unit of work is defined to be the energy expended in raising a unit of weight (1 lb.) through a unit of height (1 ft.). In England, the unit of work is the *"foot-pound,"* in France, the *"kilogrammetre."*

The work done by the heart at each contraction can be readily found by multiplying the weight of blood expelled by the ventricles by the height to which the blood rises in a tube tied into an artery. This height was found to be about 9 ft. in the horse, and the estimate is nearly correct for a large artery in man. Taking the weight of blood expelled from the left ventricle at each systole as 6 oz., *i.e.*, $\frac{3}{8}$ lb., we have $9 \times \frac{3}{8} = 3\cdot375$ foot-pounds as the work done by the left ventricle at each systole; and adding to this the work done by the right ventricle (about one-third that of the left) we have $3\cdot375 \times 1\cdot125 = 4\cdot5$ foot-pounds as the work done by the heart at each contraction. Other estimates give $\frac{1}{2}$ kilogrammetre, or about $3\frac{1}{2}$ foot-pounds. Haughton estimates the total work of the heart in 24 hours as about 124 foot-tons.

Influence of the Nervous System on the Action of the Heart.—The hearts of warm-blooded animals cease to beat almost if not quite immediately after removal from the body, and are, therefore, unfavorable for the study of the nervous mechanism which regulates their action. Observations have hitherto, therefore, been principally directed to the heart of cold-blooded animals, *e.g.*, the frog, tortoise, and snake, which will continue to beat under favorable conditions for many hours after removal from the body. Of these animals, the frog is the one mostly employed, and, indeed, until recently, it was from the study of the frog's heart that the chief part of our information was obtained. If removed from the body entire, the frog's heart will continue to beat for many hours and even days, and the beat has no apparent difference from the beat of the heart before removal from the body; it will take place without the presence of blood or other fluid within its chambers. If the beats have become infrequent, an additional beat may be induced by stimulating the heart by means of a blunt needle; but the time before the stimulus applied produces its result (the latent period) is very prolonged, and as in this way the cardiac beat is like the contraction of unstriped muscle, the method has been likened to a peristaltic contraction.

There is much uncertainty about the nervous mechanism of the beat of the frog's heart, but what has just been said shows, at any rate, two things; firstly, that as the heart will beat when removed from the body in a way differing not at all from the normal, it must contain within itself the mechanism of rhythmical contraction; and secondly, that as it can beat without the presence of fluid within its chambers, the movement cannot depend merely on reflex excitation by the entrance of blood. The nervous apparatus existing in the heart itself consists of collections of microscopic ganglia, and of nerve-fibres proceeding from them. These ganglia are

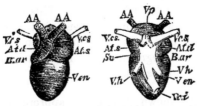

Fig. 106,—Heart of frog. (Burdon-Sanderson after Fritsche.) Front view to the left, back view to the right. *A A.* Aortæ. *V. cs.* Venæ cavæ superiores. At *s*, left auricle. At *d*, right auricle. *Ven.*, ventricle. *B. ar.*, Bulbus arteriosus. *S. v.*, Sinus venosus. *V. c. i.*, Vena cava inferior. *V. h.*, Venæ hepaticæ. *V. p.*, Venæ pulmonales.

demonstrable as being collected chiefly into three groups; one is in the wall of the sinus venosus (Remak's); a second, near the junction between the auricle and ventricle (Bidder's); and the third in the septum between the auricles.

Some very important experiments seem to identify the rhythmical contractions of the frog's heart with these ganglia. If the heart be removed entire from the body, the sequence of the contraction of its several beats will take place with rhythmical regularity, viz., of the sinus venosus, the auricles, the ventricle, and bulbus arteriosus, in order. If the heart be removed at the junction of the sinus and auricle, the former will continue to beat, but the removed portion will for a short variable time stop beating, and then resume its beats, but with a rhythm different to that of the sinus: and, further, if the ventricle be removed, it will take a still longer time before recommencing its pulsation after its removal than the larger portion consisting of the auricles and ventricle, and its rhythm is different from that of the unremoved portion, and not so regular, nor will it continue to pulsate so long: during the period of stoppage a contraction will occur if the ventricle be mechanically or otherwise stimulated. If the lower two-thirds or apex of the ventricle be removed, the remainder of the heart will go on beating regularly in the body, but this part will remain motionless, and will not beat spontaneously, although it will respond to stimuli. If the heart be divided lengthwise, its parts will continue to pulsate rhythmically, and the auricles may be cut up into pieces, and the pieces will continue their movements of contraction. It will be thus seen that the rhythmical movements appear to be more marked in the parts supplied by the ganglia, and that the apical portion of the ventricle, in which the ganglia are not found, does not possess the power of automatic movement. Although the theory that the pulsations of the rest of the heart are dependent upon that of the sinus, and to stimuli proceeding from it, when connection is maintained, and only to reflex stimuli when removal has taken place, cannot be absolutely upheld, yet it is evident that the power of spontaneous contraction is strongest in the sinus, less strong in the auricles, and less so still in the ventricle, and that, therefore, the sinus ganglia are probably important in exciting the rhythmical contraction of the whole heart. This is expressed in the following way:—"The power of independent rhythmical contraction decreases regularly as we pass from the sinus to the ventricles," and "The rhythmical power of each segment of the heart varies inversely as its distance from the sinus." (Gaskell.)

It has been recently shown that, under appropriate stimuli, even the extreme apex of the ventricle in the tortoise may take on rhythmical contractions, or in other words may be "taught to beat" rhythmically. (Gaskell.)

Inhibition of the Heart's Action.—Although, under ordinary conditions, the apparatus of ganglia and nerve-fibres in the substance of the heart forms the medium through which its action is excited and rhythmically maintained, yet they, and, through them, the heart's contractions, are regulated by nerves which pass to them from the higher nerve-centres. These nerves are branches from the *pneumogastric* or *vagus* and the *sympathetic*.

The influence of the vagi nerves over the heart-beat may be shown by stimulating one (especially the right) or both of the nerves when a record is being taken of the beats of the frog's heart. If a single induction shock be sent into the nerve, the heart, after a short interval, ceases beating, but after the suppression of several beats resumes its action. As already mentioned, the effect of the stimulus is not immediately seen, and one beat may occur before the heart stops after the application of the electric-current. The stoppage of the heart may occur apparently in one of two ways, either by diminution of the strength of the systole or by increasing the length of the diastole. The stoppage of the heart may be brought about by the application of the electrodes to any part of the vagus, but most effectually if they are applied near the position of Remak's ganglia. It is supposed that the fibres of the vagi, therefore, terminate there in *inhibitory* ganglia in the heart-walls, and that the inhibition of the heart's beats by means of the vagus, is not a simple action, but that it is produced by stimulating centres in the heart itself. These inhibitory centres are paralyzed by atropin, and then no amount of stimulation of the vagus, or of the heart itself, will produce any effect upon the cardiac beats. Urari in large doses paralyzes the vagus fibres, but in this case, as the inhibitory action can be produced by direct stimulation of the heart, it is inferred that this drug does not paralyze the ganglia themselves. Muscarin and pilocarpin appear to produce effects similar to those obtained by stimulating the vagus fibres.

If a ligature be tightly tied round the heart over the situation of the ganglia between the sinus and the auricles, the heart stops beating. This experiment (Stannius') would seem to stimulate the inhibitory ganglia, but for the remarkable fact that atropin does not interfere with its success. If the part (the ventricle) below the ligature be cut off, it will begin and continue to beat rhythmically, this may be explained by supposing that the stimulus of section induces pulsation in the part which is removed from the influence of the inhibitory ganglia.

So far, the effect of the terminal apparatus of the vagi has been considered; there is, however, reason for believing that the vagi nerves are simply the media of an *inhibitory* or restraining influence over the action of the heart, which is conveyed through them from a *centre* in the medulla oblongata which is always in operation, and, because of its restraining the heart's action, is called the *cardio-inhibitory* centre. For, on dividing these nerves, the pulsations of the heart are increased in frequency, an effect opposite to that produced by stimulation of their divided (peripheral) ends. The restraining influence of the centre in the medulla may be increased reflexly, producing slowing or stoppage of the heart, through influence passing from it down the vagi. As an example of the latter, the well-known effect on the heart of a violent blow on the epigastrium may be referred to. The stoppage of the heart's action is due to the conveyance of the stimulus by fibres of the sympathetic to the medulla oblongata, and its subsequent *reflection* through the vagi to the inhibitory ganglia of the heart. It is also believed that the power of the medullary inhibitory centre may be reflexly lessened, producing accelerated action of the heart.

Acceleration of Heart's Action.—Through certain fibres of the sympathetic, the heart receives an *accelerating* influence from the medulla oblongata. These accelerating nerve-fibres, issuing from the spinal cord in the neck, reach the inferior cervical ganglion, and pass thence to the cardiac plexus, and so to the heart. Their function is shown in the quickened pulsation which follows stimulation of the spinal cord, when the latter has been cut off from all connection with the heart, excepting that which is formed by the accelerating filaments from the inferior cervical ganglion.   Unlike the inhibitory fibres of the pneumogastric, the accelerating fibres are not continuously in action.

The accelerator nerves must not, however, be considered as direct antagonists of the vagus; for if at the moment of their maximum stimulation, the vagus be stimulated with minimum currents, inhibition is produced with the same readiness as if these were not acting.

The connection of the heart with other organs by means of the nervous system, and the influences to which it is subject through them, are shown in a striking manner by the phenomena of disease.   The influence of mental shock in arresting or modifying the action of the heart, the slow pulsation which accompanies compression of the brain, the irregularities and palpitations caused by dyspepsia or hysteria, are good evidence of the connection of the heart with other organs through the nervous system.

The action of the heart is no doubt also very materially affected by the nutrition of its walls by a sufficient supply of healthy blood sent to them, and it is not unlikely that the apparently contradictory effect of poisons may be explained by supposing that the influence of some of them is either partially or entirely directed to the muscular tissue itself, and not to the nervous apparatus alone.   As will be explained presently, the heart exercises a considerable influence upon the condition of the pressure of blood within the arteries, but in its turn the blood-pressure within the arteries reacts upon the heart, and has a distinct effect upon its contractions, increasing by its increase, and *vice versâ*, the force of the cardiac beat, although the frequency is diminished as the blood-pressure rises. The quantity (and quality?) of the blood contained in each chamber, too, has an influence upon its systole, and within normal limits the larger the quantity the stronger the contraction.   Rapidity of systole does not of necessity indicate strength, as two weak contractions often do no more work than one strong and prolonged.   In order that the heart may do its maximum work, it must be allowed free space to act; for if obstructed in its action by mechanical outside pressure, as by an excess of fluid within the pericardium, such as is produced by inflammation, or by an overloaded stomach, or what not, the pulsations become irregular and feeble.

THE ARTERIES.

Distribution.—The arterial system begins at the left ventricle in a single large trunk, the aorta, which almost immediately after its origin gives off in its course in the thorax three large branches for the supply of the head, neck, and upper extremities; it then traverses the thorax and abdomen, giving off branches, some large and some small, for the supply of the various organs and tissues it passes on its way.   In the abdomen it divides into two chief branches, for the supply of the lower extremities. The arterial branches wherever given off divide and sub-divide, until the calibre of each subdivision becomes very minute, and these minute vessels pass into capillaries. Arteries are, as a rule, placed in situations, protected from pressure and other dangers, and are, with few exceptions, straight in their course, and frequently communicate with other arteries (anastomose or inosculate). The branches are usually given off at an acute angle, and the area of the branches of an artery gen-erally exceeds that of the parent trunk; and as the distance from the origin is increased, the area of the combined branches is increased also.

After death, arteries are usually found dilated (not collapsed as the veins are) and empty, and it was to this fact that their name was given them, as the ancients believed that they conveyed air to the various parts of the body. As regards the arterial system of the lungs (pulmonary system) it begins at the right ventricle in the pulmonary artery, and is distributed much as the arteries belonging to the general systemic cir-culation.

Structure.—The walls of the arteries are composed of three principal coats, termed the *external* or *tunica adventitia,* the *middle* or *tunica media,* and the *internal* coat or *tunica intima.*

The *external coat* or *tunica adventitia* (Figs. 107 and 111, *t. a.*), the strongest and toughest part of the wall of the artery, is formed of areolar

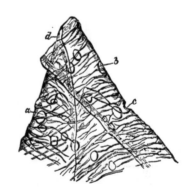

FIG. 107. FIG. 108.

FIG. 107.—Minute artery viewed in longitudinal section. *e,* Nucleated endothelial membrane, with faint nuclei in lumen, looked at from above. *i.* Thin elastic tunica intima. *m.* Muscular coat or tunica media. *a.* Tunica adventitia. (Klein and Noble Smith.) × 250.
FIG. 108.—Portion of fenestrated membrane from the femoral artery. × 200. *a, b, c.* Perfo-rations. (Henle.)

tissue, with which is mingled throughout a network of elastic fibres. At the inner part of this outer coat the elastic network forms in most arteries so distinct a layer as to be sometimes called the *external elastic coat* (Fig. 123, *e. e.*).

The *middle coat* (Fig. 107, *m*) is composed of both muscular and

elastic fibres, with a certain proportion of areolar tissue. In the larger
arteries (Fig. 110) its thickness is comparatively as well as absolutely
much greater than in the small, constituting, as it does, the greater part
of the arterial wall.

The muscular fibres, which are of the unstriped variety (Fig. 109) are
arranged for the most part transversely to the long axis of the artery
(Fig. 107, *m*); while the elastic element, taking also a transverse direction, is disposed in the form of closely interwoven and branching fibres,
which intersect in all parts the layers of muscular fibre. In arteries of

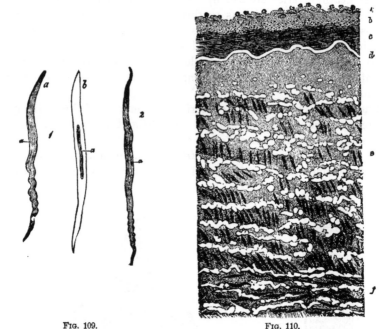

FIG. 109.   FIG. 110.

FIG. 109.—Muscular fibre-cells from human arteries, magnified 350 diameters. (Kölliker.) *a*.
Nucleus. *b*. A fibre-cell treated with acetic acid.
FIG. 110.—Transverse section of aorta through internal and about half the middle coat. *a*, Lining endothelium with the nuclei of the cells only shown. *b*. Subepithelial layer of connective tissue.
c, d. Elastic tunica intima proper, with fibrils running circularly or longitudinally. *e, f*. Middle coat,
consisting of elastic fibres arranged longitudinally, with muscle-fibres cut obliquely, or longitudinally.
(Klein.)

various size there is a difference in the proportion of the muscular and
elastic element, elastic tissue preponderating in the largest arteries, while
this condition is reversed in those of medium and small size.

The *internal coat* is formed by layers of elastic tissue, consisting in
part of coarse longitudinal branching fibres, and in part of a very thin
and brittle membrane which possesses little elasticity, and is thrown into
folds or wrinkles when the artery contracts. This latter membrane,
the striated or *fenestrated coat of Henle* (Fig. 108), is peculiar in its tendency to curl up, when peeled off from the artery, and in the perforated

and streaked appearance which it presents under the microscope. Its inner surface is lined with a delicate layer of endothelium, composed of elongated cells (Fig. 112, a), which make it smooth and polished, and furnish a nearly impermeable surface, along which the blood may flow with the smallest possible amount of resistance from friction.

Immediately external to the endothelial lining of the artery is fine connective tissue, *sub-endothelial layer*, with branched corpuscles. Thus the internal coat consists of three parts, (a) an endothelial lining, (b) the sub-endothelial layer, and (c) elastic layers.

Vasa Vasorum.—The walls of the arteries, with the possible exception of the endothelial lining and the layers of the internal coat immediately outside it, are not nourished by the blood which they convey, but are, like other parts of the body, supplied with little arteries, ending in

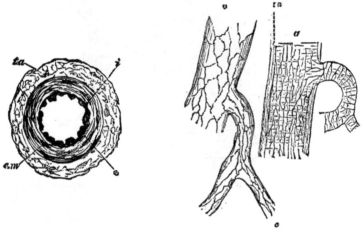

FIG. 111.   FIG. 112.

FIG. 111.—Transverse section of small artery from soft palate. *e*, endothelial lining, the nuclei of the cells are shown; *i*, elastic tissue of the intima, which is a good deal folded; *c. m.* circular muscular coat, showing nuclei of muscle cells; *t. a.* tunica adventitia. × 300. (Schofield.)

FIG. 112.—Two blood-vessels from a frog's mesentery, injected with nitrate of silver, showing the outlines of the endothelial cells. *a.* Artery. The endothelial cells are long and narrow; the transverse markings indicate the muscular coat. *t. a.* Tunica adventitia. *v.* Vein, showing the shorter and wider endothelial cells with which it is lined. *c, c.* Two capillaries entering the vein. (Schofield.)

capillaries and veins, which, branching throughout the external coat, extend for some distance into the middle, but do not reach the internal coat. These nutrient vessels are called *vasa vasorum*.

Lymphatics of Arteries and Veins.—Lymphatic spaces are present in the coats of both arteries and veins; but in the tunica adventitia or external coat of large vessels they form a distinct plexus of more or less tubular vessels. In smaller vessels they appear as sinons spaces lined by endothelium. Sometimes, as in the arteries of the omentum, mesentery, and membranes of the brain, in the pulmonary, hepatic, and splenic arteries, the spaces are continuous with vessels which distinctly ensheath

them—*perivascular lymphatic sheaths* (Fig. 121). Lymph channels are
said to be present also in the tunica media.

Nervi Vasorum.—Most of the arteries are surrounded by a plexus
of sympathetic nerves, which twine around the vessel very much like ivy
round a tree: and ganglia are found at frequent intervals. The smallest

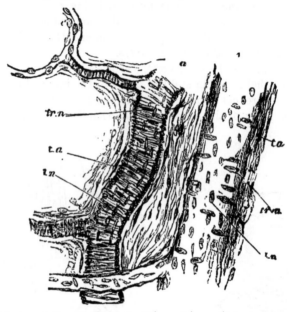

FIG. 113.—Blood-vessels from mesocolon of rabbit. *a,* Artery, with two branches, showing *tr. n.*
nuclei of transverse muscular fibres; *l. n.* nuclei of endothelial lining; *t. a.* tunica adventitia. *v.*
Vein. Here the transverse nuclei are more oval than those of the artery. The vein receives a small
branch at the lower end of the drawing; it is distinguished from the artery among other things by its
straighter course and larger calibre. *c.* Capillary, showing nuclei of endothelial cells. × 300.
(Schofield.)

arteries and capillaries are also surrounded by a very delicate network of
similar nerve-fibres, many of which appear to end in the nuclei of the
transverse muscular fibres (Fig. 122). It is through these plexuses that
the calibre of the vessels is regulated by the nervous system (p. 152).

THE CAPILLARIES.

Distribution.—In all vascular textures, except some parts of the
corpora cavernosa of the penis, and of the uterine placenta, and of the
spleen, the transmission of the blood from the minute branches of the
arteries to the minute veins is effected through a network of *microscopic*
vessels, called *capillaries*. These may be seen in all minutely injected
preparations; and during life, in any transparent vascular parts,—such
as the web of the frog's foot, the tail or external branchiæ of the tadpole,
or the wing of the bat.

The branches of the minute arteries form repeated anastomoses with each other, and give off the capillaries which, by their anastomoses, compose a continuous and uniform network, from which the venous radicles take their rise (Fig. 114). The point at which the arteries terminate and the minute veins commence, cannot be exactly defined, for the transition is gradual; but the capillary network has, nevertheless, this peculiarity, that the small vessels which compose it maintain the same diameter throughout: they do not diminish in diameter in one direction, like arteries and veins; and the meshes of the network that they compose are more uniform in shape and size than those formed by the anastomoses of the minute arteries and veins.

FIG. 114.—Blood-vessels of an intestinal villus, representing the arrangement of capillaries between the ultimate venous and arterial branches; *a, a,* the arteries; *b,* the vein.

Structure.—This is much more simple than that of the arteries or veins. Their walls are composed of a single layer of elongated or radiate, flattened and nucleated cells, so joined and dovetailed together as to form a continuous transparent membrane (Fig. 115). Outside these cells, in the larger capillaries, there is a structureless, or very finely fibrillated membrane, on the inner surface of which they are laid down.

In some cases this external membrane is nucleated, and may then be regarded as a miniature representative of the tunica adventitia of arteries.

Here and there, at the junction of two or more of the delicate endothelial cells which compose the capillary wall, *pseudo-stomata* may be seen

FIG. 115.—Capillary blood-vessels from the omentum of rabbit, showing the nucleated endothelial membrane of which they are composed. (Klein and Noble Smith.)

resembling those in serous membranes (p. 296). The endothelial cells are often continuous at various points with processes of adjacent connective-tissue corpuscles.

Capillaries are surrounded by a delicate nerve-plexus resembling, in miniature, that of the larger blood-vessels.

The *diameter* of the capillary vessels varies somewhat in the different textures of the body, the most common size being about $\frac{1}{3000}$th of an inch. Among the smallest may be mentioned those of the brain, and of the follicles of the mucous membrane of the intestines; among the largest, those of the skin, and especially those of the medulla of bones.

The *size* of capillaries varies necessarily in different animals in relation to the size of their blood corpuscles: thus, in the Proteus, the capillary circulation can just be discerned with the naked eye.

The *form* of the capillary network presents considerable variety in the different textures of the body: the varieties consisting principally of modifications of two chief kinds of mesh, the rounded and the elongated. That

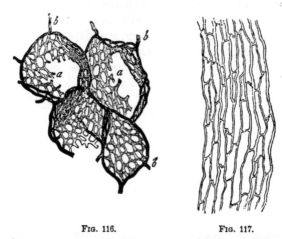

FIG. 116.   FIG. 117.

FIG. 116.—Network of capillary vessels of the air-cells of the horse's lung magnified. *a, a,* capillaries proceeding from *b, b,* terminal branches of the pulmonary artery. (Frey.)
FIG. 117.—Injected capillary vessels of muscle seen with a low magnifying power. (Sharpey.)

kind of which the meshes or interspaces have a roundish form is the most common, and prevails in those parts in which the capillary network is most dense, such as the lungs (Fig. 116), most glands, and mucous membranes, and the cutis. The meshes of this kind of network are not quite circular but more or less angular, sometimes presenting a nearly regular quadrangular or polygonal form, but being more frequently irregular. The capillary network with elongated meshes (Fig. 117) is observed in parts in which the vessels are arranged among bundles of fine tubes or fibres, as in muscles and nerves. In such parts, the meshes usually have the form of a parallelogram, the short sides of which may be from three to eight or ten times less than the long ones; the long sides always corresponding to the axis of the fibre or tube, by which it is placed. The appearance of both the rounded and elongated meshes is much varied according as the vessels composing them have a straight or tortuous form. Sometimes the capillaries have a looped arrangement, a single capillary projecting from the common network into some prominent organ, and returning after forming one or more loops, as in the papillæ of the tongue and skin.

The *number* of the capillaries and the *size of the meshes* in different parts determine in general the degree of *vascularity* of those parts. The parts in which the network of capillaries is closest, that is, in which the meshes or interspaces are the smallest, are the lungs and the choroid membrane of the eye. In the iris and ciliary body, the interspaces are somewhat wider, yet very small. In the human liver the interspaces are of the same size or even smaller than the capillary vessels themselves. In the human lung they are smaller than the vessels; in the human kidney, and in the kidney of the dog, the diameter of the injected capillaries, compared with that of the interspaces, is in the proportion of one to four, or of one to three. The brain receives a very large quantity of blood; but the capillaries in which the blood is distributed through its substance are very minute, and less numerous than in some other parts. Their diameter, according to E. H. Weber, compared with the long diameter of the meshes, being in the proportion of one to eight or ten ; compared with the transverse diameter, in the proportion of one to four or six. In the mucous membranes—for example in the conjunctiva and in the cutis vera, the capillary vessels are much larger than in the brain, and the interspaces narrower,—namely, not more than three or four times wider than the vessels. In the periosteum the meshes are much larger. In the external coat of arteries, the width of the meshes is ten times that of the vessels (Henle).

It may be held as a general rule, that the more active the functions of an organ are, the more vascular it is. Hence the narrowness of the interspaces in all glandular organs, in mucous membranes, and in growing parts; their much greater width in bones, ligaments, and other very tough and comparatively inactive tissues; and the usually complete absence of vessels in cartilage, and such parts as those in which, probably, very little *vital* change occurs after they are once formed.

THE VEINS.

. **Distribution.**—The venous system begins in small vessels which are slightly larger than the capillaries from which they spring. These vessels are gathered up into larger and larger trunks until they terminate (as regards the systemic circulation) in the two venæ cavæ and the coronary veins, which enter the right auricle, and (as regards the pulmonary circulation) in four pulmonary veins, which enter the left auricle. The capacity of the veins diminishes as they approach the heart; but, as a rule, the capacity of the veins exceeds by several times (twice or three times) that of their corresponding arteries. The pulmonary veins, however, are an exception to this rule, as they do not exceed in capacity the pulmonary arteries. The veins are found after death as a rule to be more or less collapsed, and often to contain blood. The veins are usually distributed in a superficial and a deep set which communicate frequently in their course.

Structure.—In structure the coats of veins bear a general resemblance to those of arteries (Fig. 118). Thus, they possess an *outer*,

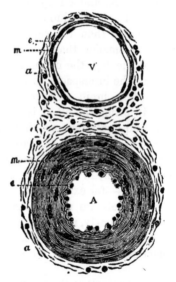

Fig. 118.—Transverse section through a small artery and vein of the mucous membrane of a child's epiglottis: the contrast between the thick-walled artery and the thin-walled vein is well shown. A. Artery, the letter is placed in the lumen of the vessel. *e.* Endothelial cells with nuclei clearly visible: these cells appear very thick from the contracted state of the vessel. Outside it, a double wavy line marks the elastic tunica intima. *m.* Tunica media forming the chief part of arterial wall and consisting of unstriped muscular fibres circularly arranged: their nuclei are well seen. *a.* Part of the tunica adventitia showing bundles of connective-tissue fibres in section, with the circular nuclei of the connective-tissue corpuscles. This coat gradually merges into the surrounding connective-tissue. V. In the lumen of the vein: The other letters indicate the same as in the artery. The muscular coat of the vein (*m*) is seen to be much thinner than that of the artery. × 350. (Klein and Noble Smith.)

middle, and *internal* coat. The *outer* coat is constructed of areolar tissue like that of the arteries, but is thicker. In some veins it contains muscular fibre-cells, which are arranged longitudinally.

The *middle* coat is considerably thinner than that of the arteries; and, although it contains circular unstriped muscular fibres or fibre-cells, these are mingled with a larger proportion of yellow elastic and white fibrous tissue. In the large veins, near the heart, namely the *venæ cavæ* and pulmonary veins, the middle coat is replaced, for some distance from the heart, by circularly arranged striped muscular fibres, continuous with those of the auricles.

The *internal* coat of veins is less brittle than the corresponding coat of an artery, but in other respects resembles it closely.

Valves.—The chief influence which the veins have in the circulation, is effected with the help of the *valves*, which are placed in all veins subject to local pressure from the muscles between or near which they run. The general construction of these valves is similar to that of the semilunar valves of the aorta and pulmonary artery, already described; but their free margins are turned in the opposite direction, *i.e.*, *toward* the heart, so as to stop any movement of blood backward in the veins. They are commonly placed in pairs, at various distances in different veins, but almost uniformly in each (Fig. 119). In the smaller veins, single valves are often met with; and three or four are sometimes placed together, or near one another, in the largest veins, such as the subclavian, and at their junction with the jugular veins. The valves are semilunar; the

Fig. 119.—Diagram showing valves of veins. A, part of a vein laid open and spread out, with two pairs of valves. B. Longitudinal section of a vein, showing the apposition of the edges of the valves in their closed state. C, portion of a distended vein, exhibiting a swelling in the situation of a pair of valves.

unattached edge being in some examples concave, in others straight. They are composed of inextensile fibrous tissue, and are covered with endothelium like that lining the veins. During the period of their inaction, when the venous blood is flowing in its proper direction, they lie by the sides of the veins; but when in action, they close together like the valves of the arteries, and offer a complete barrier to any backward movement of the blood (Figs. 119 and 120). Their situation in the superficial veins of the forearm is readily discovered by pressing along its surface, in a direction opposite to the venous current, *i.e.*, from the elbow toward the wrist; when little swellings (Fig. 119, c) appear in the position of each pair of valves. These swellings at once disappear when the pressure is relaxed.

Valves are not equally numerous in all veins, and in many they are absent altogether. They are most numerous in the veins of the extremities, and more so in those of the leg than the arm. They are commonly *absent* in veins of less than a line in diameter, and, as a general rule, there are few or none in those which are not subject to muscular pressure. Among those veins which have no valves may be mentioned the superior and inferior vena cava, the trunk and branches of the portal vein, the

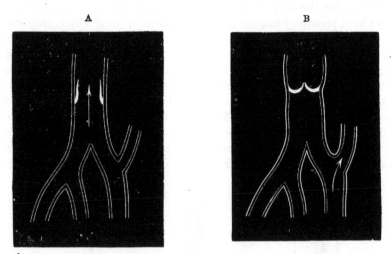

Fig. 120.—A, vein with valves open. B, vein with valves closed: stream of blood passing off by lateral channel. (Dalton.)

hepatic and renal veins, and the pulmonary veins; those in the interior of the cranium and vertebral column, those of the bones, and the trunk and branches of the umbilical vein are also destitute of valves.

CIRCULATION IN THE ARTERIES.

Functions of the External Coat of Arteries.—The external coat forms a strong and tough investment, which, though capable of extension, appears principally designed to strengthen the arteries and to guard against their excessive distension by the force of the heart's action. It is this coat which alone prevents the complete severance of an artery when a ligature is tightly applied; the internal and middle coats being divided. In it, too, the little *vasa vasorum* (p. 131) find a suitable tissue in which to subdivide for the supply of the arterial coats.

Functions of the Elastic Tissue in Arteries.—The purpose of the elastic tissue, which enters so largely into the formation of all the coats of the arteries, is, (a) to guard the arteries from the suddenly exerted pressure to which they are subjected at each contraction of the ventricles. In every such contraction, the contents of the ventricles are forced into the arteries more quickly than they can be discharged into and through the capillaries. The blood therefore, being, for an instant, resisted in its onward course, a part of the force with which it was impelled is directed against the sides of the arteries; under this force their elastic walls dilate, stretching enough to receive the blood, and as they stretch, becoming more tense and more resisting. Thus, by yielding, they break - the shock of the force impelling the blood. On the subsidence of the pressure, when the ventricles cease contracting, the arteries are able, by the same elasticity, to resume their former calibre; (*b*) It equalizes the current of the blood by maintaining pressure on it in the arteries during the periods at which the ventricles are at rest or dilating. If the arteries had .been rigid tubes, the blood, instead of flowing, as it does,. in a constant stream, would have been propelled through the arterial system in a series of jerks corresponding to the ventricular contractions, with intervals of almost complete rest during the inaction of the ventricles.. But in the actual condition of the arteries, the force of the successive contractions of the ventricles is expended partly in the direct propulsion of the blood, and partly in the dilatation of the elastic arteries; and in the intervals between the contractions of the ventricles, the force of the recoil is employed in continuing the same direct propulsion. Of course, the pressure they exercise is equally diffused in every direction, and the blood tends to move backward as well as onward, but all movement backward is pre-

FIG. 121.—Surface view of an artery from the mesentery of a frog, ensheathed in a perivascular lymphatic vessel. *a.* The artery, with its circular muscular coat (media) indicated by broad transverse markings, with an indication of the adventitia outside. *l.* Lymphatic vessel; its wall is a simple endothelial membrane. (Klein and Noble Smith.)

vented by the closure of the semilunar arterial valves (p. 114), which takes place at the very commencement of the recoil of the arterial walls.

By this exercise of the elasticity of the arteries, all the force of the ventricles is made advantageous to the circulation; for that part of their force which is expended in dilating the arteries, is restored in full when they recoil. There is thus no loss of force; but neither is there any gain, for the elastic walls of the artery cannot originate any force for the propulsion of the blood—they only restore that which they received from the ventricles. The force with which the arteries are dilated every time the ventricles contract, might be said to be received by them in store, to be all given out again in the next succeeding period of dilatation of the ventricles. It is by this equalizing influence of the successive branches of every artery that, at length, the intermittent accelerations produced in the arterial current by the action of the heart, cease to be observable, and the jetting stream is converted into the continuous and equable movement of the

blood which we see in the capillaries and veins. In the production of a
continuous stream of blood in the smaller arteries and capillaries, the
resistance which is offered to the blood-stream in these vessels (p. 158),
is a necessary agent. Were there no greater obstacle to the *escape* of
blood from the larger arteries than exists to its *entrance* into them from
the heart, the stream would be intermittent, notwithstanding the elasticity of the walls of the arteries.

(*c.*) By means of the elastic tissue in their walls (and of the muscular
tissue also), the arteries are enabled to dilate and contract readily in correspondence with any temporary increase or diminution of the total
quantity of blood in the body; and within a certain range of diminution

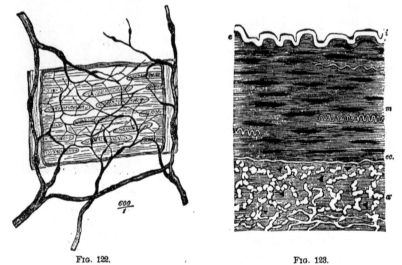

FIG. 122.   FIG. 123.

FIG. 122.—Ramification of nerves and termination in the muscular coat of a small artery of the
frog. (Arnold.)
FIG. 123.—Transverse section through a large branch of the inferior mesenteric artery of a pig.
e, endothelial membrane; *i*, tunica elastica interna, no subendothelial layer is seen; *m*, muscular tunica media, containing only a few wavy elastic fibres; *ee*, tunica elastica externa, dividing the media
from the connective tissue adventitia, *a.* (Klein and Noble Smith.) × 350.

of the quantity, still to exercise due pressure on their contents; (*d.*) The
elastic tissue assists in restoring the normal state after *diminution* of its
calibre, whether this has been caused by a contraction of the muscular
coat, or the temporary application of a compressing force from without.
This action is well shown in arteries which, having contracted by means
of their muscular element, after death, regain their average patency on
the cessation of post-mortem rigidity (p. 142). (*e.*) By means of their
elastic coat the arteries are enabled to adapt themselves to the different
movements of the several parts of the body.

Tension of Arteries.—The natural state of all arteries, in regard at least
to their length, is one of tension—they are always more or less stretched,
and ever ready to recoil by virtue of their elasticity, whenever the oppos-

ing force is removed. The extent to which the divided extremities of arteries retract is a measure of this tension, not of their elasticity. (Savory.)

Functions of the Muscular Coat.—The most important office of the *muscular* coat is, (1) that of regulating the quantity of blood to be received by each part or organ, and of adjusting it to the requirements of each, according to various circumstances, but, chiefly, according to the activity with which the functions of each are at different times performed. The amount of work done by each organ of the body varies at different times, and the variations often quickly succeed each other, so that, as in the brain, for example, during sleep and waking, within the same hour a part may be now very active and then inactive. In all its active exercise of function, such a part requires a larger supply of blood than is sufficient for it during the times when it is comparatively inactive. It is evident that the heart cannot regulate the supply to each part at different periods; neither could this be regulated by any general and uniform contraction of the arteries; but it may be regulated by the power which the arteries of each part have, in their muscular tissue, of contracting so as to diminish, and of passively dilating or yielding so as to permit an increase of the supply of blood, according to the requirements of the part to which they are distributed. And thus, while the ventricles of the heart determine the total quantity of blood, to be sent onward at each contraction, and the force of its propulsion, and while the large and merely elastic arteries distribute it and equalize its stream, the smaller arteries, in addition, regulate and determine, by means of their muscular tissue, the proportion of the whole quantity of blood which shall be distributed to each part.

It must be remembered, however, that this regulating function of the arteries is itself governed and directed by the nervous system (vaso-motor centres and fibres).

Another function of the muscular element of the middle coat of arteries is (2), to co-operate with the elastic in adapting the calibre of the vessels to the quantity of blood which they contain. For the amount of fluid in the blood-vessels varies very considerably even from hour to hour, and can never be quite constant; and were the elastic tissue only present, the pressure exercised by the walls of the containing vessels on the contained blood would be sometimes very small, and sometimes inordinately great. The presence of a muscular element, however, provides for a certain uniformity in the amount of pressure exercised; and it is by this adaptive, uniform, gentle, muscular contraction, that the normal *tone* of the blood-vessels is maintained. Deficiency of this *tone* is the cause of the soft and yielding pulse, and its unnatural excess, of the hard and tense one.

The elastic and muscular contraction of an artery may also be regarded as fulfilling a natural purpose when (3), the artery being cut, it first limits and then, in conjunction with the coagulated fibrin, arrests the escape of blood. It is only in consequence of such contraction and coagulation that we are free from danger through even very slight wounds; for it is only when the artery is closed that the processes for the more permanent and secure prevention of bleeding are established.

(4) There appears no reason for supposing that the muscular coat assists, to more than a very small degree, in propelling the onward current of blood.

(1.) When a small artery in the living subject is exposed to the air or cold, it gradually but manifestly contracts. Hunter observed that the posterior tibial artery of a dog when laid bare, became in a short time so much contracted as almost to prevent the transmission of blood; and the observation has been often and variously confirmed. Simple elasticity could not effect this.

(2.) When an artery is cut across, its divided ends contract, and the orifices may be completely closed. The rapidity and completeness of this contraction vary in different animals; they are generally greater in young than in old animals; and less, apparently, in man than in the lower animals. This contraction is due in part to elasticity, but in part, also, to muscular action; for it is generally increased by the application of cold, or of any simple stimulating substances, or by mechanically irritating the cut ends of the artery, as by picking or twisting them.

(3.) The contractile property of arteries continues many hours after death, and thus affords an opportunity of distinguishing it from their elasticity. When a portion of an artery of a recently killed animal is exposed, it gradually contracts, and its canal may be thus completely closed: in this contracted state it remains for a time, varying from a few hours to two days: then it dilates again, and permanently retains the same size.

This persistence of the contractile property after death was well shown in an observation of Hunter, which may be mentioned as proving, also, the greater degree of contractility possessed by the smaller than by the larger arteries. Having injected the uterus of a cow, which had been removed from the animal upward of twenty-four hours, he found, after the lapse of another day, that the larger vessels had become much more turgid than when he injected them, and that the smaller arteries had contracted so as to force the injection back into the larger ones.

THE PULSE.

If one extremity of an elastic tube be fastened to a syringe, and the other be so constricted as to present an obstacle to the escape of fluid, we shall have a rough model of what is present in the living body:—The syringe representing the heart, the elastic tube the arteries, and the contracted orifice the arterioles (smallest arteries) and capillaries. If the apparatus be filled with water, and if a finger-tip be placed on any part of the elastic tube, there will be felt with every action of the syringe, an impulse or beat, which corresponds exactly with what we feel in the arteries of the living body with every contraction of the heart, and call the *pulse*. The pulse is essentially caused by an expansion *wave*, which is due to the injection of blood into an already full aorta; which blood expanding the vessel produces the ·pulse in it, almost coincidently with the systole of the left ventricle. As the force of the left ventricle, however, is not expended in dilating the aorta only, the wave of blood passes on, expanding the arteries as it goes, running as it were on the surface of the more slowly traveling blood already contained in them, and producing the pulse as it proceeds.

The distension of each artery increases both its length and its diameter. In their elongation, the arteries change their form, the straight ones becoming slightly curved, and those already curved becoming more so, but they recover their previous form as well as their diameter when the ventricular contraction ceases, and their elastic walls recoil. The increase of their curves which accompanies the distension of arteries, and the succeeding recoil, may be well seen in the prominent temporal artery of an old person. In feeling the pulse, the finger cannot distinguish the sensation produced by the dilatation from that produced by the elongation and

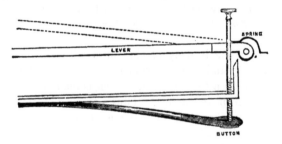

FIG. 124.—Diagram of the mode of action of the Sphygmograph.

curving; that which it perceives most plainly, however, is the dilatation, or return, more or less, to the cylindrical form, of the artery which has been partially flattened by the finger.

The pulse—due to any given beat of the heart—is not perceptible at the same moment in all the arteries of the body. Thus,—it can be felt in the carotid a very short time before it is perceptible in the radial artery, and in this vessel again before the dorsal artery of the foot. The delay in the beat is in proportion to the distance of the artery from the heart, but the difference in time between the beat of any two arteries never exceeds probably $\frac{1}{4}$ to $\frac{1}{3}$ of a second.

A distinction must be carefully made between the passage of the *wave* along the arteries and the velocity of the *stream* (p. 165) of blood. Both wave and current are present; but the rates at which they travel are very different; that of the wave 16·5 to 33 feet per second (5 to 10 metres) being twenty or thirty times as great as that of the current.

The Sphygmograph.—A great deal of light has been thrown on what may be called the form of the pulse by the sphygmograph (Figs. 124 and 125). The principle on which the sphygmograph acts is very simple (see Fig. 124). The small button replaces the finger in the act of taking the pulse, and is made to rest lightly on the artery, the pulsations of which it is desired to investigate. The up-and-down movement of the button is communicated to the lever, to the hinder end of which is attached a slight spring, which allows the lever to move up, at the same time that it is just strong enough to resist its making any sudden jerk,

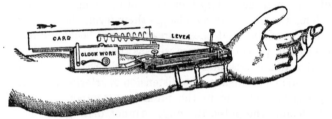

FIG. 125.—The Sphygmograph applied to the arm.

and in the interval of the beats also to assist in bringing it back to its original position. For ordinary purposes the instrument is bound on the wrist (Fig. 125).

It is evident that the beating of the pulse with the reaction of the spring will cause an up-and-down movement of the lever, the pen of which will write the effect on a smoked card, which is made to move by clockwork in the direction of the arrow. Thus a tracing of the pulse is obtained, and in this way much more delicate effects can be seen, than can be felt on the application of the finger.

The pulse-tracing differs somewhat according to the artery upon which the sphygmograph is applied, but its general characters are much the same in all cases. It consists of:—A sudden upstroke (Fig. 126, A), which

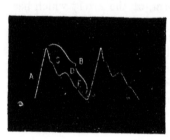

FIG. 126.— Diagram of pulse tracing. A, upstroke; B, down-stroke; C, predicrotic wave; D, dicrotic; E, post dicrotic wave.

is somewhat higher and more abrupt in the pulse of the carotid and of other arteries near the heart than in the radial and other arteries more remote; and a gradual decline (B), less abrupt, and therefore taking a longer time than (A). It is seldom, however, that the decline is an uninterrupted fall: it is usually marked about half-way by a distinct notch (C), called the *dicrotic notch*, which is caused by a second more or less marked ascent of the lever at that point by a second wave called the *dicrotic wave* (D); not unfrequently (in which case the tracing is said to have a double apex) there is also soon after the commencement of the descent a slight ascent previous to the dicrotic notch, this is called the *predicrotic wave* (C), and in addition there may be one or more slight ascents after the dicrotic, called *post dicrotic* (E).

The explanation of these tracings presents some difficulties, not, however, as regards the two primary factors, viz., the upstroke and downstroke, because they are universally taken to mean the sudden injection
of blood into the already full arteries, and that this passes through the
artery as a wave and expands them, the gradual fall of the lever signify-
ing the recovery of the arteries by their recoil. It may be demonstrated
on a system of elastic tubes, such as was described above, where a syringe
pumps in water at regular intervals, just as well as on the radial artery,
or on a more complicated system of tubes in which the heart, the arteries,
the capillaries and veins are represented, which is known as an *arterial
schema.* If we place two or more sphygmographs upon such a system
of tubes at increasing distances from the pump, we may demonstrate

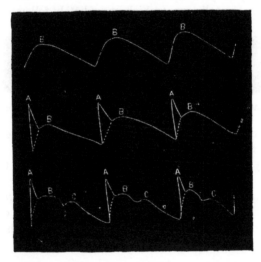

Fig., 127.—Diagram of the formation of the pulse-tracing. A, percussion wave; B, tidal wave;
C, dicrotic wave. (Mahomed.)

that the rise of the lever commences first in that nearest the pump,
and is higher and more sudden, while at a longer distance from the pump
the wave is less marked, and a little later. So in the arteries of the body
the wave of blood gradually gets less and less as we approach the periphery
of the arterial system, and is lost in the capillaries. By the sudden injection of blood two distinct waves are produced, which are called the
tidal and *percussion* waves. The tidal wave occurs whenever fluid is
injected into an elastic tube (Fig. 127, B), and is due to the expansion of
the tube and its more gradual collapse. The percussion wave occurs
(Fig. 127, A) when the impulse imparted to the fluid is more sudden;
this causes an abrupt upstroke of the lever, which then falls until it is
again caught up perhaps by the tidal wave which begins at the same time
but is not so quick.

In this way, generally speaking, the apex of the upstroke is double, the second upstroke, the so-called predicrotic elevation of the lever, representing the tidal wave. The double apex is most marked in tracings from large arteries, especially when their tone is deficient. In tracings,

FIG. 128.—Pulse-tracing of radial artery, somewhat deficient in tone. (Sanderson.)

on the other hand, from arteries of medium size, *e.g.*, the radial, the upstroke is usually single. In this case the percussion-impulse is not sufficiently strong to jerk up the lever and produce an effect distinct from that of the systolic *wave* which immediately follows it, and which

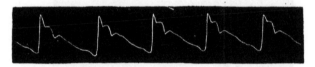

FIG. 129.—Pulse-tracing of radial artery, with double apex. (Sanderson.)

continues and completes the distension. In cases of feeble arterial tension, however, the percussion-impulse may be traced by the sphygmograph, not only in the carotid pulse, but to a less extent in the radial also (Fig. 129).

The interruptions in the downstroke are called the *katacrotic* waves, to distinguish them from an interruption in the upstroke, called the *anacrotic* wave, which is occasionally met with in cases in which the predicrotic or tidal wave is higher than the percussion wave.

FIG. 130.—Anacrotic pulse from a case of aortic aneurism. A, anacrotic wave (or percussion wave). B, tidal or predicrotic wave, continued rise in tension (or higher tidal wave).

There is considerable difference of opinion as to whether the dicrotic wave is present in health generally, and also as to its cause. The balance of opinion appears to be in favor of the belief of its presence in health, although it may be very faint; while, at any rate, in certain conditions not necessarily diseased, it becomes so marked as to be quite plain to the unaided finger. Such a pulse is called *dicrotic*. Sometimes the dicrotic rise exceeds the initial upstroke, and the pulse is then called *hyperdicrotic*.

As to the cause of dicrotism, one opinion is that it is due to a recovery of pressure during the elastic recoil, in consequence of a rebound from the periphery, and it may indeed be produced on a schema by obstructing the tube at a little distance beyond the spot where the sphygmograph is placed. Against this view, however, is the fact that the notch appears at about the same point in the downstroke in tracings from the carotid and from the radial, and not first in the radial tracing, as it should do, since that artery is nearer the periphery than the carotid, and as it does in the corresponding experiment with the arterial schema when the tube is obstructed. The generally accepted notion among clinical observers, is that the dicrotic wave is due to the rebound from the aortic valves causing a second wave; but the question cannot be considered settled, and the presence of marked dicrotism in cases of hæmorrhage, of anæmia, and of other weakening conditions, as well as its presence in cases of diminished pressure within the arteries, would imply that it might, at any rate sometimes, be due to the altered specific gravity of the blood within the vessels, either directly or through the indirect effect of these conditions on the tone ôf the arterial walls. Waves may be produced in any elastic tube when a fluid is being driven through it with an intermittent force, such waves being called *waves of oscillation* (M. Foster). They have received various explanations. In an arterial schema they vary with the specific gravity of the fluid

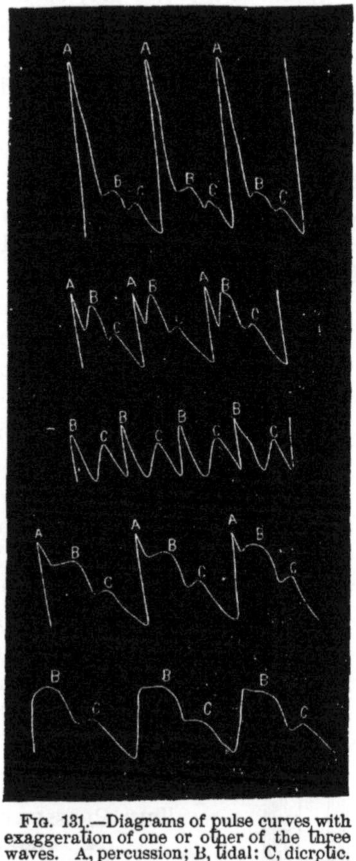

Fig. 131.—Diagrams of pulse curves with exaggeration of one or other of the three waves. A, percussion; B, tidal: C, dicrotic. 1, percussion wave very marked; 2, tidal wave sudden; 3, dicrotic pulse curve; 4 and 5, the tidal wave very exaggerated, from high tension. (Mahomed.)

used, and with the kind of tubing, and may be therefore supposed to vary in the body with the condition of the blood and of the arteries.

Some consider the secondary waves in the downstroke of a normal wave to be due to oscillation; but, as just mentioned, even if this be the case, as is most likely, with post-dicrotic waves, the dicrotic wave itself is almost certainly due to the rebound from the aortic valves.

The anacrotic notch is usually associated with disease of the arteries, *e.g.*, in atheroma and aneurism. The dicrotic notch is called diastolic or aortic, and indicates closure of the aortic valves.

Of the three main parts then of a pulse-tracing, viz., the percussion wave, the tidal, and the dicrotic, the percussion wave is produced by sudden and forcible contraction of the heart, perhaps exaggerated by an excited action, and may be transmitted much more rapidly than the tidal wave, and so the two may be distinct; frequently, however, they are inseparable. The dicrotic wave may be as great or greater than the other two.

According to Mahomed, the distinctness of the three waves depends upon the following conditions:—

The *percussion wave* is increased by:—1. Forcible contraction of the Heart; 2. Sudden contraction of the Heart; 3. Large volume of blood; 4. Fulness of vessel; and diminished by the reversed conditions.

The *tidal wave* is increased by:—1. Slow and prolonged contraction of the Heart; 2. Large volume of blood; 3. Comparative emptiness of vessels; 4. Diminished outflow or slow capillary circulation; and diminished by the reversed conditions.

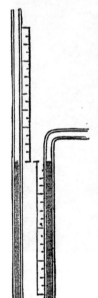

The *dicrotic wave* is increased by:—1. Sudden contraction of the Heart; 2. Comparative emptiness of vessels; 3. Increased outflow or rapid capillary circulation; 4. Elasticity of the aorta; 5. Relaxation of muscular coat; and diminished by the reversed conditions.

One very important precaution in the use of the sphygmograph lies in the careful regulation of the pressure. If the pressure be too great, the characters of the pulse may be almost entirely obscured, or the artery may be entirely obstructed, and no tracing is obtained; and on the other hand, if the pressure be too slight, a very small part of the characters may be represented on the tracing.

THE PRESSURE OF THE BLOOD WITHIN THE ARTERIES (PRODUCING ARTERIAL TENSION).

Fig. 132.—Diagram of mercurial manometer.

It will be understood from the foregoing that the arteries in a normal condition, are continually on the stretch during life, and in consequence of the injection of more blood at each systole of the ventricle into the elastic aorta, this stretched condition is exaggerated each time the ventricle empties itself. This condition of the arteries is due to the pressure of blood within them, because of the resistance presented by the smaller arteries and capillaries (peripheral resistance) to the emptying of the arterial system in the intervals between the contractions of the ventricle, and is called the condition of *arterial tension*. On the other hand, it must be equally clear that, as the blood is forcibly injected into the already full

arteries against their elasticity, it must be subjected to the pressure of
the arterial walls, the elastic recoil sending on the blood after the immediate effect of the systole has passed; so that, when an artery is cut across,
the blood is projected forward by this force for a considerable distance;
at each ventricular systole, a jet of blood escaping, although the stream
does not cease flowing during the diastole.

The relations which exist between the arteries and their contained
blood are obviously of the utmost importance to the carrying on of the
circulation, and it therefore becomes necessary to be able to gauge the

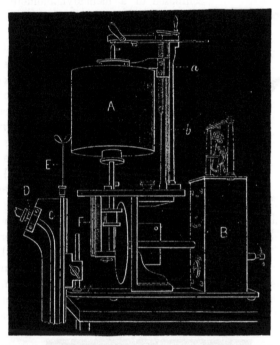

Fig. 133.—Diagram of mercurial kymograph. A, revolving cylinder, worked by a clockwork arrangement contained in the box (B), the speed being regulated by a fan above the box; cylinder supported by an upright (b), and capable of being raised or lowered by a screw (a), by a handle attached
to it; D, C, E, represent mercurial manometer, a somewhat different form of which is shown in next
figure.

alterations in blood-pressure very accurately.   This may be done by
means of a *mercurial manometer* in the following way:—The short horizontal limb of this (Fig. 132, 1) is connected, by means of an elastic tube
and cannula, with the interior of an artery; a solution of sodium or potassium carbonate being previously introduced into this part of the apparatus to prevent coagulation of the blood.   The blood-pressure is thus
communicated to the upper part of the mercurial column (2); and the
depth to which the latter sinks, added to the height to which it rises in
the other (3), will give the height of the mercurial column which the

blood-pressure balances; the weight of the soda solution being sub-tracted.

For the estimation of the arterial tension at any given moment, no further apparatus than this, which is called Poiseuille's *hæmadynamometer*, is necessary; but for noting the *variations* of pressure in the arterial sys-tem, as well as its absolute amount, the instrument is usually combined

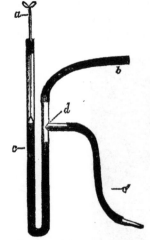

with a *registering* apparatus and in this form is called a *kymograph*.

The kymograph, invented by Ludwig, is composed of a hæmadynamometer, the open mercurial column of which supports a floating piston and vertical rod, with short horizontal pen (Fig. 134). The pen is adjusted in con-tact with a sheet of paper, which is caused to move at a uniform rate by clockwork; and thus the up-and-down movements of the mer-curial column, which are communicated to the rod and pen, are marked or *registered* on the moving paper, as in the registering apparatus of the sphygmograph, and minute variations are graphically recorded (Fig. 135).

For some purposes the *spring kymograph* of Fick (Fig. 136) is preferable to the mercurial kymograph. It consists of a hollow C-shaped spring, filled with fluid, the interior of which is brought into connection with the interior

FIG. 134.—Diagram of mercu-rial manometer. *a.* Floating rod and pen. *b.* Tube, which commu-nicates with a bottle containing an alkaline solution. *c'.* Elastic tube and cannula, the latter being intended for insertion in an artery.

of an artery, by means of a flexible metallic tube and cannula. In response to the pressure transmitted to its interior, the spring, *c*, tends to straighten itself, and the movement thus produced is communicated by means of a lever, *b*, to a writing-needle and registering apparatus.

FIG. 135.—Normal tracing of arterial pressure in the rabbit obtained with the mercurial kymo-graph. The smaller undulations correspond with the heart beats; the larger curves with the respir-atory movements. (Burdon-Sanderson.)

Fig. 137 exhibits an ordinary arterial pulse-tracing, as obtained by the spring-kymograph.

From observations which have been made by means of the mercurial manometer, it has been found that the pressure of blood in the carotid of a rabbit is capable of supporting a column of 2 to 3½ inches (50 to 90 mm.) of mercury, in the dog 4 to 7 inches (100 to 175 mm.), in the horse 5 to 8 inches (150 to 200 mm.), and in man about the same.

To measure the absolute amount of this pressure in any artery, it is necessary merely to multiply the area of its transverse section by the height of the column of mercury which is already known to be supported by the blood-pressure in any part of the arterial system. The weight of a column of mercury thus found will represent the pressure of the blood. Calculated in this way, the blood-pressure in the human aorta is

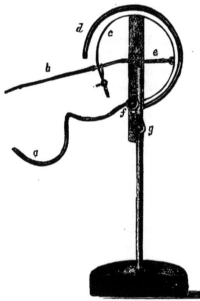

FIG. 136.—A form of Fick's Spring Kymograph. *a*, tube to be connected with artery; *c*, hollow spring, the movement of which moves *b*, the writing lever; *e*, screw to regulate height of *b*; *d*, outside protective spring; *g*, screw to fix on the upright of the support.

equal to 4 lb. 4 oz. avoirdupois; that in the aorta of the horse being 11 lb. 9 oz.; and that in the radial artery at the human wrist only 4 drs. Supposing the muscular power of the right ventricle to be only one-half that of the left, the blood-pressure in the pulmonary artery will be only 2 lb. 2 oz. avoirdupois. The amounts above stated represent the arterial tension at the time of the ventricular contraction.

The blood-pressure is greatest in the left ventricle and at the beginning of the aorta, and decreases toward the capillaries. It is greatest in the arteries at the period of the ventricular systole, and is least in the auricles, during diastole, when the pressure there and in the great veins becomes, as we have seen, negative. The mean arterial pressure equals the average of the pressures in all the arteries. The pressure in the veins is never more than one-tenth of the pressure in the corresponding arteries and is greatest at the time of auricular systole. There is no periodic variation in venous pressure, as there is in the arterial, except in the great veins.

FIG. 137.—Normal arterial tracing obtained with Fick's kymograph in the dog. (Burdon-Sanderson.)

Variations of Blood Pressure.—Many circumstances cause considerable variations in the amount of the blood-pressure. The following are the chief:—(1) *Changes in the beat of the Heart;* (2) *Changes in the Arteries and Capillaries;* (3) *Changes due to Nerve Action;* (4) *Changes in the Blood;* (5) *Respiratory Changes.*

1. *Changes in the Beat of the Heart.*—The systole and diastole of the muscular chambers. The arterial tension increases during systole and diminishes during diastole. The greater the frequency, moreover, of the heart's contractions, the greater is the blood-pressure, *cæteris paribus;* although this effect is not constant, as it may be compensated for by the delivery into the arteries at each beat of a comparatively small quantity of blood. The greater the quantity of blood expelled from the heart at each contraction the greater is the blood-pressure.

The quantity and quality of the blood nourishing the heart's substance through the coronary arteries must exercise also a very considerable influence upon its action, and therefore upon the blood-pressure.

2. *Changes in the Arteries and Capillaries.*—Variations in the degree of contraction of the smaller arteries modify the blood-pressure by favoring or impeding the accumulation of blood in the arterial system which follows every contraction of the heart; the contraction of the arterial walls increasing the blood-pressure, while their relaxation lowers it.

3. *Changes due to Nerve Action.*—As with the heart, so with the blood-vessels, the action of the nervous system is very important in relation to the blood-pressure; regulating, as it does, not only the force, frequency, and length of the heart's systole, but also the condition of the arteries, both through the central and peripheral vaso-motor centres. As this subject has not yet been fully considered it will be as well to treat of it here.

It is upon the muscular coat of the arteries that the nervous system exercises its influence; the elastic element possessing, as must be obvious, rather physical than vital properties. The muscular tissue in the walls of the vessels increases relatively to the other coats as the arteries grow smaller, so that in the smallest arteries it is developed out of all proportion to the other elements; in fact, in passing from capillary vessels, made up as we have seen of endothelial cells with a ground substance, the first change which occurs as the vessels become larger (on the side of the arteries) is the appearance of muscular fibres. Thus the nervous system is more powerful in regulating the calibre of the smaller than of the larger arteries.

It has been shown that if the cervical sympathetic nerve be divided in a rabbit, the blood-vessels of the corresponding side become dilated. The effect is best seen in the ear, which if held up to the light is seen to become redder, and the arteries to become larger. The whole ear is distinctly warmer than the opposite one. This effect is produced by removing the arteries from the influence of the central nervous system, which

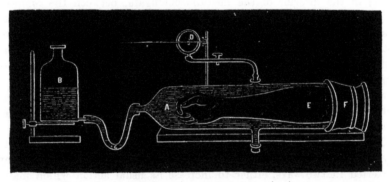

FIG. 138.—Plethysmograph. By means of this apparatus, the alteration in volume of the arm, E, which is enclosed in a glass tube, A, filled with fluid, the opening through which it passes being firmly closed by a thick gutta percha band, F, is communicated to the lever, D, and registered by a recording apparatus. The fluid in A communicates with that in B, the upper limit of which is above that in A. The chief alterations in volume are due to alteration in the blood contained in the arm. When the volume is increased, fluid passes out of the glass cylinder, and the lever, D, also is raised, and when a decrease takes place the fluid returns again from B to A. It will therefore be evident that the apparatus is capable of recording alterations of blood-pressure in the arm. Apparatus founded upon the same principle have been used for recording alterations in the volume of the spleen and kidney.

influence usually passes down the divided nerve; for if the peripheral end of the divided nerve (*i.e.*, that farthest from the brain) be stimulated, the arteries which were before dilated return to their natural size, and the parts regain their primitive condition. And, besides this, if the stimulus which is applied be too strong or too long continued, the point of normal constriction is passed, and the vessels become much more contracted than normal. The natural condition, which is somewhere about midway between extreme contraction and extreme dilatation, is called the natural *tone* of an artery, and if this be not maintained, the vessel is said to have lost tone, or if it be exaggerated, the tone is said to be too great. The influence of the nervous system upon the vessels consists in maintaining a natural tone. The effects described as having been produced by section of the cervical sympathetic and by subsequent stimulation are not peculiar to that nerve, as it has been found that for every part of the body there exists a nerve the division of which produces the same effects, viz., dilatation of the arteries; such may be cited as the case with the sciatic, the splanchnic nerves, and the nerves of the brachial plexus: when divided, dilatation of the blood-vessels in the parts supplied by them taking place. It appears, therefore, that nerves exist which have a distinct control over the vascular supply of a part.

These nerves are called *vaso-motor;* or, since they seem to run now in cerebro-spinal nerves, now in the sympathetic, we speak of those nerves as containing vaso-motor fibres, in addition to the fibres which have other functions.

Vaso-motor centres.—Experiments by Ludwig and others show that the vaso-motor fibres come primarily from grey matter (vaso-motor *centre*) in the interior of the medulla oblongata, between the *calamus scriptorius* and the *corpora quadrigemina*. Thence the vaso-motor fibres pass down in the interior of the spinal cord, and issuing with the anterior roots of the spinal nerves, traverse the various ganglia on the præ-vertebral cord of the sympathetic, and, accompanied by branches from these ganglia, pass to their destination.

Secondary or subordinate centres exist in the spinal cord, and local centres in various regions of the body, and through these, directly under ordinary circumstances, vaso-motor changes are also effected.

The influence exerted by the chief vaso-motor centre is called into play in several ways, but chiefly by afferent (sensory) stimuli, and it may be exerted in two ways, either to increase its usual action which maintains a medium tone of the arteries or to diminish such action. This afferent influence upon the centre may be extremely well shown by the action of a nerve the existence of which was demonstrated by Cyon and Ludwig, and which is called the *depressor,* because of its characteristic influence on the blood-pressure.

Depressor Nerve.—This small nerve arises, in the rabbit, from the superior laryngeal branch, or from this and the trunk of the pneumogastric nerve, and after communicating with filaments of the inferior cervical ganglion proceeds to the heart.

If during an observation of the blood-pressure of a rabbit this nerve be divided, and the central end (*i.e.,* that nearest the brain) be stimulated, a remarkable fall of blood-pressure ensues (Fig. 139).

The cause of the fall of blood-pressure is found to proceed from the dilatation of the vascular district supplied by the splanchnic nerves, in consequence of which it holds a much larger quantity of blood than usual, and this very greatly diminishes the blood in the vessels elsewhere, and so materially affects the blood-pressure. This effect of the depressor nerve is presumed to prove that the nerve is a means of conveying to the vaso-motor centre indications of such conditions of the heart as require a diminution of the tension in the blood-vessels; as, for example, when the heart cannot, with sufficient ease, propel blood into the already too full or too tense arteries.

The action of the depressor nerve illustrates the effect of afferent impulses in causing an inhibition of the vaso-motor centre as regards its action upon certain arteries. There exist other nerves, however, the stimulation of the central end of which causes a reverse action of the centre, or, in other words, increases its tonic influence, and by causing

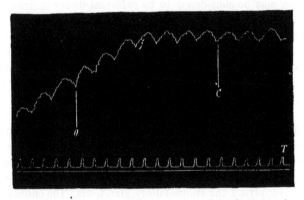

FIG. 139.—Tracing showing the effect on blood pressure of stimulating the central end of the Depressor nerve in the rabbit. To be read from right to left. T, indicates the rate at which the recording-surface was traveling, the intervals correspond to seconds; C, the moment of entrance of current; O, moment at which it was shut off. The effect is some time in developing and lasts after the current has been taken off. The larger undulations are the respiratory nerves; the pulse oscillations are very small. (M. Foster.)

considerable constriction of certain arterioles, either locally or generally, increases the blood-pressure. Moreover, the effect of stimulating an afferent nerve may be to dilate or constrict the arteries either generally or in the part supplied by the afferent nerve; and it is said that stimulation of an afferent nerve may produce a kind of paradoxical effect, causing *general* vascular constriction and so general increase of blood-pressure but at the same time *local* dilatation. This must evidently have an immense influence in increasing the flow of blood through a part.

Not only may the vaso-motor centre be reflexly affected, but it may also be affected by impulses proceeding to it from the cerebrum, as in the case of blushing from mind disturbance, or of pallor from sudden fear. It will be shown, too, in the chapter on Respiration that the circulation of deoxygenated blood may directly stimulate the centre itself.

Local Tonic Centres.—Although the tone of the arteries is influenced by the centres in the cerebro-spinal axis, certain experiments point out that this is not the only way in which it may be affected. Thus the dilatation which occurs after section of the cervical sympathetic in the first experiment cited above, only remains for a short time, and is soon followed—although a portion of the nerve may have been removed entirely—by the vessels regaining their ordinary calibre; and afterward local stimulation, *e.g.*, the application of heat or cold, will cause dilatation or constriction. From this it is probable that there exists a local mechanism distinct for each vascular area, and that the effect produced by the central nervous system acts through it much in the same way as the cardio-inhibitory centre in the medulla acts upon the heart through the ganglia contained within its muscular substance.

Central impulses may inhibit or increase the action of these local centres, which may be considered to be sufficient under ordinary circum-stances to maintain the local tone of the vessels. The observations upon the functions of the vaso-motor nerves appear to divide them into four classes: (1) those on division of which dilatation occurs for some time, and which on stimulation of their peripheral end produce constriction; (2) those on division of which momentary dilatation followed by constric-tion occurs, with dilatation on stimulation; (3) those on division of which dilatation is caused, which lasts for a limited time, with constriction if stimulated at once, but dilatation if some time is allowed to elapse before the stimulation is applied; (4) a class, division of which produces no effect but which, on stimulation, cause according to their function either dilatation or constriction. A good example of this fourth class is afforded by the nerves supplying the submaxillary gland, viz., the chorda tympani and the sympathetic. When either of these nerves is simply divided, no change takes place in the vessels of the gland; but on stimulating the chorda tympani the vessels dilate, and, on the other hand, when the sympathetic is stimulated the vessels contract. The nerves acting like the chorda tympani in this case are called *vaso-dilators,* and those like the sympathetic *vaso-constrictors.* The third class, which produce at one time dilatation, at another time constriction, are believed to contain both kinds of vaso-motor nerve-fibres, or to act as dilators or contractors according to the condition of the local apparatus. It is probable that these nerves act by inhibiting or augmenting the action of the local nerv-ous mechanism already referred to; and as they are in connection with the central nervous system, it is through this arrangement that that sys-tem is capable of influencing or of maintaining the normal local tone.

It may also be supposed that the local nerve-centres themselves may be directly affected by the condition of blood nourishing them.

The following table may serve as a summary of the effect of the nerv-ous system upon the arteries and so upon the blood-pressure —

A. An increase of the blood-pressure may be produced:—

(1.) By stimulation of the vaso-motor centre in medulla, either
α. *Directly,* as by carbonated or deoxygenated blood.
β. *Indirectly,* by impressions descending from the cerebrum, *e.g.*, in sudden pallor.
γ. *Reflexly,* by stimulation of sensory nerves anywhere.

(2.) By stimulation of the centres in spinal cord.
Possibly directly or indirectly, certainly reflexly.

(3.) By stimulation of the local centres for each vascular area, by the vaso-constrictor nerves, or directly by means of altered blood.

B. A decrease of the blood pressure may be produced:—

(1.) By stimulation of the vaso-motor centre in medulla, either

(α.) *Directly*, as by oxygenated or aërated blood.

(β.) *Indirectly*, by impressions descending from the cerebrum —*e.g.*, in blushing.

(γ.) *Reflexly*, by stimulation of the *depressor* nerve, and consequent dilatation of vessels of splanchnic area, and possibly by stimulation of other sensory nerves, the sensory impulse being interpreted as an indication for diminished blood-pressure.

(2.) By stimulation of the centres in spinal cord. Possibly directly, indirectly, or reflexly.

(3.) By stimulation of local centres for each vascular area by the vaso-dilator nerve, or directly by means of altered blood.

4. *Changes in the blood.*—*a.* As regards *quantity*. At first sight it would appear that one of the easiest ways to diminish the blood-pressure would be to remove blood from the vessels by bleeding; it has been found by experiment, however, that although the blood-pressure sinks whilst large abstractions of blood are taking place, as soon as the bleeding ceases it rises rapidly, and speedily becomes normal; that is to say, unless so large an amount of blood has been taken as to be positively dangerous to life, abstraction of blood has little effect upon the blood-pressure. The rapid return to the normal pressure is due not so much to the withdrawal of lymph and other fluids from the body into the blood, as was formerly supposed, as to the regulation of the peripheral resistance by the vaso-motor nerves; in other words, the small arteries contract, and in so doing maintain pressure on the blood and favor its accumulation in the arterial system. This is due to the stimulation of the vaso-motor centre from diminution of the supply of blood, and therefore of oxygen. The failure of the blood-pressure to return to normal in the too great abstraction must be taken to indicate a condition of exhaustion of the centre, and consequently of want of regulation of the peripheral resistance. In the same way it might be thought that injection of blood into the already pretty full vessels would be at once followed by rise in the blood-pressure, and this is indeed the case up to a certain point—the pressure does rise, but there is a limit to the rise. Until the amount of blood injected equals about 2 to 3 per cent. of the body weight the pressure continues to rise gradually; but if the amount exceed this proportion, the rise does not continue. In this case therefore, as in the opposite when blood is ab-

stracted, the vaso-motor apparatus must counteract the great increase of pressure by dilating the small vessels, and so diminishing the peripheral resistance, for after each rise there is a partial fall of pressure; and after the limit is reached the whole of the injected blood displaces, as it were, an equal quantity which passes into the small veins, and remains within them. It should be remembered that the veins are capable of holding the whole of the blood of the body.

The amount of blood supplied to the heart both to its substance and to its chambers, has a marked effect upon the blood-pressure.

b. As regards *quality.* The quality of the blood supplied to the heart has a distinct effect upon its contraction, as too watery or too little oxygenated blood must interfere with its action. Thus it appears that blood containing certain substances affects the peripheral resistance by acting upon the muscular fibres of the arterioles themselves or upon the local centres, and so altering directly, as it were, the calibre of the vessels.

5. *Respiratory changes* affecting the blood-pressure will be considered in the next Chapter.

CIRCULATION IN THE CAPILLARIES.

When seen in any transparent part of a living adult animal by means of the microscope (Fig. 140) the blood flows with a constant equable motion; the red blood-corpuscles moving along, mostly in single file, and bending in various ways to accommodate themselves to the tortuous course

of the capillary, but instantly recovering their normal outline on reaching a wider vessel.

It is in the capillaries that the chief resistance is offered to the progress of the blood; for in them the friction of the blood is greatly increased by the enormous multiplication of the surface with which it is brought in contact.

At the circumference of the stream in the larger capillaries, but chiefly in the small arteries and veins, in contact with the walls of the vessel, and adhering to them, there is a layer of liquor sanguinis which appears to be motionless. The existence of this *still layer*, as it is termed, is inferred both from the general fact that such an one exists in all fine tubes traversed by fluid, and from what can be seen in watching the movements of the blood-corpuscles. The red corpuscles occupy the middle of the stream and move with comparative rapidity; the colorless lymph-corpuscles run much more slowly by the walls of the vessel; while next to the wall there is often a transparent space in which the fluid appears to be at rest; for if any of the corpuscles happen to be forced within it, they move more slowly than before, rolling lazily along the side of the vessel, and often adhering to its wall. Part of this slow movement of the pale corpuscles aǹd their occasional stoppage may be due to their having a natural tendency to adhere to the walls of the vessels. Sometimes, indeed, when the motion of the blood is not strong, many of the white corpuscles collect in a capillary vessel, and for a time entirely prevent the passage of the red corpuscles.

Intermittent flow in the Capillaries.—When the peripheral resistance is greatly diminished by the dilatation of the small arteries and capillaries, so much blood passes on from the arteries into the capillaries at each stroke of the heart, that there is not sufficient remaining in the arteries to distend them. Thus, the intermittent current of the ventricular systole is not converted into a continuous stream by the elasticity of the arteries before the capillaries are reached; and so intermittency of the flow occurs in capillaries and veins and a pulse is produced. The same phenomenon may occur when the arteries become rigid from disease, and when the beat of the heart is so slow or so feeble that the blood at each cardiac systole has time to pass on to the capillaries before the next stroke occurs, the amount of blood sent at each stroke being insufficient to properly distend the elastic arteries.

Diapedesis of Blood Corpuscles.—Until within the last few years it has been generally supposed that the occurrence of any transudation from the interior of the capillaries into the midst of the surrounding tissues was confined, in the absence of injury, strictly to the fluid part of the blood; in other words, that the corpuscles could not escape from the circulating stream, unless the wall of the containing blood-vessel were ruptured. It is true that an English physiologist, Augustus Waller, affirmed, in 1846, that he had seen blood-corpuscles, both red and white, pass bodily through the wall of the capillary vessel in which they were contained (thus confirming what had been stated a short time previously by Addison); and that, as no opening could be seen before their escape, so none could be observed afterward—so rapidly was the part healed.

FIG. 141.—A large capillary from the frog's mesentery, eight hours after irritation had been set up, showing emigration of leucocytes. *a*, Cells in the act of traversing the capillary wall; *b*, some already escaped. (Frey.)

But these observations did not attract much notice until the phenomena of escape of the blood-corpuscles from the capillaries and minute veins, apart from mechanical injury, were rediscovered by Professor Cohnheim in 1867.

Cohnheim's experiment demonstrating the passage of the corpuscles through the wall of the blood-vessel, is performed in the following man-

ner. A frog is urarized, that is to say, paralysis is produced by injecting under the skin a minute quantity of the poison called urari; and the
abdomen having been opened, a portion of small intestine is drawn out,
and its transparent mesentery spread out under a microscope. After a
variable time, occupied by dilatation, following contraction of the minute
vessels and accompanying quickening of the blood-stream, there ensues a
retardation of the current, and blood-corpuscles, both red and white,
begin to make their way through the capillaries and small veins.

"Simultaneously with the retardation of the blood-stream, the leucocytes, instead of loitering here and there at the edge of the axial current, begin to crowd in numbers against the vascular wall. In this way
the vein becomes lined with a continuous pavement of these bodies, which
remain almost motionless, notwithstanding that the axial current sweeps
by them as continuously as before, though with abated velocity. Now is
the moment at which the eye must be fixed on the outer contour of the
vessel, from which, here and there, minute, colorless, button-shaped elevations spring, just as if they were produced by budding out of the wall
of the vessel itself. The buds increase gradually and slowly in size, until
each assumes the form of a hemispherical projection, of width corresponding to that of the leucocyte. Eventually the hemisphere is converted into
a pear-shaped body, the small end of which is still attached to the surface
of the vein, while the round part projects freely. Gradually the little
mass of protoplasm removes itself further and further away, and, as it
does so, begins to shoot out delicate prongs of transparent protoplasm from
its surface, in nowise differing in their aspect from the slender thread by
which it is still moored to the vessel. Finally the thread is severed and
the process is complete." (Burdon Sanderson.)

The process of *diapedesis* of the red corpuscles, which occurs under
circumstances of impeded venous circulation, and consequently increased blood-pressure, resembles closely the migration of the leucocytes, with the exception that they are squeezed through the wall of
the vessel, and do not, like them, work their way through by amœboid
movement.

Various explanations of these remarkable phenomena have been suggested. Some believe that minute openings (*stigmata* or *pseudo stomata*)
between contiguous endothelial cells (p. 133) provide the means of escape
for the blood-corpuscles. But the chief share in the process is to be found
in the vital endowments with respect to mobility and contraction of the
parts concerned—both of the corpuscles (Bastian) and the capillary wall
(Stricker). Burdon-Sanderson remarks, "the capillary is not a dead
conduit, but a tube of living protoplasm. There is no difficulty in understanding how the membrane may open to allow the escape of leucocytes,
and close again after they have passed out; for it is one of the most striking peculiarities of contractile substance that when two parts of the same

mass are separated, and again brought into contact, they melt together as if they had not been severed." ·

Hitherto, the escape of the corpuscles from the interior of the blood-vessels into the surrounding tissues has been studied chiefly in connection with pathology. But it is impossible to say, at present, to what degree the discovery may not influence all present notions regarding the nutrition of the tissues, even in health.

Vital Capillary Force.—The circulation through the capillaries must, of necessity, be largely influenced by that which occurs in the vessels on either side of them—in the arteries or the veins; their intermediate position causing them to feel at once, so to speak, any alteration in the size or rate of the arterial or venous blood-stream. Thus, the apparent contraction of the capillaries, on the application of certain irritating substances, and during fear, and their dilatation in blushing, may be referred to the action of the small arteries, rather than to that of the capillaries themselves. But largely as the capillaries are influenced by these, and by the conditions of the parts which surround and support them, their own endowments must not be disregarded. They must be looked upon, not as mere passive channels for the passage of blood, but as possessing endowments of their own (vital capillary force), in relation to the circulation. The capillary wall is actively living and contractile; and there is no reason to doubt that, as such, it must have an important influence in connection with the blood-current.

Blood-Pressure in the Capillaries.—From observations upon the web of the frog's foot, the tongue and mesentery of the frog, the tails of newts, and small fishes (Roy and Brown), as well as upon the skin of the finger behind the nail (Kries), by careful estimation of the amount of pressure required to empty the vessels of blood under various conditions, it appears that the blood-pressure is subject to variations in the capillaries, apparently following the variations of that of the arteries; and that up to a certain point, as the extravascular pressure is increased, so does the pulse in the arterioles, capillaries, and venules become more and more evident. The pressure in the first case (web of the frog's foot) has been found to be equal to about 14 to 20 mm. of mercury; in other experiments to be equal to about $\frac{1}{3}$ to $\frac{1}{2}$ of the ordinary arterial pressure.

The Circulation in the Veins.

The blood-current in the veins is maintained by the slight vis a tergo remaining of the contraction of the left ventricle. Very effectual assistance, however, to the flow of blood is afforded by the action of the muscles capable of pressing on such veins as have valves.

The effect of such muscular pressure may be thus explained. When pressure is applied to any part of a vein, and the current of blood in it is obstructed, the portion behind the seat of pressure becomes swollen and distended as far back as to the next pair of valves. These, acting like the semilunar valves of the heart, and being, like them, inextensible both in themselves and at their margins of attachment, do not follow the vein in its distension, but are drawn out toward the axis of the canal. Then, if the pressure continues on the vein, the compressed blood, tending to move equally in all directions, presses the valves down into contact at their free edges, and they close the vein and prevent regurgitation of the blood. Thus, whatever force is exercised by the pressure of the muscles on the veins, is distributed partly in pressing the blood onward in the proper course of the circulation, and partly in pressing it backward and closing the valves behind (Fig. 128, A and B).

The circulation might lose as much as it gains by such compression of the veins, if it were not for the numerous anastomoses by which they communicate, one with another; for through these, the closing up of the venous channel by the backward pressure is prevented from being any serious hindrance to the circulation, since the blood, of which the onward course is arrested by the closed valves, can at once pass through some anastomosing channel, and proceed on its way by another vein. Thus, therefore, the effect of muscular pressure upon veins which have valves, is turned almost entirely to the advantage of the circulation; the pressure of the blood onward is all advantageous, and the pressure of the blood backward is prevented from being a hindrance by the closure of the valves and the anastomoses of the veins.

The effects of such muscular pressure are well shown by the acceleration of the stream of blood when, in venesection, the muscles of the forearm are put in action, and by the general acceleration of the circulation during active exercise: and the numerous movements which are continually taking place in the body while awake, though their single effects may be less striking, must be an important auxiliary to the venous circulation. Yet they are not essential; for the venous circulation continues unimpaired in parts at rest, in paralyzed limbs, and in parts in which the veins are not subject to any muscular pressure.

Rhythmical Contraction of Veins.—In the web of the bat's wing, the veins are furnished with valves, and possess the remarkable property of rhythmical contraction and dilatation, whereby the current of blood within them is distinctly accelerated. (Wharton Jones.) The contraction occurs, on an average, about ten times in a minute; the existence of valves preventing regurgitation, the entire effect of the contractions was auxiliary to the onward current of blood. Analogous phenomena have been frequently observed in other animals.

Blood-Pressure in the Veins.—The blood-pressure gradually falls as we proceed from the heart to the arteries, from these to the capillaries, and thence along the veins to the right auricle. The blood-pressure in the veins is nowhere very great, but is greatest in the small veins, while in the large veins toward the heart the pressure becomes *negative*, or, in other words, when a vein is put in connection with a mercurial manometer the mercury will fall in the area furthest away from the vein and will rise in the area nearest the vein, having a tendency to suck in rather than to push forward. In the veins in the neck this tendency to suck in air is especially marked, and is the cause of death in some operations in that region. The amount of pressure in the brachial vein is said to support 9 mm. of mercury, whereas the pressure in the veins of the neck is about equal to a negative pressure of −3 to −8 mm.

The variations of venous pressure during systole and diastole of the heart are very slight, and a distinct pulse is seldom seen in veins except under very extraordinary circumstances.

The formidable obstacle to the upward current of the blood in the veins of the trunk and extremities in the erect posture supposed to be presented by the gravitation of the blood, has no real existence, since the pressure exercised by the column of blood in the arteries, will be always sufficient to support a column of venous blood of the same height as itself: the two columns mutually balancing each other. Indeed, so long as both arteries and veins contain continuous columns of blood, the force of gravitation, whatever be the position of the body, can have no power to move or resist the motion of any part of the blood in any direction. The lowest blood-vessels have, of course, to bear the greatest amount of pressure; the pressure on each part being directly proportionate to the height of the column of blood above it: hence their liability to distension. But this pressure bears equally on both arteries and veins, and cannot either move, or resist the motion of, the fluid they contain, so long as the columns of fluid are of equal height in both, and continuous.

Velocity of the Circulation.

The velocity of the blood-current at any given point in the various divisions of the circulatory system is inversely proportional to their sectional area at that point. If the sectional area of all the branches of a vessel united were always the same as that of the vessel from which they arise, and if the aggregate sectional area of the capillary vessels were equal to that of the aorta, the mean rapidity of the blood's motion in the capillaries would be the same as in the aorta and largest arteries; and if a similar correspondence of capacity existed in the veins and arteries, there would be an equal correspondence in the rapidity of the circulation in them. But the arterial and venous systems may be represented by two truncated cones with their apices directed toward the heart; the area of their united base (the sectional area of the capillaries) being 400—800 times as great as that of the truncated apex representing the aorta. Thus the velocity of blood in the capillaries is at least $\frac{1}{400}$ of that in the aorta.

Velocity in the Arteries.—The velocity of the stream of blood is greater in the arteries than in any other part of the circulatory system, and in them it is greatest in the neighborhood of the heart, and during the ventricular systole; the rate of movement diminishing during the diastole of the ventricles, and in the parts of the arterial system most distant from the heart. Chauveau has estimated the rapidity of the blood-stream in the carotid of the horse at over 20 inches per second during the heart's systole, and nearly 6 inches during the diastole (520—150 mm.).

Estimation of the Velocity.—Various instruments have been devised for measuring the velocity of the blood-stream in the arteries. Ludwig's

FIG. 142.—Ludwig's Stromuhr.

"*Stromuhr*" (Fig. 142) consists of a U-shaped glass tube dilated at a and a', and whose extremities, h and i, are of known calibre. The bulbs can be filled by a common opening at k. The instrument is so contrived that at b and b' the glass part is firmly fixed into metal cylinders, which are fixed into a circular horizontal table, $c\ c'$, capable of horizontal movement on a similar table $d\ d'$ about the vertical axis marked in figure by a dotted line. The opening in $c\ c'$, when the instrument is in position, as in Fig., corresponds exactly with those in $d\ d'$; but if $c\ c'$ be turned at right angles to its present position, there is no communication between h and a, and i and a', but h communicates directly with i; and if turned through two right angles c' communicates with d, and c with d', and there is no direct connection between h and i. The experiment is performed in the following way:—The artery to be experimented upon is divided and connected with two cannulæ and tubes which fit it accurately with h and i—h the central end, and i the peripheral; the bulb a is filled with olive oil up to a point rather lower than k, and a' and the remainder of a is filled with defibrinated blood; the tube on k is then carefully clamped; the tubes d and d' are also filled with defibrinated blood. When everything is ready, the blood is allowed to flow into a through h, and it pushes before it the oil, and that the defibrinated blood into the artery through i, and replaces it in a'; when the blood reaches the former level of the oil in a, the disc $c\ c'$ is turned rapidly through two right angles, and the blood flowing through d into a' again displaces the oil which is driven into a. This is repeated several times, and the duration of the experiment noted. The capacity of a and a' is known; the diameter of the artery is also known by its corresponding with the cannulæ of known diameter, and as the number of times a has been filled in a given time is known, the velocity of the current can be calculated.

Chauveau's instrument (Fig. 143) consists of a thin brass tube, a, in one side of which is a small perforation closed by thin vulcanized india-rubber. Passing through the rubber is a fine lever, one end of which, slightly flattened, extends.into the *lumen* of the tube, while the other moves over the face of a dial. The tube is inserted into the interior of

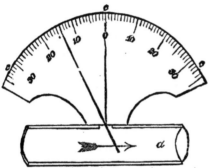

Fig. 143.—Diagram of Chauveau's Instrument. *a.* Brass tube for introduction into the lumen of the artery, and containing an index-needle, which passes through the elastic membrane in its side, and moves by the impulse of the blood-current. *c.* Graduated scale, for measuring the extent of the oscillations of the needle.

an artery, and ligatures applied to fix it, so that the movement of the blood may, in flowing through the tube, be indicated by the movement of the outer extremity of the lever on the face of the dial.

The *Hæmatochometer* of Vierordt, and the instrument of Lortet, resemble in principle that of Chauveau.

Velocity in the Capillaries.—The observations of Hales, E. H. Weber, and Valentin agree very closely as to the rate of the blood-current in the capillaries of the frog; and the mean of their estimates gives the velocity of the *systemic* capillary circulation at about one inch (25 mm.) per minute. The velocity in the capillaries of warm-blooded animals is greater. In the dog $\frac{1}{50}$ to $\frac{3}{100}$ inch ('5 to '75 mm.) a second. This may seem inconsistent with the facts which show that the whole circulation is accomplished in about half a minute. But the whole length of capillary vessels, through which any given portion of blood has to pass, probably does not exceed from $\frac{1}{30}$th to $\frac{1}{50}$th of an inch ('5 mm.); and therefore the time required for each quantity of blood to traverse its own appointed portion of the general capillary system will scarcely amount to a second.

Velocity in the Veins.—The *velocity* of the blood is greater in the veins than in the capillaries, but less than in the arteries: this fact depending upon the relative capacities of the arterial and venous systems. If an accurate estimate of the proportionate areas of arteries and the veins corresponding to them could be made, we might, from the velocity of the arterial current, calculate that of the venous. A usual estimate is, that the capacity of the veins is about twice or three times as great as that of the arteries, and that the velocity of the blood's motion is, therefore,

about twice or three times as great in the arteries as in the veins, 8 inches
(about 200 mm.) a second. The rate at which the blood moves in the
veins gradually increases the nearer it approaches the heart, for the sectional area of the venous trunks, compared with that of the branches
opening into them, becomes gradually less as the trunks advance toward
the heart.

Velocity of the Circulation as a whole.—It would appear that a
portion of blood can traverse the entire course of the circulation, in the
horse, in half a minute. Of course it would require longer to traverse
the vessels of the most distant part of the extremities than to go through
those of the neck: but taking an average length of vessels to be traversed,
and assuming, as we may, that the movement of blood in the human
subject is not slower than in the horse, it may be concluded that half a
minute represents the average rate.

Satisfactory data for these estimates are afforded by the results of
experiments to ascertain the rapidity with which poisons introduced into
the blood are transmitted from one part of the vascular system to
another. The time required for the passage of a solution of potassium
ferrocyanide, mixed with the blood, from one jugular vein (through the
right side of the heart, the pulmonary vessels, the left cavities of the
heart, and the general circulation) to the jugular vein of the opposite
side, varies from twenty to thirty seconds. The same substance was
transmitted from the jugular vein to the great saphena in twenty seconds;
from the jugular vein to the masseteric artery, in between fifteen and
thirty seconds; to the facial artery, in one experiment, in between ten
and fifteen seconds; in another experiment in between twenty and twenty-
five seconds; in its transit from the jugular vein to the metatarsal artery,
it occupied between twenty and thirty seconds, and in one instance more
than forty seconds. The result was nearly the same whatever was the
rate of the heart's action.

In all these experiments, it is assumed that the substance injected
moves with the blood, and at the same rate, and does not move from one
part of the organs of circulation to another by diffusing itself through the
blood or tissues more quickly than the blood moves. The assumption is
sufficiently probable, to be considered nearly certain, that the times above
mentioned, as occupied in the passage of the injected substances, are
those in which the portion of blood, into which each was injected, was
carried from one part to another of the vascular system.

Another mode of estimating the general velocity of the circulating
blood, is by calculating it from the quantity of blood supposed to be contained in the body, and from the quantity which can pass through the
heart in each of its actions. But the conclusions arrived at by this
method are less satisfactory. For the estimates both of the total quantity
of blood, and of the capacity of the cavities of the heart, have as yet only

approximated to the truth. Still the most careful of the estimates thus made accord very nearly with those already mentioned; and it may be assumed that the blood may all pass through the heart in from twenty-five to fifty seconds.

Peculiarities of the Circulation in Different Parts.—The most remarkable peculiarities attending the circulation of blood through different organs are observed in the cases of the *brain,* the *erectile organs,* the *lungs,* the *liver,* and the *kidney.*

1. *In the Brain.*—For the due performance of its functions, the brain requires a large supply of blood. This object is effected through the number and size of its arteries, the two *internal carotids,* and the two *vertebrals.* It is further necessary that the force with which this blood is sent to the brain should be less, or at least should be subject to less variation from external circumstances than it is in other parts, and so the large arteries are very tortuous and anastomose freely in the circle of Willis, which thus insures that the supply of blood to the brain is uniform, though it may by an accident be diminished, or in some way changed, through one or more of the principal arteries. The transit of the large arteries through bone, especially the carotid canal of the temporal bone, may prevent any undue distension; and uniformity of supply is further insured by the arrangement of the vessels in the pia mater, in which, previous to their distribution to the substance of the brain, the large arteries break up and divide into innumerable minute branches ending in capillaries, which, after frequent communications with one another, enter the brain, and carry into nearly every part of it uniform and equable streams of blood. The arteries are also enveloped in a special lymphatic sheath. The arrangement of the *veins* within the cranium is also peculiar. The large venous trunks or sinuses are formed so as to be scarcely capable of change of size; and composed, as they are, of the tough tissue of the dura mater, and, in some instances, bounded on one side by the bony cranium, they are not compressible by any force which the fulness of the arteries might exercise through the substance of the brain; nor do they admit of distension when the flow of venous blood from the brain is obstructed.

The general uniformity in the supply of blood to the brain, which is thus secured, is well adapted, not only to its functions, but also to its condition as a mass of nearly incompressible substance placed in a cavity with unyielding walls. These conditions of the brain and skull have appeared, indeed, to some, enough to justify the opinion that the quantity of blood in the brain must be at all times the same. It was found that in animals bled to death, without any aperture being made in the cranium, the brain became pale and anæmic like other parts. And in death from strangling or drowning, congestion of the cerebral vessels; while in death by prussic acid, the quantity of blood in the cavity of the cranium was determined by the position in which the animal was placed after death, the cerebral vessels being congested when the animal was suspended with its head downward, and comparatively empty when the animal was kept suspended by the ears. ˙ That, it was concluded, although the total volume of the contents of the cranium is probably nearly always the same, yet the quantity of blood in it is liable to variation, its increase or diminution being accompanied by a simultaneous diminution or increase in the quantity of the cerebro-spinal fluid, which, by readily admitting of being removed from one part of the brain and spinal cord to another, and of being rapidly absorbed, and as readily effused, would serve as a kind of supplemental fluid to the other contents of the cranium, to keep it uniformly filled in case of variations in their quantity (Burrows). And there can be no doubt that, although the arrangements of the blood-vessels, to which reference has been made, ensure to the brain an amount of blood which is tolerably uniform, yet, inasmuch as with every beat of the heart and every act of respiration, and under many other circumstances, the quantity of blood in the cavity of the cranium is constantly varying, it is plain that, were there not provision made for the possible displacement of some of the contents of the unyielding bony case in which the brain is contained, there would be often alternations of excessive pressure with insufficient supply of blood. Hence we may consider that the cerebro-spinal fluid in the interior of the skull not only subserves the mechanical functions of fat in other parts as a *packing* material, but by the readiness with which it can be displaced into the spinal canal, provides the means whereby undue pressure and insufficient supply of blood are equally prevented.

Chemical Composition of Cerebro-spinal Fluid.—The cerebro-spinal fluid is transparent, colorless, not viscid, with a saline taste and alkaline reaction, and is not affected by heat or acids. It contains 981–984 parts water, sodium chloride, traces of potassium chloride, of sulphates, carbonates, alkaline and earthy phosphates, minute traces of urea, sugar, sodium lactate, fatty matter, cholesterin, and albumen (Flint).

2. *In Erectile Structures.*—The instances of greatest variation in the quantity of blood contained, at different times, in the same organs, are found in certain structures which, under ordinary circumstances, are soft and flaccid, but, at certain times, receive an unusually large quantity of blood, become distended and swollen by it, and pass into the state which has been termed *erection.* Such structures are the *corpora cavernosa* and *corpus spongiosum* of the penis in the male, and the *clitoris* in the female; and, to a less degree, the *nipple* of the mammary gland in both sexes. The corpus cavernosum penis, which is the best example of an erectile structure, has an external fibrous membrane or sheath; and from the inner surface of the latter are prolonged numerous fine lamellæ which

divide its cavity into small compartments looking like cells when they are inflated. Within these is situated the plexus of veins upon which the peculiar erectile property of the organ mainly depends. It consists of short veins which very closely interlace and anastomose with each other in all directions, and admit of great variation of size, collapsing in the passive state of the organ, but, for erection, capable of an amount of dilatation which exceeds beyond comparison that of the arteries and veins which convey the blood to and from them. The strong fibrous tissue lying in the intervals of the venous plexuses, and the external fibrous membrane or sheath with which it is connected, limit the distension of the vessels, and, during the state of erection, give to the penis its condition of tension and firmness. The same general condition of vessels exists in the corpus spongiosum urethræ, but around the urethra the fibrous tissue is much weaker than around the body of the penis, and around the glans there is none. The venous blood is returned from the plexuses by comparatively small veins; those from the glans and the fore part of the urethra empty themselves into the dorsal veins of the penis; those from the cavernosum pass into deeper veins which issue from the corpora cavernosa at the crura penis; and those from the rest of the urethra and bulb pass more directly into the plexus of the veins about the prostate. For all these veins one condition is the same; namely, that they are liable to the pressure of muscles when they leave the penis. The muscles chiefly concerned in this action are the erector penis and accelerator urinæ. Erection results from the distension of the venous plexuses with blood. The principal exciting cause in the erection of the penis is nervous irritation, originating in the part itself, or derived from the brain and spinal cord. The nervous influence is communicated to the penis by the pudic nerves, which ramify in its vascular tissue: and after their division in the horse, the penis is no longer capable of erection.

This influx of the blood is the first condition necessary for erection, and through it alone much enlargement and turgescence of the penis may ensue. But the erection is probably not complete, nor maintained for any time except when, together with this influx, the muscles already mentioned contract, and by compressing the veins, stop the efflux of blood, or prevent it from being as great as the influx.

It appears to be only the most perfect kind of erection that needs the help of muscles to compress the veins; and none such can materially assist the erection of the nipples, or that amount of turgescence, just falling short of erection, of which the spleen and many other parts are capable. For such turgescence nothing more seems necessary than a large plexiform arrangement of the veins, and such arteries as may admit, upon occasion, augmented quantities of blood.

(3, 4, 5.) *The circulation in the Lungs, Liver, and Kidneys* will be described under those heads.

Agents concerned in the circulation.—Before quitting this subject it will be as well to bring together in a tabular form the various agencies concerned in maintaining the circulation.

1. The *Systole and Diastole of the Heart,* the former pumping into the aorta and so into the arterial system a certain amount of blood, and the latter to some extent sucking in the blood from the veins.

2. *The elastic and muscular coats of the arteries,* which serve to keep up an equable and continuous stream.

3. The so-called *vital capillary force.*

4. The pressure of the *muscles on veins with valves,* and the slight rhythmic contraction of the veins.

5. *Aspiration of the Thorax* during inspiration, by means of which the blood is drawn from the large veins into the thorax (to be treated of in next Chapter).

DISCOVERY OF THE CIRCULATION.

Up to nearly the close of the sixteenth century it was generally believed that the blood passed from one ventricle to the other through foramina in the "septum ventriculorum." These foramina are of course purely imaginary, but no one ventured to dispute their existence till Servetus boldly stated that he could not succeed in finding them. He further asserted that the blood passed from the Right to the Left side of the heart by way of the lungs, and also advanced the hypothesis that it is thus "revivified," remarking that the Pulmonary Artery is too large to serve merely for the nutrition of the lungs (a theory then generally accepted).

Realdus, Columbo, and Cæsalpinus added several important observations. The latter showed that the blood is slightly cooled by passing through the lungs, also that the veins swell up on the distal side of a ligature. The existence of valves in the veins had previously been discovered by Fabricius of Aquapendente, the teacher of Harvey.

The honor of first demonstrating the general course of the circulation belongs by right to Harvey, who made his grand discovery about 1618. He was the first to establish the muscular structure of the heart, which had been denied by many of his predecessors; and by careful study of its action both in the body and when excised, ascertained the order of contraction of its cavities. He did not content himself with inferences from the anatomy of the parts, but employed the experimental method of injection, and made an extensive and accurate series of observations on the circulation in cold-blooded animals. He forced water through the Pulmonary Artery till it trickled out through the Left Ventricle, the tip of which had been cut off. Another of his experiments was to fill the Right side of the heart with water, tie the Pulmonary Artery and the Venæ Cavæ and then squeeze the Right ventricle: not a drop could be forced through into the Left ventricle, and thus he conclusively disproved the existence of foramina in the septum ventriculorum. "I have sufficiently proved," says he, "that by the beating of the heart the blood passes from the veins into the arteries through the ventricles, and is distributed over the whole body."

"In the warmer animals, such as man, the blood passes from the Right

Ventricle of the Heart through the Pulmonary Artery into the Lungs, and thence through the Pulmonary Veins into the Left Auricle, thence into the Left Ventricle."

Proofs of the Circulation of the Blood.—The following are the main arguments by which Harvey established the fact of the circulation:—

1. The heart in half an hour propels more blood than the whole mass of blood in the body.

2. The great force and jetting manner with which the blood spurts from an opened artery, such as the carotid, with every beat of the heart.

3. If true, the normal course of the circulation explains why after death the arteries are commonly found empty and the veins full.

4. If the large *veins* near the heart were tied in a fish or snake, the heart became pale, flaccid, and bloodless; on removing the ligature, the blood again flowed into the heart. If the *artery* were tied, the heart became distended; the distension lasting until the ligature was removed.

5. The evidence to be derived from a ligature round a limb. If it be drawn very tight, no blood can enter the limb, and it becomes pale and cold. If the ligature be somewhat relaxed, blood can enter but cannot leave the limb; hence it becomes swollen and congested. If the ligature be removed, the limb soon regains its natural appearance.

6. The existence of valves in the veins which only permit the blood to flow toward the heart.

7. The general constitutional disturbance resulting from the introduction of a poison at a single point, *e. g.*, snake poison.

To these may now be added many further proofs which have accumulated since the time of Harvey, *e. g.*:—

8. Wounds of arteries and veins. In the former case hæmorrhage may be almost stopped by pressure between the heart and the wound, in the latter by pressure beyond the seat of injury.

9. The direct observation of the passage of blood corpuscles from small arteries through capillaries into veins in all transparent vascular parts, as the mesentery, tongue or web of the frog, the tail or gills of a tadpole, etc.

10. The results of injecting certain substances into the blood.

Further, it is obvious that the mere fact of the existence of a hollow muscular organ (the heart) with valves so arranged as to permit the blood to pass only in one direction, of itself suggests the course of the circulation. The only part of the circulation which Harvey could not follow is that through the capillaries, for the simple reason that he had no lenses sufficiently powerful to enable him to see it. Malpighi (1661) and Leeuwenhoek (1668) demonstrated it in the tail of the tadpole and lung of the frog.

THE maintenance of animal life necessitates the continual absorption of oxygen and excretion of carbonic acid; the blood being, in all animals which possess a well developed blood-vascular system, the medium by which these gases are carried. By the blood, oxygen is absorbed from without and conveyed to all parts of the organism, and, by the blood, carbonic acid, which comes from within, is carried to those parts by which it may escape from the body. The two processes,—absorption of oxygen and excretion of carbonic acid,—are complementary, and their sum is termed the process of *Respiration.*

In all Vertebrata, and in a large number of Invertebrata, certain parts, either *lungs or gills,* are specially constructed for bringing the blood into proximity with the aërating medium (atmospheric air, or water containing air in solution). In some of the lower Vertebrata (frogs and other naked Amphibia) the skin is important as a respiratory organ, and is capable of supplementing, to some extent, the functions of the *proper breathing* apparatus; but in all·the higher animals, including man, the respiratory capacity of the skin is so infinitesimal that it may be practically disregarded.

Essentially, a lung or gill is constructed of a fine transparent membrane, one surface of which is exposed to the air or water, as the case may be, while, on the other, is a network of blood-vessels,—the only separation between the blood and aërating medium being the thin wall of the blood-vessels, and the fine membrane on one side of which vessels are distributed. The difference between the simplest and the most complicated respiratory membrane is one of degree only.

The various complexity of the respiratory membrane, and the kind of aërating medium, are not, however, the only conditions which cause a difference in the respiratory capacity of different animals. The number and size of the red blood-corpuscles, the mechanism of the breathing apparatus, the presence or absence of a *pulmonary* heart, physiologically distinct from the *systemic,* are, all of them, conditions scarcely second in importance.

In the heart of man and all other Mammalia, the *right* side from which the blood is propelled into and through the lungs may be termed the "pulmonary" heart; while the *left* side is "systemic" in function. In many of the lower animals, however, no such distinction can be drawn. Thus, in Fish the heart propels the blood to the respiratory organ (gills); but there is no contractile sac corresponding to the left side of the heart, to propel the blood directly into the systemic vessels.

It may be well to state here that the lungs are only the medium for the *exchange,* on the part of the blood, of carbonic acid for oxygen. They are not the seat, in any special manner, of those combustion-processes of which the production of carbonic acid is the final result. These occur in all parts of the body—more in one part, less in another: chiefly in the substance of the tissues, but in part in the capillary blood-vessels contained in them.

THE RESPIRATORY PASSAGES AND TISSUES.

The object of respiration is the interchange of gases in the lungs; for this purpose it is necessary that the atmospheric air shall pass into them

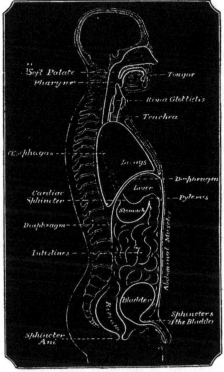

FIG. 144.

and be expelled from them. The lungs are contained in the *chest* or *thorax,* which is a closed cavity having no communication with the outside, except by means of the respiratory passages. The air enters these passages through the nostrils or through the mouth, thence it passes through the *larynx* into the *trachea* or windpipe, which about the middle of the chest divides into two tubes, *bronchi*, one to each (right and left) lung.

The Larynx is the upper part of the passage which leads exclusively to the lung; it is formed by the thyroid, cricoid, and arytenoid cartilages (Fig. 145), and contains the *vocal cords*, by the vibration of which the voice is chiefly produced. These vocal cords are ligamentous bands attached to certain cartilages capable of movement by muscles. By their approximation the cords can entirely close the entrance into the larynx; but under the ordinary conditions, the entrance of the larynx is formed by a more or less triangular chink between them, called the *rima glottidis*. Projecting at an acute angle between the base of the tongue and the larynx to which it is attached, is a leaf-shaped cartilage, with its larger extremity free, called the *epiglottis* (Fig. 145, *e*). The whole of the larynx is lined by mucous membrane, which, however, is extremely thin over the cords. At its lower extremity the larynx joins the trachea.[1] With the exception of the epiglottis and the so-called cornicula laryngis, the cartilages of the larynx are of the *hyaline* variety.

Structure of Epiglottis.—The supporting cartilage is composed of yellow elastic cartilage, enclosed in a fibrous sheath (perichondrium), and covered on both sides with mucous membrane. The *anterior* surface, which looks toward the base of the tongue, is covered with mucous membrane, the basis of which is fibrous tissue, elevated toward both surfaces in the form of rudimentary papillæ, and covered with several layers of squamous epithelium. In it ramify capillary blood-vessels, and in its meshes are a large number of lymphatic channels. Under the mucous membrane, in the less dense fibrous tissue of which it is composed, are a number of tubular glands. The *posterior* or laryngeal surface of the epiglottis is covered by a mucous membrane, similar in structure to that on the other surface, but that the epithelial coat is thinner, the number of strata of cells being less, and the papillæ few and less distinct. The fibrous tissue which constitutes the mucous membrane is in great part of the adenoid variety, and this is here and there collected into distinct masses or follicles. The glands of the posterior surface are smaller but more numerous than those on the other surface. In many places the glands which are situated nearest to the perichondrium are directly continuous through apertures in the cartilage with those on the other side, and often the ducts of the glands from one side of the cartilage pass through and open on the mucous surface of the other side. *Taste goblets* have been

---

[1] A detailed account of the structure and function of the Larynx will be found in Chapter XVI.

found in the epithelium of the posterior surface of the epiglottis, and in several other situations in the laryngeal mucous membrane.

The Trachea and Bronchial Tubes.—The *trachea* or wind-pipe extends from the cricoid cartilage, which is on a level with the fifth cervi-

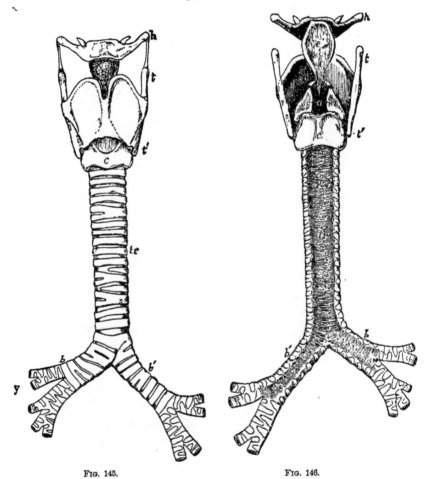

FIG. 145.   FIG. 146.

FIG. 145.—Outline showing the general form of the larynx, trachea, and bronchi, as seen from before. *h*, the great cornu of the hyoid bone; *e*, epiglottis; *t*, superior, and *t'*, inferior cornu of the thyroid cartilage; *c*, middle of the cricoid cartilage; *tr*, the trachea, showing sixteen cartilaginous rings; *b*, the right, and *b'*, the left bronchus. (Allen Thomson.) × ½.

FIG. 146.—Outline showing the general form of the larynx, trachea, and bronchi, as seen from behind. *h*, great cornu of the hyoid bone; *t*, superior, and *t'*, the inferior cornu of the thyroid cartilage; *e*, the epiglottis; *a*, points to the back of both the arytenoid cartilages which are surmounted by the cornicula; *c*, the middle ridge on the back of the cricoid cartilage; *tr*, the posterior membranous part of the trachea; *b*, *b'*, right and left bronchi. (Allen Thomson.) ½.

cal vertebra, to a point opposite the third dorsal vertebra, where it divides into the two bronchi, one for each lung (Fig. 146). It measures, on an average, four or four-and-a-half inches in length, and from three-quarters of an inch to an inch in diameter.

Structure.—The trachea is essentially a tube of fibro-elastic membrane, within the layers of which are enclosed a series of cartilaginous rings, from sixteen to twenty in number. These rings extend only around the front and sides of the trachea (about two-thirds of its circumference), and are deficient behind; the interval between their posterior extremities being bridged over by a continuation of the fibrous membrane in which they are enclosed (Fig. 145). The cartilages of the trachea and bronchial tubes are of the *hyaline* variety.

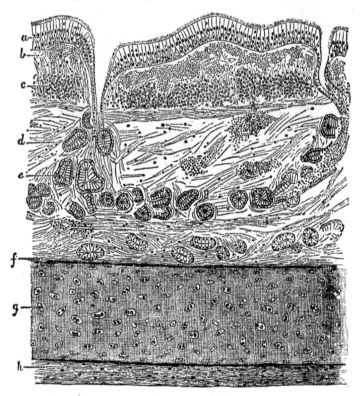

FIG. 147.—Section of trachea. *a*, columnar ciliated epithelium; *b* and *c*, proper structure of the mucous membrane, containing elastic fibres cut across transversely; *d*, submucous tissue containing mucous glands, *e*, separated from the hyaline cartilage, *g*, by a fine fibrous tissue, *f*; *h*, external investment of fine fibrous tissue. (S. K. Alcock.)

Immediately within this tube, at the back, is a layer of unstriped muscular fibres, which extends, *transversely*, between the ends of the cartilaginous rings to which they are attached, and opposite the intervals between them, also; their evident function being to diminish, when required, the calibre of the trachea by approximating the ends of the cartilages. Outside these are a few *longitudinal* bundles of muscular tissue which, like the preceding, are attached both to the fibrous and cartilaginous framework.

The mucous membrane consists of adenoid tissue, separated from the stratified columnar epithelium which lines it by a homogeneous basement membrane. This is penetrated here and there by channels which connect the adenoid-tissue of the *mucosa* with the intercellular substance of the epithelium. The stratified columnar epithelium is formed of several layers of cells (Fig. 147), of which the most superficial layer is ciliated, and is often branched downward to join connective-tissue corpuscles; while between these branched cells are smaller elongated cells prolonged up toward the surface and down to the basement membrane. Beneath these are one or more layers of more irregularly shaped cells. In the deeper part of the mucosa are many elastic fibres between which lie connective-tissue corpuscles and capillary blood-vessels.

Numerous mucous glands are situate on the exterior and in the substance of the fibrous framework of the trachea; their ducts perforating the various structures which form the wall of the trachea, and opening through the mucous membrane into the interior.

The two bronchi into which the trachea divides, of which the right is shorter, broader, and more horizontal than the left (Fig. 145), resemble the trachea exactly in structure, and in the arrangement of their cartilaginous rings. On entering the substance of the lungs, however, the rings, although they still form only larger or smaller segments of a circle, are no longer confined to the front and sides of the tubes, but are distributed impartially to all parts of their circumference.

The bronchi divide and subdivide, in the substance of the lungs, into a number of smaller and smaller branches, which penetrate into every part of the organ, until at length they end in the smaller subdivisions of the lungs, called *lobules*.

All the larger branches still have walls formed of tough membrane, containing portions of cartilaginous rings, by which they are held open, and unstriped muscular fibres, as well as longitudinal bundles of elastic tissue. They are lined by mucous membrane, the surface of which, like that of the larynx and trachea, is covered with ciliated epithelium (Fig. 148). The mucous membrane is abundantly provided with mucous glands.

As the bronchi become smaller and smaller, and their walls thinner, the cartilaginous rings become scarcer and more irregular, until, in the smaller bronchial tubes, they are represented only by minute and scattered cartilaginous flakes. And when the bronchi, by successive branches, are reduced to about $\frac{1}{40}$ of an inch in diameter, they lose their cartilaginous element altogether, and their walls are formed only of a tough fibrous elastic membrane, with circular muscular fibres; they are still lined, however, by a thin mucous membrane, with ciliated epithelium, the length of the cells bearing the cilia having become so far diminished, that the cells are now almost cubical. In the smaller bronchi the circular muscular

fibres are more abundant than in the trachea and larger bronchi, and form
a distinct circular coat.

The Lungs and Pleura.—The Lungs occupy the greater portion of
the thorax. They are of a spongy elastic texture, and on section appear
to the naked eye as if they were in great part solid organs, except here
and there, at certain points, where branches of the bronchi or air-tubes
may have been cut across, and show, on the surface of the section, their

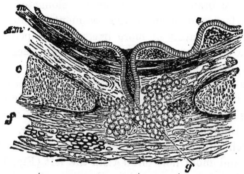

FIG. 148.—Transverse section of a bronchus, about one-fourth of an inch in diameter. *e*, Epithelium (ciliated); immediately beneath it is the mucous membrane or internal fibrous layer, of varying
thickness; *m*, muscular layer; *s, m*, submucous tissue; *f*, fibrous tissue; *c*, cartilage enclosed within
the layers of fibrous tissue; *g*, mucous gland. (F. E. Schulze.)

tubular structure. In fact, however, the lungs are hollow organs, each
of which communicates by a separate orifice with a common air-tube, the
trachea.

The Pleura.—Each lung is enveloped by a serous membrane—the
pleura, one layer of which adheres closely to the surface of the lung,

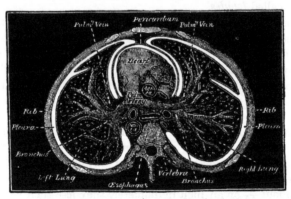

FIG. 149.—Transverse section of the chest (after Gray).

and provides it with its smooth and slippery covering, while the other
adheres to the inner surface of the chest-wall. The continuity of the
two layers, which form a closed sac, as in the case of other serous membranes, will be best understood by reference to Fig. 149. The appearance

of a space, however, between the pleura which covers the lung (*visceral* layer), and that which lines the inner surface of the chest (*parietal* layer), is inserted in the drawing only for the sake of distinctness. These layers are, in health, everywhere in contact, one with the other; and between them is only just so much fluid as will ensure the lungs gliding easily, in their expansion and contraction, on the inner surface of the parietal layer, which lines the chest-wall. While considering the subject of normal respiration, we may discard altogether the notion of the existence of any space or cavity between the lungs and the wall of the chest.

If, however, an opening be made so as to permit air or fluid to enter the pleural sac, the lung, in virtue of its elasticity, recoils, and a considerable space is left between the lung and the chest-wall. In other words, the natural elasticity of the lungs would cause them at all times to contract away from the ribs, were it not that the contraction is resisted by atmospheric pressure which bears only on the *inner* surface of the air-tubes and air-cells. On the admission of air into the pleural sac, atmospheric pressure bears alike on the inner and outer surfaces of the lung, and their elastic recoil is thus no longer prevented.

Structure of the Pleura and Lung.—The pulmonary pleura consists of an outer or denser layer and an inner looser tissue. The former or *pleura proper* consists of dense fibrous tissue with elastic fibres, covered by endothelium, the cells of which are large, flat, hyaline, and transparent when the lung is expanded, but become smaller, thicker, and granular when the lung collapses. In the pleura is a lymph-canalicular system; and connective tissue corpuscles are found in the fibres and tissue which forms its groundworr. The inner, looser, or subpleural tissue contains lamellæ of fibrous connective tissue and connective tissue corpuscles between them. Numerous lymphatics are to be met with, which form a dense plexus of vessels, many of which contain valves. They are simple endothelial tubes, and take origin in the lymph-canalicular system of the pleura proper. Scattered bundles of unstriped muscular fibre occur in the pulmonary pleura. They are especially strongly developed on those parts (anterior and internal surfaces of lungs) which move most freely in respiration: their function is doubtless to aid in expiration. The structure of the *parietal* portion of the pleura is very similar to that of the visceral layer.

Each lung is partially subdivided into separate portions called *lobes;* the right lung into three lobes, and the left into two. Each of these lobes, again, is composed of a large number of minute parts, called *lobules.* Each pulmonary lobule may be considered a lung in miniature, consisting, as it does, of a branch of the bronchial tube, of air-cells, blood vessels, nerves, and lymphatics, with a sparing amount of areolar tissue.

On entering a lobule, the small bronchial tube, the structure of which has been just described (*a*, Fig. 150), divides and subdivides; its walls at the same time becoming thinner and thinner, until at length they are formed only of a thin membrane of areolar and elastic tissue, lined by a layer of *squamous* epithelium, not provided with cilia. At the same time, they are altered in shape; each of the minute terminal branches

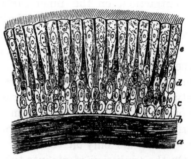

FIG. 150.—Ciliary epithelium of the human trachea. *a*, Layer of longitudinally arranged elastic fibres; *b*, basement membrane; *c*, deepest cells, circular in form; *d*, intermediate elongated cells; *e*, outermost layer of cells fully developed and bearing cilia. × 350. (Kölliker.)

widening out funnel-wise, and its walls being pouched out irregularly into small saccular dilatations, called *air-cells* (Fig. 151, *b*). Such a funnel-shaped terminal branch of the bronchial tube, with its group of pouches or air-cells, has been called an *infundibulum* (Figs. 151, 152),

FIG. 151.   FIG. 152.

FIG. 151.—Terminal branch of a bronchial tube, with its infundibula and air-cells, from the margin of the lung of a monkey, injected with quicksilver. *a*, terminal bronchial twig; *b b*, infundibula and air-cells, × 10. (F. E. Schulze.)

FIG. 152.—Two small infundibula or groups of air-cells, *a a*, with air-cells, *b b*, and the ultimate bronchial tubes, *c c*, with which the air-cells communicate. From a new-born child. (Kölliker.)

and the irregular oblong space in its centre, with which the air-cells communicate, an *intercellular passage*.

The air-cells, or air-vesicles, may be placed singly, like recesses from the intercellular passage, but more often they are arranged in groups or even in rows, like minute sacculated tubes; so that a short series of vesicles, all communicating with one another, open by a common orifice into the tube. The vesicles are of various forms, according to the mutual pressure to which they are subject; their walls are nearly in contact, and they vary from $\frac{1}{60}$ to $\frac{1}{70}$ of an inch in diameter. Their walls are formed of fine membrane, similar to that of the intercellular passages, and continuous with it, which membrane is folded on itself so as to form a sharp-edged border at each circular orifice of communication between contiguous air-vesicles, or between the vesicles and the bronchial passages. Numerous fibres of elastic tissue are spread out between contiguous air-

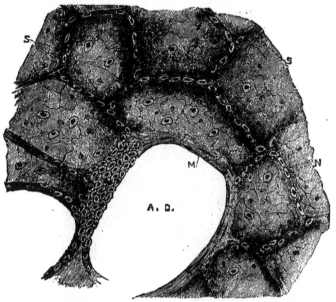

Fig. 153.—From a section of lung of a cat, stained with silver nitrate. A. D. Alveolar duct or intercellular passage. S. Alveolar septa. N. Alveoli or air-cells, lined with large flat, nucleated cells, with some smaller polyhedral nucleated cells. Circular muscular fibres are seen surrounding the interior of the alveolar duct, and at one part is seen a group of small polyhedral cells continued from the bronchus. (Klein and Noble Smith.)

cells, and many of these are attached to the outer surface of the fine membrane of which each cell is composed, imparting to it additional strength, and the power of recoil after distension. The cells are lined by a layer of epithelium (Fig. 153), not provided with cilia. Outside the cells, a network of pulmonary capillaries is spread out so densely (Fig. 154), that the interspaces or meshes are even narrower than the vessels, which are, on an average, $\frac{1}{3000}$ of an inch in diameter. Between the atmospheric air in the cells and the blood in these vessels, nothing intervenes but the thin walls of the cells and capillaries; and the exposure of the blood to the air is the more complete, because the folds of membrane between contiguous cells, and often the spaces between the walls of the same, contain only a single layer of capillaries, both sides of which are thus at once exposed to the air.

The air-vesicles situated nearest to the centre of the lung are smaller and their networks of capillaries are closer than those nearer to the circumference. The vesicles of adjacent lobules do not communicate; and those of the same lobule or proceeding from the same intercellular passage, do so as a general rule only near angles of bifurcation; so that, when any bronchial tube is closed or obstructed, the supply of air is lost for all the cells opening into it or its branches.

Blood-supply.—The lungs receive blood from two sources, (*a*) the pulmonary artery, (*b*) the bronchial arteries. The former conveys *venous* blood to the lungs for its *arterialization,* and this blood takes no share in the nutrition of the pulmonary tissues through which it passes. (*b*) The

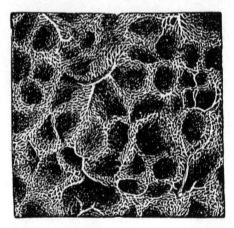

Fig. 154.—Capillary network of the pulmonary blood-vessels in the human lung. × 60. (Kölliker.)

branches of the bronchial arteries ramify for nutrition's sake in the walls of the bronchi, of the larger pulmonary vessels, in the interlobular connective tissue, etc.; the blood of the bronchial vessels being returned chiefly through the bronchial and partly through the pulmonary veins.

Lymphatics.—The lymphatics are arranged in three sets:—1. Irregular lacunæ in the walls of the alveoli or air-cells. The lymphatic vessels which lead from these accompany the pulmonary vessels toward the root of the lung. 2. Irregular anastomosing spaces in the walls of the bronchi. 3. Lymph-spaces in the pulmonary pleura. The lymphatic vessels from all these irregular sinuses pass in toward the root of the lung to reach the bronchial glands.

Nerves.—The nerves of the lung are to be traced from the anterior and posterior pulmonary plexuses, which are formed by branches of the vagus and sympathetic. The nerves follow the course of the vessels and bronchi, and in the walls of the latter many small ganglia are situated.

Respiration consists of the alternate expansion and contraction of the thorax, by means of which air is drawn into or expelled from the lungs. These acts are called **Inspiration** and **Expiration** respectively.

For the *inspiration* of air into the lungs it is evident that all that is necessary is such a movement of the side-walls or floor of the chest, or of both, that the capacity of the interior shall be enlarged. By such increase of capacity there will be of course a diminution of the pressure of the air in the lungs, and a fresh quantity will enter through the larynx and trachea to equalize the pressure on the inside and outside of the chest.

For the *expiration* of air, on the other hand, it is also evident that, by an opposite movement which shall diminish the capacity of the chest, the pressure in the interior will be increased, and air will be expelled, until the pressures within and without the chest are again equal. In both cases the air passes through the trachea and larynx, whether in entering or leaving the lungs, there being no other communication with the exterior of the body; and the lung, for the same reason, remains under all the circumstances described closely in contact with the walls and floor of the chest. To speak of expansion of the chest, is to speak also of expansion of the lung.

We have now to consider the means by which the respiratory movements are effected.

RESPIRATORY MOVEMENTS.

A. Inspiration.—The enlargement of the chest in *inspiration* is a muscular act; the effect of the action of the inspiratory muscles being an increase in the size of the chest-cavity (a) in the vertical, and (b) in the lateral and antero-posterior diameters. The muscles engaged in *ordinary* inspiration are the diaphragm; the external intercostals; parts of the internal intercostals; the levatores costarum; and serratus posticus superior.

(*a.*) The *vertical diameter* of the chest is increased by the contraction and consequent descent of the diaphragm,—the sides of the muscle descending most, and the central tendon remaining comparatively unmoved; while the intercostal and other muscles, by acting at the same time, prevent the diaphragm, during its contraction, from drawing in the sides of the chest.

(*b.*) The increase in the *lateral* and *antero-posterior diameters* of the chest is effected by the raising of the ribs, the greater number of which are attached very obliquely to the spine and sternum (see Figure of Skeleton in frontispiece).

The elevation of the ribs takes place both in front and at the sides—

the hinder ends being prevented from performing any upward movement
by their attachment to the spine.   The movement of the front extremities
of the ribs is of necessity accompanied by an upward and forward movement of the sternum to which they are attached, the movement being
greater at the lower end than at the upper end of the latter bone.

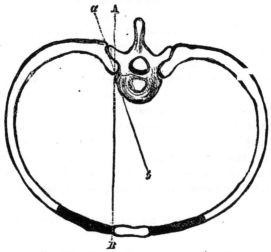

FIG. 155.—Diagram of axes of movement of ribs.

The axes of rotation in these movements are two; one corresponding
with a line drawn through the two articulations which the rib forms with
the spine (*a b,* Fig. 155); and the other, with a line drawn from one of
these (head of rib) to the sternum (A B, Fig. 155, and Fig. 156); the

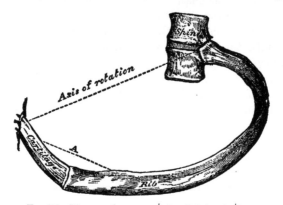

FIG. 156.—Diagram of movement of a rib in inspiration.

motion of the rib around the latter axis being somewhat after the fashion
of raising the handle of a bucket.
    The elevation of the ribs is accompanied by a slight opening out of the

angle which the bony part forms with its cartilage (Fig. 156, A); and thus an additional means is provided for increasing the antero-posterior diameter of the chest.

. The muscles by which the ribs are raised, in *ordinary* quiet inspiration, are the *external intercostals,* and that portion of the *internal intercostals* which is situate between the costal cartilages; and these are assisted by the *levatores costarum,* and the *serratus posticus superior.* The action of the *levatores* and the *serratus* is very simple. Their fibres, arising from the spine as a fixed point, pass obliquely downward and forward to the ribs, and necessarily raise the latter when they contract. The action of the intercostal muscles is not quite so simple, inasmuch as, passing merely from rib to rib, they seem at first sight to have no fixed point toward which they can pull the bones to which they are attached.

A very simple apparatus will explain this apparent anomaly and make their action plain. Such an apparatus is shown in Fig. 157. A B is an upright bar, representing the spine, with which are jointed two parallel bars, C and D, which represent two of the ribs, and are connected in front by movable joints with another upright, representing the sternum.

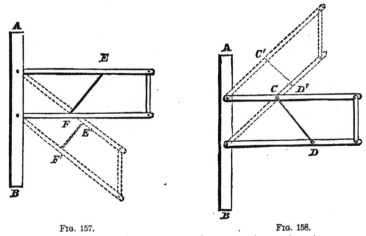

FIG. 157.   FIG. 158.

FIG. 157.--Diagram of apparatus showing the action of the external intercostal muscles.
FIG. 158.—Diagram of apparatus showing the action of the internal intercostal muscles.

If with such an apparatus elastic bands be connected in imitation of the intercostal muscles, it will be found that when stretched on the bars after the fashion of the *external* intercostal fibres (Fig. 157, C D), *i.e.,* passing downward and forward, they raise them (Fig. 157, C' D'); while on the other hand, if placed in imitation of the position of the *internal* intercostals (Fig. 158, E F), *i.e.,* passing downward and backward, they depress them (Fig. 158, E' F').

. The explanation of the foregoing facts is very simple. The intercostal muscles, in contracting, merely do that which all other contracting fibres do, viz., bring nearer together the points to which they are attached; and in order to do this, the external intercostals must raise the ribs, the points C and D (Fig. 157) being nearer to each other when the parallel bars are in the position of the dotted lines. The limit of the movement in the apparatus is reached when the elastic band extends at right angles to the two bars which it connects—the points of attachment C' and D' being then at the smallest possible distance one from the other.

The *internal* intercostals (excepting those fibres which are attached to the cartilages of the ribs), have an opposite action to that of the external. In contracting they must pull down the ribs, because the points E and F (Fig. 158) can only be brought nearer one to another (Fig. 158, E' F') by such an alteration in their position.

On account of the oblique position of the *cartilages* of the ribs with reference to the sternum, the action of the *inter-cartilaginous* fibres of the internal intercostals must, of course, on the foregoing principles, resemble that of the external intercostals.

In tranquil breathing, the expansive movements of the lower part of the chest are greater than those of the upper. In forced inspiration, on the other hand, the greatest extent of movement appears to be in the upper antero-posterior diameter.

Muscles of Extraordinary Inspiration.—In *extraordinary* or forced inspiration, as in violent exercise, or in cases in which there is some interference with the due entrance of air into the chest, and in which, therefore, strong efforts are necessary, other muscles than those just enumerated, are pressed into the service. It is very difficult or impossible to separate by a hard and fast line, the so-called muscles of *ordinary* from those of *extraordinary* inspiration; but there is no doubt that the following are but little used as *respiratory* agents, except in cases in which unusual efforts are required—the *scaleni* muscles, the *sternomastoid*, the *serratus magnus*, the *pectorales*, and the *trapezius*.

Types of Respiration.—The expansion of the chest in inspiration presents some peculiarities in different persons. In young children, it is effected chiefly by the diaphragm, which being highly arched in expiration, becomes flatter as it contracts, and, descending, presses on the abdominal viscera, and pushes forward the front walls of the abdomen. The movement of the abdominal walls being here more manifest than that of any other part, it is usual to call this the *abdominal* type of respiration. In men, together with the descent of the diaphragm, and the pushing forward of the front wall of the abdomen, the chest and the sternum are subject to a wide movement in inspiration (*inferior costal* type). In women, the movement appears less extensive in the lower, and more so in the upper, part of the chest (*superior costal* type). (See Figs. 159, 160.)

B. Expiration.—From the enlargement produced in inspiration, the chest and lungs return in ordinary tranquil expiration, by their elasticity; the force employed by the inspiratory muscles in distending the chest and overcoming the elastic resistance of the lungs and chest-walls, being returned as an expiratory effort when the muscles are relaxed. This elastic recoil of the lungs is sufficient, in ordinary quiet breathing, to expel air from the chest in the intervals of inspiration, and no muscular power is required. In all voluntary expiratory efforts, however, as in speaking, singing, blowing, and the like, and in many involuntary actions also, as sneezing, coughing, etc., something more than merely passive elastic power is necessary, and the proper expiratory muscles are brought into action. By far the chief of these are the abdominal muscles, which, by

FIG. 159.   FIG. 160.

FIG. 159.—The changes of the thoracic and abdominal walls of the male during respiration. The back is supposed to be fixed, in order to throw forward the respiratory movement as much as possible. The outer black continuous line in front represents the ordinary breathing movement; the anterior margin of it being the boundary of inspiration, the posterior margin the limit of expiration. The line is thicker over the abdomen, since the ordinary respiratory movement is chiefly abdominal; thin over the chest, for there is less movement over that region. The dotted line indicates the movement on deep inspiration, during which the sternum advances while the abdomen recedes.

FIG. 160.—The respiratory movement in the female. The lines indicate the same changes as in the last figure. The thickness of the continuous line over the sternum shows the larger extent of the ordinary breathing movement over that region in the female than in the male. (John Hutchinson.) The posterior continuous line represents in both figures the limit of forced expiration.

pressing on the viscera of the abdomen, push up the floor of the chest formed by the diaphragm, and by thus making pressure on the lungs, expel air from them through the trachea and larynx. All muscles, however, which depress the ribs, must act also as muscles of expiration, and therefore we must conclude that the abdominal muscles are assisted in their action by the greater part of the *internal* intercostals, the *triangularis sterni*, the *serratus posticus inferior*, and *quadratus lumborum*. When by the efforts of the expiratory muscles, the chest has been squeezed to less than its average diameter, it again, on relaxation of the muscles, returns to the normal dimensions by virtue of its elasticity. The construction of the chest-walls, therefore, admirably adapts them for recoiling against and resisting as well undue contraction as undue dilatation.

In the natural condition of the parts, the lungs can never contract to the utmost, but are always more or less "on the stretch," being kept closely in contact with the inner surface of the walls of the chest by atmospheric pressure, and can contract away from these only when, by some means or other, as by making an opening into the pleural cavity, or by the effusion of fluid there, the pressure on the exterior and interior of the lungs becomes equal. Thus, under ordinary circumstances, the degree of contraction or dilatation of the lungs is dependent on that of the boundary walls of the chest, the outer surface of the one being in close contact with the inner surface of the other, and obliged to follow it in all its movements.

Respiratory Rhythm.—The acts of expansion and contraction of the chest, take up, under ordinary circumstances, a nearly equal time. The act of inspiring air, however, especially in women and children, is a little shorter than that of expelling it, and there is commonly a very slight pause between the end of expiration and the beginning of the next inspiration. The respiratory rhythm may be thus expressed:—

Inspiration   6
Expiration   .   .   .   . 7 or 8
                  A very slight **pause.**

Respiratory Sounds.—If the ear be placed in contact with the wall of the chest, or be separated from it only by a good conductor of sound, a faint *respiratory murmur* is heard during inspiration. This sound varies somewhat in different parts—being loudest or coarsest in the neighborhood of the trachea and large bronchi (tracheal and bronchial breathing), and fading off into a faint sighing as the ear is placed at a distance from these (vesicular breathing). It is best heard in children, and in them a faint murmur is heard in expiration also. The cause of the vesicular murmur has received various explanations. Most observers hold that the sound is produced by the friction of the air against the walls of the alveoli of the lungs when they are undergoing distension (Laennec, Skoda), others that it is due to an oscillation of the current of air as it enters the alveoli (Chauveau), whilst others believe that the sound is produced in the glottis, but that it is modified in its passage to the pulmonary alveoli (Beau, Gee).

Respiratory Movements of the Nostrils and of the Glottis.— During the action of the muscles which directly draw air into the chest, those which guard the opening through which it enters are not passive. In hurried breathing the instinctive dilatation of the nostrils is well seen, although under ordinary conditions it may not be noticeable. The opening at the upper part of the larynx, however, or *rima glottidis* (Fig. 297), is dilated at each inspiration, for the more ready passage of air, and becomes smaller at each expiration; its condition, therefore, corresponding during respiration with that of the walls of the chest. There is a further likeness between the two acts in that, under ordinary circumstances, the dilatation of the rima glottidis is a muscular act, and its contraction chiefly an elastic recoil; although, under various conditions, to be hereafter mentioned, there may be, in the contraction of the glottis, considerable muscular power exercised.

Terms used to express Quantity of Air breathed.—*Breathing* or *tidal air,* is the quantity of air which is habitually and almost uniformly changed in each act of breathing. In a healthy adult man it is about 30 cubic inches.

Complemental air, is the quantity over and above this which can be drawn into the lungs in the deepest inspiration; its amount is various, as will be presently shown.

Reserve air. After ordinary expiration, such as that which expels the breathing or tidal air, a certain quantity of air remains in the lungs, which may be expelled by a forcible and deeper expiration. This is termed reserve air.

Residual air is the quantity which still remains in the lungs after the most violent expiratory effort. Its amount depends in great measure on the absolute size of the chest, but may be estimated at about 100 cubic inches.

The total quantity of air which passes into and out of the lungs of an adult, at rest, in 24 hours, is about 686,000 cubic inches. This quantity, however, is largely increased by exertion; the average amount for a hard-working laborer in the same time, being 1,568,390 cubic inches.

Respiratory Capacity.—The greatest respiratory capacity of the chest is indicated by the quantity of air which a person can expel from his lungs by a forcible expiration after the deepest inspiration that he can make; it expresses the power which a person has of breathing in the emergencies of active exercise, violence, and disease. The average capacity of an adult (at 60° F. or 15·4° C.) is about 225 cubic inches.

The *respiratory* capacity, or as Hutchinson called it, *vital* capacity, is usually measured by a modified gasometer (*spirometer* of Hutchinson), into which the experimenter breathes,—making the most prolonged expiration possible after the deepest possible inspiration. The quantity of air which is thus expelled from the lungs is indicated by the height to which the air chamber of the spirometer rises; and by means of a scale placed in connection with this, the number of cubic inches is read off.

In healthy men, the respiratory capacity varies chiefly with the stature, weight, and age.

It was found by Hutchinson, from whom most of our information on this subject is derived, that at a temperature of 60° F., 225 cubic inches is the average *vital* or respiratory capacity of a healthy person, five feet seven inches in height

Circumstances affecting the amount of respiratory capacity.—For every inch of height above this standard the capacity is increased, on an average, by eight cubic inches; and for every inch below, it is diminished by the same amount.

The influence of *weight* on the capacity of respiration is less manifest and considerable than that of height; and it is difficult to arrive at any definite conclusions on this point, because the natural average weight of a healthy man in relation to stature has not yet been determined. As a general statement, however, it may be said that the capacity of respiration is not affected by weights under 161 pounds, or 11½ stones; but that, above this point, it is diminished at the rate of one cubic inch for every additional pound up to 196 pounds, or 14 stones.

By *age*, the capacity appears to be increased from about the fifteenth to the thirty-fifth year, at the rate of five cubic inches per year; from thirty-five to sixty-five it diminishes at the rate of about one and a half cubic inch per year; so that the capacity of respiration of a man of sixty years old would be about 30 cubic inches less than that of a man forty years old, of the same height and weight. (John Hutchinson.)

Number of Respirations, and Relation to the Pulse.—The

number of respirations in a healthy adult person usually ranges from fourteen to eighteen per minute. It is greater in infancy and childhood. It varies also much according to different circumstances,.such as exercise or rest, health, or disease, etc. Variations in the number of respirations correspond ordinarily with similar variations in the pulsations of the heart. In health the proportion is about 1 to 4, or 1 to 5, and when the rapidity of the heart's action is increased, that of the chest movement is commonly increased also; but not in every case in equal proportion. It happens occasionally in disease, especially of the lungs or air-passages, that the number of *respiratory* acts increases in quicker proportion than the beats of the *pulse;* and, in other affections, much more commonly, that the number of the pulses is greater in proportion than that of the respirations.

There can be no doubt that the number of respirations of any given animal is largely affected by its size. Thus, comparing animals of the same kind, in a tiger (lying quietly) the number of respirations was 20 per minute, while in a small leopard (lying quietly) the number was 30. In a small monkey 40 per minute; in a large baboon, 20.

The rapid, panting respiration of mice, even when quite still, is familiar, and contrasts strongly with the slow breathing of a large animal such as the elephant (eight or nine times per minute). These facts may be explained as follows:—The heat-producing power of any given animal depends largely on its bulk, while its loss of heat depends to a great extent upon the surface area of its body. If of two animals of similar shape, one be ten times as long as the other, the area of the large animal (representing its loss of heat) is 100 times that of the small one, while its bulk (representing production of heat) is about 1000 times as great. Thus in order to balance its much greater relative loss of heat, the smaller animal must have all its vital functions, circulation, respiration, etc., carried on much more rapidly.

Force of Inspiratory and Expiratory Muscles.—The force with which the inspiratory muscles are capable of acting is greatest in individuals of the height of from five feet seven inches to five feet eight inches, and will elevate a column of three inches of mercury. Above this height, the force decreases as the stature increases; so that the average of men of six feet can elevate only about two and a half inches of mercury. The force manifested in the strongest expiratory acts is, on the average, one-third greater than that exercised in inspiration. But this difference is in great measure due to the power exerted by the elastic reaction of the walls of the chest; and it is also much influenced by the disproportionate strength which the expiratory muscles attain, from their being called into use for other purposes than that of simple expiration. The force of the inspiratory act is, therefore, better adapted than that of the expiratory for testing the muscular strength of the body. (John Hutchinson.)

The instrument used by Hutchinson to gauge the inspiratory and expiratory power was a mercurial manometer, to which was attached a tube fitting the nostrils, and through which the inspiratory or expiratory effort was made. The following table represents the results of numerous experiments:

Power of Inspiratory Muscles.						Power of Expiratory Muscles.
1·5 in.	.	.	.	.	Weak . . .	2·0 in.
2·0 "	.	.	.	.	Ordinary . .	2·5 "
2· "	.	.	.	.	Strong . .	3·5 "
3 "	.	.	.	.	Very strong .	4·5 "
4 "	.	.	.	.	Remarkable .	5·8 "
5 "	.	.	.	.	Very remarkable .	7·0 "
6 "	.	.	.	.	Extraordinary .	8·5 "
7·5 "	.	.	.	.	Very extraordinary .	10·0 "

The greater part of the force exerted in deep inspiration is employed in overcoming the resistance offered by the elasticity of the walls of the chest and of the lungs.

The amount of this elastic resistance was estimated by observing the elevation of a column of mercury raised by the return of air forced, after death, into the lungs, in quantity equal to the known capacity of respiration during life; and Hutchinson calculated, according to the well-known hydrostatic law of equality of pressures (as shown in the Bramah press), that the total force to be overcome by the muscles in the act of inspiring 200 cubic inches of air is more than 450 lbs.

The elastic force overcome in ordinary inspiration is, according to the same authority, equal to about 170 lbs.

Douglas Powell has shown that within the limits of *ordinary tranquil respiration*, the elastic resilience of the *walls of the chest* favors *inspiration;* and that it is only in deep inspiration that the ribs and rib-cartilages offer an opposing force to their dilatation. ` In other words, the elastic resilience of the lungs, at the end of an act of ordinary breathing, has drawn the chest-walls within the limits of their normal degree of expansion. Under all circumstances, of course, the elastic tissue of the *lungs* opposes inspiration, and favors expiration.

Functions of Muscular Tissue of Lungs.—It is possible that the contractile power which the bronchial tubes and air-vesicles possess, by means of their *muscular fibres* may (1) assist in expiration; but it is more likely that its chief purpose is (2) to regulate and adapt, in some measure, the quantity of air admitted to the lungs, and to each part of them, according to the supply of blood; (3) the muscular tissue contracts upon and gradually expels collections of mucus, which may have accumulated within the tubes, and cannot be ejected by forced expiratory efforts, owing to collapse or other morbid conditions of the portion of lung connected with the obstructed tubes (Gairdner). (4) Apart from any of the before-mentioned functions, the presence of muscular fibre in the walls of a hollow viscus, such as a lung, is only what might be expected from analogy with other organs. Subject as the lungs are to such great variation in size it might be anticipated that the elastic tissue, which enters so largely into their composition, would be supplemented by the presence of much muscular fibre also.

RESPIRATORY CHANGES IN THE AIR AND IN THE BLOOD.

A. In the Air.

Composition of the Atmosphere.—The *atmosphere* we breathe has, in every situation in which it has been examined in its natural state, a nearly uniform composition. It is a mixture of oxygen, nitrogen, carbonic acid, and watery vapor, with, commonly, traces of other gases, as ammonia, sulphuretted hydrogen, etc. Of every 100 *volumes* of pure atmospheric air, 79 volumes (on an average) consist of nitrogen, the remaining 21 of oxygen. By weight the proportion is N. 75, O. 25. The proportion of carbonic acid is extremely small; 10,000 volumes of atmospheric air contain only about 4 or 5 of carbonic acid.

The quantity of watery vapor varies greatly according to the temperature and other circumstances, but the atmosphere is never without some. In this country, the average quantity of watery vapor in the atmosphere is 1·40 per cent.

Composition of Air which has been breathed.—The changes effected by respiration in the atmospheric air are: 1, an increase of temperature; 2, an increase in the quantity of carbonic acid; 3, a diminution in the quantity of oyxgen; 4, a diminution of volume; 5, an increase in the amount of watery vapor; 6, the addition of a minute amount of organic matter and of free ammonia.

1. The expired air, heated by its contact with the interior of the lungs, is (at least in most climates) hotter than the inspired air. Its temperature varies between 97° and 99.5° F. (36°—37·5° C.), the lower temperature being observed when the air has remained but a short time in the lungs. Whatever may be the temperature of the air when inhaled, it nearly acquires that of the blood before it is expelled from the chest.

2. The Carbonic Acid in respired air is always increased; but the quantity exhaled in a given time is subject to change from various circumstances. From every volume of air inspired, about 4·8 per cent. of oxygen is abstracted; while a rather smaller quantity, 4·3, of carbonic acid is added in its place: the air will contain, therefore, 434 vols. of carbonic acid in 10,000. Under ordinary circumstances, the quantity of carbonic acid exhaled into the air breathed by a healthy adult man amounts to 1346 cubic inches, or about 636 grains per hour. According to this estimate, the weight of carbon excreted from the lungs is about 173 grains per hour, or rather more than 8 ounces in twenty-four hours. These quantities must be considered approximate only, inasmuch as various circumstances, even in health, influence the amount of carbonic acid excreted, and, correlatively, the amount of oxygen absorbed.

Circumstances influencing the amount of carbonic acid excreted.—The following are the chief:—Age and sex. Respiratory movements. External temperature. Season of year. Condition of respired air. Atmospheric conditions. Period of the day. Food and drink. Exercise and sleep.

a. Age and Sex.—The quantity of carbonic acid exhaled into the air breathed by males, regularly increases from eight to thirty years of age; from thirty to fifty the quantity, after remaining stationary for awhile, gradually diminishes, and from fifty to extreme age it goes on diminishing, till it scarcely exceeds the quantity exhaled at ten years old. In females (in whom the quantity exhaled is always less than in males of the same age) the same regular increase in quantity goes on from the eighth year to the age of puberty, when the quantity abruptly ceases to increase, and remains stationary so long as they continue to menstruate. When menstruation has ceased, it soon decreases at the same rate as it does in old men.

b. Respiratory Movements.—The more quickly the movements of respiration are performed, the smaller is the proportionate quantity of carbonic acid contained in each volume of the expired air. Although, however, the proportionate quantity of carbonic acid is thus diminished during frequent respiration, yet the absolute amount exhaled into the air within a given time is increased thereby, owing to the larger quantity of air which is breathed in the time. The last half of a volume of expired air contains more carbonic acid than the half first expired; a circumstance ·which is explained by the one portion of air coming from the remote part of the lungs, where it has been in more immediate and prolonged contact with the blood than the other has, which comes chiefly from the larger bronchial tubes.

c. *External temperature.*—The observation made by Vierordt at vari᠎ous temperatures between 38° F. and 75° F. (3·4°—23·8° C.) show, for warm-blooded animals, that within this range, every rise equal to 10° F. ·causes a diminution of about two cubic inches in the quantity of carbonic ·acid exhaled per minute.

d. *Season of the Year.*—The season of the year, independently of temperature, materially influences the respiratory phenomena; spring being the season of the greatest, and autumn of the least activity of the respiratory and other functions. (Edward Smith.)

e. *Purity of the Respired Air.*—The average quantity of carbonic acid given out by the lungs constitutes about 4·3 per cent. of the expired air; but if the air which is breathed be previously impregnated with carbonic acid (as is the case when the same air is frequently respired), then the quantity of carbonic acid exhaled becomes much less.

f. *Hygrometric State of Atmosphere.*—The amount of carbonic acid exhaled is considerably influenced by the degree of moisture of the atmosphere, much more being given off when the air is moist than when it is dry. (Lehmann.)

g. *Period of the Day.*—During the daytime more carbonic acid is exhaled than corresponds to the oxygen absorbed; while, on the other hand, at night very much more oxygen is absorbed than is exhaled in carbonic acid. There is, thus, a *reserve fund* of oxygen absorbed by night to meet the requirements of the day. If the total quantity of carbonic acid exhaled in 24 hours be represented by 100, 52 parts are exhaled during the day, and 48 at night. While, similarly, 33 parts of the oxygen are absorbed during the day, and the remaining 67 by night. (Pettenkofer and Voit.)

h. *Food and Drink.*—By the use of *food* the quantity is increased, whilst by fasting it is diminished; it is greater when animals are fed on farinaceous food than when fed on meat. The effects produced by spirituous drinks depend much on the kind of drink taken. Pure alcohol tends rather to increase than to lessen respiratory changes, and the amount therefore of carbonic acid expired; rum, ale, and porter, also sherry, have very similar effects. On the other hand, brandy, whisky, and gin, particularly the latter, almost always lessened the respiratory changes, and consequently the amount of carbonic acid exhaled. (Edward Smith.)

i. *Exercise—Bodily exercise,* in moderation, increases the quantity to about one-third more than it is during rest: and for about an hour after exercise the volume of the air expired in the minute is increased about 118 cubic inches: and the quantity of carbonic acid about 7·8 cubic inches per minute. Violent exercise, such as full labor on the treadwheel, still further increases the amount of the acid exhaled. (Edward Smith.)

A larger quantity is exhaled when the barometer is low than when it is high.

3. The oxygen is diminished, and its diminution is generally proportionate to the increase of the carbonic acid.

For every volume of carbonic acid exhaled into the air, 1·17421 volumes of oxygen are absorbed from it, and 1346 cubic inches, or 636 grains, being exhaled in the hour, the quantity of oxygen absorbed in the same time is 1584 cubic inches, or 542 grains. According to this estimate, there is more oxygen absorbed than is exhaled with carbon to form carbonic acid.

4. The volume of air expired in a given time is less than that of the air inspired (allowance being made for the expansion in being heated), and that the loss is due to a portion of oxygen absorbed and not returned in the exhaled carbonic acid, all observers agree, though as to the actual quantity of oxygen so absorbed, they differ even widely. The amount of oxygen absorbed is on an average 4·8 per cent., so that the expired air contains 16·2 volumes per cent. of that gas.

The quantity of oxygen that does not combine with the carbon given off in carbonic acid from the lungs is probably disposed of in forming some of the carbonic acid and water given off from the skin, and in combining with sulphur and phosphorus to form part of the acids of the sulphates and phosphates excreted in the urine, and probably also, with the nitrogen of the decomposing nitrogenous tissues. (Bence Jones.)

The quantity of oxygen in the atmosphere surrounding animals, appears to have very little influence on the amount of this gas absorbed by them, for the quantity consumed is not greater even though an excess of oxygen be added to the atmosphere experimented with.

It has often been discussed whether *Nitrogen* is absorbed by or exhaled from the lungs during respiration. At present, all that can be said on the subject is that, under most circumstances, animals appear to expire a very small quantity above that which exists in the inspired air. During prolonged fasting, on the contrary, a small quantity appears to be absorbed.

5. The watery vapor is increased. The quantity emitted is, as a general rule, sufficient to saturate the expired air, or very nearly so. Its absolute amount is, therefore, influenced by the following circumstances, (1), by the quantity of air respired; for the greater this is, the greater also will be the quantity of moisture exhaled. (2), by the quantity of watery vapor contained in the air previous to its being inspired; because the greater this is, the less will be the amount required to complete the saturation of the air; (3), by the temperature of the expired air; for the higher this is, the greater will be the quantity of watery vapor required to saturate the air; (4), by the length of time which each volume of inspired air is allowed to remain in the lungs; for although, during ordinary respiration, the expired air is always saturated with watery vapor, yet when respiration is performed very rapidly the air has scarcely time to be raised to the highest temperature, or be fully charged with moisture ere it is expelled.

The quantity of water exhaled from the lungs in twenty-four hours ranges (according to the various modifying circumstances already mentioned) from about 6 to 27 ounces, the ordinary quantity being about 9 or 10 ounces. Some of this is probably formed by the chemical combination of oxygen with hydrogen in the system; but the far larger proportion of it is water which has been absorbed, as such, into the blood from the alimentary canal, and which is exhaled from the surface of the air-passages and cells, as it is from the free surfaces of all moist animal membranes, particularly at the high temperature of warm-blooded animals.

6. A small quantity of ammonia is added to the ordinary constituents of expired air. It seems probable, however, both from the fact that this substance cannot be always detected, and from its minute amount when present, that the whole of it may be derived from decomposing particles of food left in the mouth, or from carious teeth or the like; and that it is, therefore, only an accidental constituent of expired air.

7. The quantity of organic matter in the breath is about 3 grains in twenty-four hours. (Ransome.)

The following represents the kind of experiment by which the foregoing facts regarding the excretion of carbonic acid, water, and organic matter, have been established.

A bird or mouse is placed in a large bottle, through the stopper of which two tubes pass, one to supply fresh air, and the other to carry off that which has been expired. Before entering the bottle, the air is made to bubble through a strong solution of caustic potash, which absorbs the carbonic acid, and then through lime-water, which by remaining limpid, proves the absence of carbonic acid. The air which has been breathed by the animal is made to bubble through lime water, which at once becomes turbid and soon quite milky from the precipitation of calcium carbonate; and it finally passes through strong sulphuric acid, which, by turning brown, indicates the presence of organic matter. The watery vapor in the expired air will condense inside the bottle if the surface be kept cool.

By means of an apparatus sufficiently large and well constructed, experiments of the kind have been made extensively on man.

METHODS BY WHICH THE RESPIRATORY CHANGES IN THE AIR ARE EFFECTED.

The method by which fresh air is inhaled and expelled from the lungs has been considered. It remains to consider how it is that the blood absorbs oxygen from, and gives up carbonic acid to, the air of the alveoli. In the first place, it must be remembered that the tidal air only amounts to about 25—30 cubic inches at each inspiration, and that this is of course insufficient to fill the lungs, but it mixes with the stationary air by *diffusion*, and so supplies to it new oxygen. The amount of oxygen in expired air, which may be taken as the average composition of the mixed air in the lungs, is about 16 to 17 per cent.; in the pulmónary alveoli it may be rather less than this. From this air the venous blood has to take up oxygen in the proportion of 8 to 12 vols. in every hundred volumes of blood, as the differènce between the. amount of oxygen in arterial and venous blood is no less than that. It seems therefore somewhat difficult to understand how this can be accomplished at the low oxygen tension of the pulmonary air. But as was pointed out in a previous Chapter (IV.), the oxygen is not simply dissolved in the blood, but is to a great extent chemically combined with the hæmoglobin of the red corpuscles; and when a fluid contains a body which enters into loose chemical combination in this way with a gas, the tension of the gas in the fluid is not directly proportional to the total quantity of the gaś taken up by the fluid, but to the excess above the total quantity which the substance dissolved in the fluid is capable of taking up (a known quantity in the case of hæmoglobin, viz., 1·59 cm. for one grm. hæmoglobin). On the other hand, if the substance be not saturated, i.e., if it be not combined with as much of the gas as it is capable of taking up, further combination leads to no increase of its tension. However, there is a point at which the hæmoglobin gives up its oxygen when it is exposed to a low partial pressure of oxygen, and there is also a point at which it neither takes up nor gives out oxygen; in the case of arterial blood of the dog, this is found to be when the oxygen tension of the atmosphere is equal to 3·9 per cent. (or 29·6 mm. of mercury), which is equivalent to saying that the oxygen tension of arterial blood is 3·9 per cent.; venous blood, in a similar manner, has been found to have an oxygen tension of 2·8 per cent. At a higher temperature, the tension is raised, as there is a greater tendency at a high temperature for the chemical compound to undergo dissociation. It is therefore easy to see that the oxygen tension of the air of the pulmonary alveoli is quite sufficient, even supposing it much less than that of the expired air, to enable the venous blood to take up oxygen, and what is more, it will take it up until the hæmoglobin is very nearly saturated with the gas.

As regards the elimination of carbonic acid from the blood, there is evidence to show that it is given up by a process of simple diffusion, the only condition necessary for the process being that the tension of the carbonic acid of the air in the pulmonary alveoli should be less than the tension of the carbonic acid in venous blood. The carbonic acid tension of the alveolar air probably does not exceed in the dog 3 or 4 per cent., while that of the venous blood is 5·4 per cent., or equal to 41 mm. of mercury.

B. Respiratory Changes in the Blood.

Circulation of Blood in the Respiratory Organs.—To be exposed to the air thus alternately moved into and out of the air cells and minute bronchial tubes, the blood is propelled from the right ventricle through the pulmonary capillaries in steady streams, and slowly enough to permit every minute portion of it to be for, a few seconds exposed to the air, with only the thin walls of the capillary vessels and the air-cells intervening. The pulmonary circulation is of the simplest kind: for the pulmonary artery branches regularly; its successive branches run in straight lines, and do not anastomose: the capillary plexus is uniformly spread over the air-cells and intercellular passages; and the veins derived from it proceed in a course as simple and uniform as that of the arteries, their branches converging but not anastomosing. The veins have no valves, or only small imperfect ones prolonged from their angles of junc- tion, and incapable of closing the orifice of either of the veins between which they are placed. The pulmonary circulation also is unaffected by changes of atmospheric pressure, and is not exposed to the influence of the pressure of muscles: the force by which it is accomplished, and the course of the blood, are alike simple.

Changes produced in the Blood by Respiration.—The most obvious change which the blood of the pulmonary artery undergoes in its passage through the lungs is 1st, that of color, the dark crimson of venous blood being exchanged for the bright scarlet of arterial blood; 2nd, and in connection with the preceding change, it gains oxygen; 3rd, it loses carbonic acid; 4th, it becomes slightly cooler (p. 193); 5th, it coagu- lates sooner and more firmly, and, apparently, contains more fibrin (see p. 87). The oxygen absorbed into the blood from the atmospheric air in the lungs is combined chemically with the hæmoglobin of the red blood-corpuscles. In this condition it is carried in the arterial blood to the various parts of the body, and brought into near relation or contact with the tissues. In these tissues, and in the blood which circulates in them, a certain portion of the oxygen, which the arterial blood contains, disappears, and a proportionate quantity of carbonic acid and water is formed. The venous blood, containing the new-formed carbonic acid, returns to the lungs, where a portion of the carbonic acid is exhaled, and a fresh supply of oxygen is taken in.

Mechanism of Various Respiratory Actions.—It will be well here, perhaps, to explain some respiratory acts, which appear at first sight somewhat complicated, but cease to be so when the mechanism by, which they are performed is clearly understood. The accompanying dia- gram (Fig. 161) shows that the cavity of the chest is separated from that of the abdomen by the diaphragm, which, when acting, will lessen its curve, and thus descending, will push *downward and forward* the ab- dominal viscera; while the abdominal muscles have the opposite effect, and in acting will push the viscera *upward and backward*, and with them the diaphragm, supposing its ascent to be not from any cause inter- fered with. From the same diagram it will be seen that the lungs com- municate with the exterior of the body through the glottis, and further on through the mouth and nostrils—through either of them separately, or through both at the same time, according to the position of the soft palate. The stomach communicates with the exterior of the body through the œsophagus, pharynx, and mouth; while below the rectum opens at the anus, and the bladder through the urethra. All these openings, through which the hollow viscera communicate with the exterior of the body, are guarded by muscles, called sphincters, which can act independently of each other. The position of the latter is indicated in the diagram.

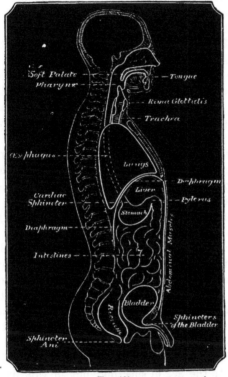

FIG. 161.

Sighing.—In sighing there is a rather prolonged inspiration; the air almost noiselessly passing in through the glottis, and by the elastic recoil of the lungs and chest-walls, and probably also of the abdominal walls, being rather suddenly expelled again.

Now, in the first, or *inspiratory* part of this act, the descent of the diaphragm presses the abdominal viscera downward, and of course this pressure tends to evacuate the contents of such as communicate with the exterior of the body. Inasmuch, however, as their various openings are guarded by sphincter muscles, in a state of constant tonic contraction, there is no escape of their contents, and air simply enters the lungs. In the second, or *expiratory* part of the act of sighing, there is also pressure made on the abdominal viscera in the opposite direction, by the .elastic or muscular recoil of the abdominal walls; but the pressure is relieved by the escape of air through the open glottis, and the relaxed diaphragm is pushed up again into its original position. The sphincters of the stomach, rectum, and bladder, act as before.

Hiccough resembles sighing in that it is an inspiratory act; but the inspiration is sudden instead of gradual, from the diaphragm acting suddenly and spasmodically; and the air, therefore, suddenly rushing through the unprepared rima glottidis, causes vibration of the vocal cords, and the peculiar sound.

Coughing.—In the act of coughing, there is most often first an inspiration, and this is followed by an expiration; but when the lungs have been filled by the preliminary inspiration, instead of the air being easily let out again through the glottis, the latter is momentarily closed by the approximation of the vocal cords, .and then the abdominal muscles, strongly acting, push up the viscera against the diaphragm, and thus make pressure on the air in the lungs until its tension is sufficient to burst open noisily the vocal cords which oppose its outward passage. In this way a considerable force is exercised, and mucus or any other matter that may need expulsion from the lungs or trachea is quickly and sharply expelled by the outstreaming current of air.

Now it is evident on reference to the diagram (Fig. 161), that pressure exercised by the abdominal muscles in the act of coughing, acts as forcibly on the abdominal viscera as on the lungs, inasmuch as the viscera form the medium by which the upward pressure on the diaphragm is made, and of necessity there is quite as great a tendency to the expulsion of their contents as of the air in the lungs. The instinctive, and if necessary, voluntarily increased contraction of the sphincters, however, prevents any escape at the openings guarded by them, and the pressure is effective at one part only, namely, the rima glottidis.

Sneezing.—The same remarks that apply to coughing, are almost exactly applicable to the act of sneezing; but in this instance the blast of air, on escaping from the lungs, is directed, by an instinctive contraction of the pillars of the fauces and descent of the soft palate, chiefly through the nose, and any offending matter is thence expelled.

Speaking.—In speaking, there is a voluntary expulsion of air through the glottis by means of the expiratory muscles; and the vocal cords are put, by the muscles of the larynx, in a proper position and state of tension for vibrating as the air passes over them, and thus producing sound. The sound is moulded into words by the tongue, teeth, lips, etc.—the vocal cords producing the sound only, and having nothing to do with *articulation*.

Singing.—Singing resembles speaking in the manner of its production; the laryngeal muscles, by variously altering the position and degree of tension of the vocal cords, producing the different notes. Words used in the act of singing are of course framed, as in speaking, by the tongue, teeth, lips, etc.

Sniffing.—Sniffing is produced by a somewhat quick action of the diaphragm and other inspiratory muscles. The mouth is, however, closed, and by these means the whole stream of air is made to enter by the nostrils. The alæ nasi are, commonly, at the same time, instinctively dilated.

Sobbing.—Sobbing consists in a series of convulsive inspirations, at the moment of which the glottis is usually more or less closed.

Laughing.—Laughing is a series of short and rapid expirations.

Yawning.—Yawning is an act of inspiration, but is unlike most of the preceding actions in being always more or less involuntary. It is attended by a stretching of various muscles about the palate and lower jaw, which is probably analogous to the stretching of the muscles of the limbs in which a weary man finds relief, as a voluntary act, when they have been some time out of action. The involuntary and reflex character of yawning depends probably on the fact that the muscles concerned are themselves at all times more or less involuntary, and require, therefore, something beyond the exercise of the will to set them in action. For the same reason, yawning, like sneezing, cannot be well performed voluntarily.

Sucking.—Sucking is not properly a respiratory act, but it may be most conveniently considered in this place. It is caused chiefly by the depressor muscles of the os hyoides. These, by drawing downward and backward the tongue and floor of the mouth, produce a partial vacuum in the latter: and the weight of the atmosphere then acting on all sides tends to produce equilibrium on the inside and outside of the mouth as best it may. The communication between the mouth and pharynx is completely shut off by the contraction of the pillars of the soft palate and descent of the latter so as to touch the back of the tongue; and the equilibrium, therefore, can be restored only by the entrance of something through the mouth. The action, indeed, of the tongue and floor of the mouth in sucking may be compared to that of the piston in a syringe, and the muscles which pull down the os hyoides and tongue, to the power which draws the handle.

Influence of the Nervous System in Respiration.—Like all other functions of the body, the discharge of which is necessary to life, respiration must be essentially an involuntary act. Else, life would be in constant danger, and would cease on the loss of consciousness for a few moments, as in sleep. But it is also necessary that respiration should be to some extent under the control of the will. For were it not so, it would be impossible to perform those voluntary respiratory acts which have been just enumerated and explained, as speaking, singing, and the like.

The respiratory movements and their rhythm, so far as they are involuntary and independent of consciousness (as on all ordinary occasions) are under the governance of a *nerve-centre* in the *medulla oblongata* corresponding with the origin of the pneumogastric nerves; that is to say, the motor nerves, and through them the muscles concerned in the respiratory movements, are excited by a stimulus which issues from this part of the nervous system. How far the medulla acts *automatically, i.e.,* how far the stimulus originates in it, or how far it is merely a nerve-centre for *reflex* action, is not certainly known. Probably, as will be seen, both events happen; and, in both cases, the stimulus is the result of the condition of the blood.

The respiratory centre is bilateral or double, since the respiratory movements continue after the medulla at this point is divided in the middle line.

As regards its supposed automatic action, it has been shown that if the spinal cord be divided below the medulla, and both vagi be divided so that no afferent impulses can reach it from below, the nasal and laryngeal respiration continues, and the only possible course of the afferent impulses would be through the cranial nerves; and when the cord and medulla are intact the division of these produces no effect upon respiration, so that it appears evident that the afferent stimuli are not absolutely necessary for maintaining the respiratory movements. But although automatic in its action the respiratory centre may be reflexly excited, and the chief channel of this reflex influence is the vagus nerve; for when the nerve of one side is divided, respiration is slowed, and if both vagi be cut the respiratory action is still slower.

The influence of the vagus trunk upon it is twofold, for if the nerve be divided below the origin of the superior laryngeal branch and the central end be stimulated, respiratory movements are increased in rapidity, and indeed follow one another so quickly if the stimuli be increased in number, that after a time cessation of respiration in inspiration follows from a tetanus of the respiratory muscles (diaphragm). Whereas if the superior laryngeal branch be divided, although no effect, or scarcely any, follows the mere division, on stimulation of the central end respiration is slowed, and after a time, if the stimulus be increased, stops, but not in inspiration as in the other case, but in expiration. Thus the vagus trunk contains fibres which slow and fibres which accelerate respiration. If we adopt the theory of a doubly acting respiratory centre in the floor of the medulla, one tending to produce inspiration and the other expiration, and acting in antagonism as it were, so that there is a gradual increase in the tendency to produce respiratory action, until it culminates in an inspiratory effort, which is followed by a similar action of the expiratory part of the centre, producing an expiration, we must look upon the main trunk of the vagus as aiding the inspiratory, and of the superior laryngeal as aiding the expiratory part of the centre, the first nerve possibly inhibiting the action of the expiratory centre, whilst it aids the inspiratory, and the latter nerve having the very opposite effect. But inasmuch as the respiration is slowed on division of the vagi, and not quickened or affected manifestly on simple division of the superior laryngeal, it must be supposed that the vagi fibres are always in action, whereas the superior laryngeal fibres are not.

It appears, however, that there are, in some animals at all events, subordinate centres in the spinal cord which are able, under certain conditions, to discharge the function of the chief medullary centre.

The centre in the medulla may be influenced not only by afferent impulses proceeding along the vagus and laryngeal nerves but also by those proceeding from the cerebrum, as well as by impressions made upon the nerves of the skin, or upon part of the fifth nerve distributed to the nasal mucous membrane, or upon other sensory nerves, as is exemplified by the deep inspiration which follows the application of cold to the surface of the skin, and by the sneezing which follows the slightest irritation of the nasal mucous membrane.

At the time of birth, the separation of the placenta, and the consequent non-oxygenation of the fœtal blood, are the circumstances which immediately lead to the issue of automatic impulses to action from the respiratory centre in the medulla oblongata. But the quickened action which ensues on the application of cold air or water, or other sudden stimulus, to the skin, shows well the intimate connection which exists between this centre and other parts which are not ordinarily connected with the function of respiration.

Methods of Stimulation of the Respiratory Centre.—It is now necessary to consider the method by which the centre or centres are stimulated themselves, as well as the manner in which the afferent vagi impulses are produced.

The more venous the blood, the more marked are the inspiratory impulses, and if the air is prevented from entering the chest, in a short time the respiration becomes very labored. Its cessation is followed by an abnormal rapidity of the inspiratory acts, which make up even in depth for the previous stoppage. The condition caused by obstruction to the entrance of air, or by any circumstance by which the oxygen of the blood is used up in an abnormally quick manner, is known as *dyspnœa,* and as the aëration of the blood becomes more and more interfered with, not only are the ordinary respiratory muscles employed, but also those extraordinary muscles which have been previously enumerated (p. 186), so that as the blood becomes more and more venous the action of the medullary centre becomes more and more active. The question arises as to what condition of the venous blood causes this increased activity, whether it is due to deficiency of oxygen or excess of carbonic acid in the blood. This has been answered by the experiments, which show on the one hand that dyspnœa occurs when there is no obstruction to the exit of carbonic acid, as when an animal is placed in an atmosphere of nitrogen, and therefore cannot be due to the accumulation of carbonic acid, and secondly, that if plenty of oxygen be supplied, dyspnœa proper does not. occur, although the carbonic acid of the blood is in excess. The respiratory centre is evidently stimulated to action by the absence of sufficient oxygen in the blood circulating in it.

The method by which the vagus is stimulated to conduct afferent impulses, influencing the action of the respiratory centre, appears to be by the venous blood circulating in the lungs, or as some say by the condition of the air in the pulmonary alveoli. And if either of these be the stimuli it will be evident that as the condition of venous blood stimulates the peripheral endings of the vagus in the lungs, the vagus action which tends to help on the discharge of inspiratory impulses from the centre, must tend also to increase the activity of the centre, when the blood in the lungs becomes more and more venous. No doubt the venous condition of the blood will affect all the sensory nerves in a similar manner, but it has been shown that the circulation of too little blood through the centre is quite sufficient by itself for the purpose; as when its blood supply is cut off increased inspiratory actions ensue.

Effects of Vitiated Air.—Ventilation.—We have seen that the air expired from the lungs contains a large proportion of carbonic acid and a minute amount of organic putrescible matter.

Hence it is obvious that if the same air be breathed again and again, the proportion of carbonic acid and organic matter will constantly increase till fatal results are produced; but long before this point is reached, uneasy sensations occur, such as headache, languor, and a sense of oppression. It is a remarkable fact that the organism after a time adapts itself to such a vitiated atmosphere, and that a person soon comes to breathe, without sensible inconvenience, an atmosphere which, when he first entered it, felt intolerable. Such an adaptation, however, can only take place at the expense of a depression of all the vital functions, which must be injurious if long continued or often repeated.

This power of adaptation is well illustrated by the experiments of Claude Bernard. A sparrow is placed under a bell-glass of such a size that it will live for three hours. If now at the end of the second hour (when it could have survived another hour) it be taken out and a fresh healthy sparrow introduced, the latter will perish instantly.

The adaptation above spoken of is a gradual and continuous one: thus a bird which will live one hour in a pint of air will live three hours in two pints; and if two birds of the same species, age, and size, be placed in a quantity of air in which either, separately, would survive three hours, they will not live 1½ hour, but only 1¼ hour.

From what has been said it must be evident that provision for a constant and plentiful supply of fresh air, and the removal of that which is vitiated, is of far greater importance than the actual cubic space per head of occupants. Not less than 2000 cubic feet per head should be allowed in sleeping apartments (barracks, hospitals, etc.), and with this allowance the air can only be maintained at the proper standard of purity by such a system of ventilation as provides for the supply of 1500 to 2000 cubic feet of fresh air per head per hour. (Parkes.)

THE EFFECT OF RESPIRATION ON THE CIRCULATION.

Inasmuch as the heart and great vessels are situated in the air-tight thorax, they are exposed to a certain alteration of pressure when the

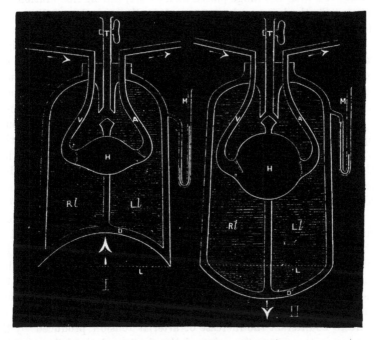

FIG. 162.—Diagram of an apparatus illustrating the effect of inspiration upon the heart and great vessels within the thorax.—I, the thorax at rest; II, during inspiration; D, represents the diaphragm when relaxed; D′ when contracted (it must be remembered that this position is a mere diagram), i.e., when the capacity of the thorax is enlarged; H, the heart; V, the veins entering it, and A, the aorta; Rl, Ll, the right and left lung; T, the trachea; M, mercurial manometer in connection with the pleura. The increase in the capacity of the box representing the thorax is seen to dilate the heart as well as the lungs, and so to pump in blood through V, whereas the valve prevents reflex through A. The position of the mercury in M shows also the suction which is taking place. (Landois.)

capacity of the latter is increased; for although the expansion of the lungs during inspiration tends to counterbalance this increase of area, it never quite does so, since part of the pressure of the air which is drawn into the chest through the trachea is expended in overcoming the elasticity of the lungs themselves. The amount thus used up increases as the lungs become more and more expanded, so that the pressure inside the thorax during inspiration as far as the heart and great vessels are concerned, never quite equals that outside, and at the conclusion of inspiration is considerably less than the atmospheric pressure. It has been ascertained that the amount of the pressure used up in the way above described, varies from 5 or 7 mm. of mercury during the pause, and to 30 mm. of mercury when the lungs are expanded at the end of a deep inspiration, so that it will be understood that the pressure to which the heart and great vessels are subjected diminishes as inspiration progresses. It will be understood from the accompanying diagram how, if there were no lungs in the chest, but if its capacity were increased, the effect of the increase would be expended in pumping blood into the heart from the veins, but even with the lungs placed as they are, during inspiration the pressure outside the heart and great vessels is diminished, and they have therefore a tendency to expand and to diminish the intra-vascular pressure. The diminution of pressure within the veins passing to the right auricle and within the right auricle itself, will draw the blood into the thorax, and so assist the circulation: this suction action aiding, though independently, the suction power of the diastole of the auricle about which we have previously spoken (p. 124). The effect of sucking more blood into the right auricle will, *cæteris paribus*, increase the amount passing through the right ventricle, which also exerts a similar suction action, and through the lungs into the left auricle and ventricle and thus into the aorta, and this tends to increase the arterial tension. The effect of the diminished pressure upon the pulmonary vessels will also help toward the same end, *i.e.*, an increased flow through the lungs, so that as far as the heart and its veins are concerned inspiration increases the blood pressure in the arteries. The effect of inspiration upon the aorta and its branches within the thorax would be, however, contrary; for as the pressure outside is diminished the vessels would tend to expand, and thus to diminish the tension of the blood within them, but inasmuch as the large arteries are capable of little expansion beyond their natural calibre, the diminution of the arterial tension caused by this means would be insufficient to counteract the increase of arterial tension produced by the effect of inspiration upon the veins of the chest, and the balance of the whole action would be in favor of an increase of arterial tension during the inspiratory period. But if a tracing of the variation be taken at the same time that the respiratory movements are recorded, it will be found that, although speaking generally, the arterial tension is increased during inspiration, the maximum of arterial tension does not correspond with the acme of inspiration (Fig. 163).

As regards the effect of expiration, the capacity of the chest is diminished, and the intra-thoracic pressure returns to the normal, which is not exactly equal to the atmospheric, pressure. The effect of this on the veins is to increase their intra-vascular pressure, and so to diminish the flow of blood into the left side of the heart, and with it the arterial tension, but this is almost exactly balanced by the necessary increase of arterial tension caused by the increase of the extra-vascular pressure of the aorta and large arteries, so that the arterial tension is not much affected during expiration either way. Thus, ordinary expiration does

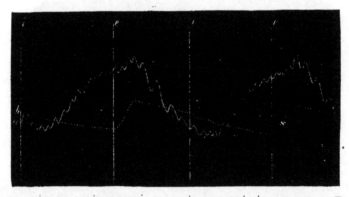

FIG. 163.—Comparison of blood-pressure curve with curve of intra-thoracic pressure. (To be read from left to right.) *a* is a curve of blood-pressure with its respiratory undulations, the slower beats on the descent being very marked; *b* is the curve of intra-thoracic pressure obtained by connecting one limb of a manometer with the pleural cavity. Inspiration begins at *i* and expiration at *e*. The intra-thoracic pressure rises very rapidly after the cessation of the inspiratory effort, and then slowly falls as the air issues from the chest; at the beginning of the inspiratory effort the fall becomes more rapid. (M. Foster.)

not produce a distinct obstruction to the circulation, as even when the expiration is at an end the intra-thoracic pressure is less than the extra-thoracic.

The effect of violent expiratory efforts, however, has a distinct action in preventing the current of blood through the lungs, as seen in the blueness of the face from congestion in straining; this condition being produced by pressure on the small pulmonary vessels.

We may summarize this mechanical effect, therefore, and say that inspiration aids the circulation and so increases the arterial tension, and that although expiration does not materially aid the circulation, yet under ordinary conditions neither does it obstruct. Under extraordinary conditions, as in violent expirations, the circulation is decidedly obstructed. But we have seen that there is no exact correspondence between the points of extreme arterial tension and the end of inspiration, and we must look to the nervous system for an explanation of this apparently contradictory result.

The effect of the nervous system in producing a rhythmical alteration of the blood pressure is twofold. In the first place the *cardio-inhibitory* centre is believed to be stimulated during the fall of blood pressure, producing a slower rate of heart-beats during expiration, which will be noticed in the tracing (Fig. 163), the undulations during the decline of blood-pressure being longer but less frequent. This effect disappears when, by section of the vagi, the effect of the centre is cut off from the heart. In the second place, the *vaso-motor centre* is also believed to send out rhythmical impulses, by which undulations of blood pressure are produced independently of the mechanical effects of respiration.

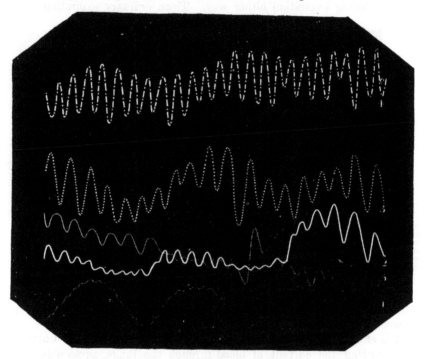

Fig. 164.—Traube-Hering's curves. (To be read from left to right.) The curves 1, 2, 3, 4, and 5 are portions selected from one continuous tracing forming the record of a prolonged observation, so that the several curves represent successive stages of the same experiment. Each curve is placed in its proper position relative to the base line, which is omitted; the blood-pressure rises in stages from 1, to 2, 3, and 4, but falls again in stage 5. Curve 1 is taken from a period when artificial respiration was being kept up, but the vagi having been divided. the pulsations on the ascent and descent of the undulations do not differ; when artificial respiration ceased these undulations for a while disappeared, and the blood-pressure rose steadily while the heart-beats became slower. Soon, as at 2, new undulations appeared; a little later, the blood-pressure was still rising, the heart-beats still slower, but the undulations still more obvious (3); still later (4), the pressure was still higher, but the heart-beats were quicker, and the undulations flatter, the pressure then began to fall rapidly (5), and continued to fall until some time after artificial respiration was resumed. (M. Foster.)

The action of the vaso-motor centre in taking part in producing rhythmical changes of blood-pressure which are called respiratory, is shown in the following way:—In an animal under the influence of urari, record of whose blood-pressure is being taken, and where artificial respiration has been stopped, and both vagi cut, the blood-pressure curve rises at first almost in a straight line; but after a time new rhythmical undulations occur very like the original respiratory undulations, only somewhat larger. These are called *Traube's* or *Traube-Hering's curves.* They continue whilst the blood-pressure continues to rise, and only cease when the vaso-motor centre and the heart are exhausted, when the pressure speedily falls. These ιcurves must be dependent upon the vaso-motor centre, as the mechanical effects of respiration have been eliminated by the poison and by the cessation of artificial respiration, and the effect of the cardio-inhibitory centre be the division of the vagi. It may be presumed therefore that the vaso-motor centre, as well as the cardio-inhibitory, must be considered to take part with the mechanical changes of inspiration and expiration in producing the so-called respiratory undulations of blood-pressure.

Cheyne-Stokes's breathing.—This is a rhythmical irregularity in respirations which has been observed in various diseases, and is especially connected with fatty degeneration of the heart. Respirations occur in groups, at the beginning of each group the inspirations are very shallow, but each successive breath is deeper than the preceding until a climax is reached, then comes in a prolonged sighing expiration, succeeded by a pause, after which the next group begins.

APNŒA.—DYSPNŒA.—ASPHYXIA.

As blood which contains a normal proportion of oxygen excites the respiratory centre (p. 204), and as the excitement and consequent respiratory muscular movements are greater (*dyspnœa*) in proportion to the deficiency of this gas, so an abnormally large proportion of oxygen in the blood leads to diminished breathing movements, and, if the proportion be large enough, to their temporary cessation. This condition of absence of breathing is termed *apnœa*,[1] and it can be demonstrated, in one of the lower animals, by performing artificial respiration to the extent of saturating the blood with oxygen.

When, on the other hand, the respiration is stopped, by, *e.g.*, interference with the passage of air to the lungs, or by supplying air devoid of oxygen, a condition ensues, which passes rapidly from the state of *dyspnœa* (difficult breathing) to what is termed *asphyxia;* and the latter quickly ends in death.

The ways by which this condition of asphyxia may be produced are very numerous; as, for example, by the prevention of the due entry of oxygen into the blood, either by direct obstruction of the trachea or other part of the respiratory passages, or by introducing instead of ordinary air a gas devoid of oxygen, or, again, by interference with the due interchange of gases between the air and the blood.

Symptoms of Asphyxia.—The most evident symptoms of asphyxia or suffocation are well known. Violent action of the respiratory muscles

---

[1] This term has been, unfortunately, often applied to conditions of *dyspnœa* or *asphyxia;* but the modern application of the term, as in the text, is the more convenient.

and, more or less, of all the muscles of the body; lividity of the skin and all other vascular parts, while the veins are also distended, and the tissues seem generally gorged with blood; convulsions, quickly followed by insensibility, and death.

The conditions which accompany these symptoms are—

(1) More or less interference with the passage of the blood through the pulmonary blood-vessels.

(2) Accumulation of blood in the right side of the heart and in the systemic veins.

(3) Circulation of impure (non-aërated) blood in all parts of the body.

Cause of Death from Asphyxia.—The causes of these conditions and the manner in which they act, so as to be incompatible with life, may be here briefly considered.

(1) The obstruction to the passage of blood through the lungs is not so great as it was once supposed to be; and such as there is occurs chiefly in the later stages of asphyxia, when, by the violent and convulsive action of the expiratory muscles, pressure is indirectly made on the lungs, and the circulation through them is proportionately interfered with.

(2) Accumulation of blood, with consequent distension of the right side of the heart and systemic veins, is the direct result, at least in part, of the obstruction to the pulmonary circulation just referred to. Other causes, however, are in operation. (*a*) The vaso-motor centres stimulated by blood deficient in oxygen, causes contraction of all the small arteries with increase of arterial tension, and as an immediate consequence the filling of the systemic veins. (*b*) The increased arterial tension is followed by inhibition of the action of the heart, and, thus, the latter, contracting less frequently, and gradually enfeebled also by deficient supply of oxygen, becomes over-distended by blood which it cannot expel. At this stage the left as well as the right cavities are distended with blood.

The ill effects of these conditions are to be looked for partly in the heart, the muscular fibres of which, like those of the urinary bladder or any other hollow muscular organ, may be paralyzed by over-stretching; and partly in the venous congestion, and consequent interference with the function of the higher nerve-centres, especially the medulla oblongata.

(3) The passage of non-aërated blood through the lungs and its distribution over the body are events incompatible with life, in one of the higher animals, for more than a few minutes; the rapidity with which death ensues in asphyxia being due, more particularly, to the effect of non-oxygenized blood on the medulla oblongata, and, through the coronary arteries, on the muscular substance of the heart. The excitability of both nervous and muscular tissue is dependent on a constant and large supply of oxygen, and, when this is interfered with, is rapidly lost. The diminution of oxygen, it may be here remarked, has a more direct in-

fluence in the production of the usual symptoms of asphyxia than the increased amount of carbonic acid. Indeed, the fatal effect of a gradual accumulation of the latter in the blood, if a due supply of oxygen be maintained, resembles rather that of a narcotic poison.

In some experiments performed by a committee appointed by the Medico-Chirurgical Society to investigate the subject of *Suspended Animation*, it was found that, in the dog, during simple asphyxia, *i.e.*, by simple privation of air, as by plugging the trachea, the average duration of the respiratory movements after the animal had been deprived of air, was 4 minutes 5 seconds; the extremes being 3 minutes 30 seconds, and 4 minutes 40 seconds. The average duration of the heart's action, on the other hand, was 7 minutes 11 seconds; the extremes being 6 minutes 40 seconds, and 7 minutes 45 seconds. It would seem, therefore, that on an average, the heart's action continues for 3 minutes 15 seconds after the animal has ceased to make respiratory efforts. A very similar relation was observed in the rabbit. Recovery never took place after the heart's action had ceased.

The results obtained by the committee on the subject of *drowning* were very remarkable, especially in this respect, that whereas an animal may recover, after simple deprivation of air for nearly four minutes, yet, after submersion in water for 1½ minute, recovery seems to be impossible. This remarkable difference was found to be due, not to the mere submersion, nor directly to the struggles of the animal, nor to depression of temperature, but to the two facts, that in drowning, a free passage is allowed to air out of the lungs, and a free entrance of water into them. It is probably to the entrance of water into the lungs that the speedy death in drowning is mainly due. The results of *post-mortem* examination strongly support this view. On examining the lungs of animals deprived of air by plugging the trachea, they were found simply congested; but in the animals drowned, not only was the congestion much more intense, accompanied with ecchymosed points on the surface and in the substance of the lung, but the air tubes were completely choked up with a sanious foam, consisting of blood, water, and mucus, churned up with the air in the lungs by the respiratory efforts of the animal. The lung-substance, too, appeared to be saturated and sodden with water, which, stained slightly with blood, poured out at any point where a section was made. The lung thus sodden with water was heavy (though it floated), doughy, pitted on pressure, and was incapable of collapsing. It is not difficult to understand how, by such infraction of the tubes, air is debarred from reaching the pulmonary cells; indeed the inability of the lungs to collapse on opening the chest is a proof of the obstruction which the froth occupying the air-tubes offers to the transit of air.

We must carefully distinguish the asphyxiating effect of an insufficient supply of oxygen from the directly poisonous action of such a gas as carbonic oxide, which is present to a considerable amount in common coal-gas. The fatal effects often produced by this gas (as in accidents from burning charcoal stoves in small, close rooms), are due to its entering into combination with the hæmoglobin of the blood-corpuscles (p. 95), and thus expelling the oxygen.

CHAPTER VII.

FOOD.

In order that life may be maintained it is necessary that the body should be supplied with food in proper quality and quantity.

The food taken in by the animal body is used for the purpose of replacing the waste of the tissues. And to arrive at a reasonable estimation of the proper diet in twenty-four hours it is necessary to consider the amount of the excreta daily eliminated from the body. The excreta contain chiefly carbon, hydrogen, oxygen, and nitrogen, but also to a less extent, sulphur, phosphorus, chlorine, potassium, sodium, and certain other of the elements. Since this is the case it must be evident that, to balance this waste, foods must be supplied containing all these elements to a certain degree, and some of them, viz., those which take the principal part in forming the excreta, in large amount. We have seen in the last Chapter that carbonic acid and ammonia, *i.e.*, the elements carbon, oxygen, nitrogen, hydrogen, are given off from the lungs. By the excretion of the kidneys—the urine—many elements are discharged from the blood, especially nitrogen, hydrogen, and oxygen. In the sweat, the elements chiefly represented are carbon, hydrogen, and oxygen, and also in the fæces. By all the excretions large quantities of water are got rid of daily, but chiefly by the urine.

The relations between the amounts of the chief elements contained in these various excreta in twenty-four hours may be represented in the following way (Landois):

	Water.	C.	H.	N.	O.
By the lungs	330	248·8	—	?	651.15
By the skin	660	2.6	—	—	7·2
By the urine	1700	9·8	3·3	15·8	11·1
By the fæces	128	20·	3·	3·	12
Grammes	2818	281·2	6·3	18·8	681·41

To this should be added 296· grammes water, which are produced by the union of hydrogen and oxygen in the body during the process of oxidation (*i.e.*, 32·89 hydrogen and 263.41 oxygen). There are twenty-six grammes of salts got rid of by the urine and six by the fæces. As the water can be supplied as such, the losses of carbon, nitrogen, and oxygen are those to which we should direct our attention in supplying food.

For the sake of example, we may now take only two elements, carbon and nitrogen, and, if we discover what amount of these is respectively discharged in a given time from the body, we shall be in a position to judge what kind of food will most readily and economically replace their loss.

The quantity of carbon daily lost from the body amounts to about 281·2 grammes or nearly 4,500 grains, and of nitrogen 18·8 grammes or nearly 300 grains; and if a man could be fed by these elements, as such, the problem would be a very simple one; a corresponding weight of charcoal, and, allowing for the oxygen in it, of atmospheric air, would be all that is necessary. But an animal can live only upon these elements when they are arranged in a particular manner with others, in the form of an organic compound, as albumen, starch, and the like; and the relative proportion of carbon to nitrogen in either of these compounds alone, is, by no means, the proportion required in the diet of man. Thus, in albumen, the proportion of carbon to nitrogen is only as 3·5 to 1. If, therefore, a man took into his body, as food, sufficient albumen to supply him with the needful amount of carbon, he would receive more than four times as much nitrogen as he wanted; and if he took only sufficient to supply him with nitrogen, he would be starved for want of carbon. It is plain, therefore, that he should take with the albuminous part of his food, which contains so large a relative amount of nitrogen in proportion to the carbon he needs, substances in which the nitrogen exists in much smaller quantities relatively to the carbon.

It is therefore evident that the diet must consist of several substances, not of one alone, and we must therefore turn to the available food-stuffs. For the sake of convenience they may be classified as follows:

A. ORGANIC.

    I. **Nitrogenous,** consisting of *Proteids, e.g.* albumen, casein, syntonin, gluten, legumin and their allies; and *Gelatins,* which include gelatin, elastin, and chondrin. All of these contain carbon; hydrogen, oxygen, and nitrogen, and some in addition, phosphorus and sulphur.

    II. **Non-Nitrogenous,** comprising:

    (1.) *Amyloid or saccharine bodies,* chemically known as carbohydrates, since they contain carbon, hydrogen, and oxygen, with the last two elements in the proportion to form water, *i.e.*, H_2O. To this class belong starch and sugar.

    (2.) *Oils and fats.*—These contain carbon, hydrogen, and oxygen; but the oxygen is less in amount than in the amyloids and saccharine bodies.

B. INORGANIC.

    I. **Mineral and saline matter.**
    II. **Water.**

To supply the loss of nitrogen and carbon, it is found by experience that it is necessary to combine substances which contain a large amount of nitrogen with others in which carbon is in considerable amount; and although, without doubt, if it were possible to relish and digest one or other of the above-mentioned proteids when combined with a due quantity of an amyloid to supply the carbon, such a diet, together with salt and water, ought to support life; yet we find that for the purposes of ordinary life this system does not answer, and instead of confining our nitrogenous foods to one variety of substance we obtain it in a large number of allied substances, for example, in flesh, of bird, beast, or fish; in eggs; in milk; and in vegetables. And, again, we are not content with one kind of material to supply the carbon necessary for maintaining life, but seek more, in bread, in fats, in vegetables, in fruits. Again, the fluid diet is seldom supplied in the form of pure water, but in beer, in wines, in tea and coffee, as well as in fruits and succulent vegetables.

Man requires that his food should be *cooked*. Very few organic substances can be properly digested without previous exposure to heat and to other manipulations which constitute the process of cooking. It will be well, therefore, to consider the composition of the various substances employed as food, and then to consider how they are affected by cooking.

A. Foods containing principally Nitrogenous bodies.

I.—*Flesh of Animals*, especially of the ox (beef, veal), sheep (mutton, lamb), pig (pork, bacon, ham).

Of these, beef is richest in nitrogenous matters, containing about 20 per cent., whereas mutton contains about 18 per cent., veal, 16·5, and pork, 10; the flesh is also firmer, more satisfying, and is supposed to be more strengthening than mutton, whereas the latter is more digestible. The flesh of young animals, such as lamb and veal, is less digestible and less nutritious. Pork is comparatively indigestible, and contains a large amount of fat.

Flesh contains:—(1) Nitrogenous bodies: *myosin, serum-albumin, gelatin* (from the interstitial fibrous connective tissue); *elastin* (from the elastic tissue), as well as *hæmoglobin*. (2) Fatty matters, including *lecithin* and *cholesterin*. (3) Extractive matters, some of which are agreeable to the palate, *e.g.*, *osmazome*, and others which are weakly stimulating, *e.g.*, *kreatin*. Besides, there are *sarcolactic* and *inositic acids, taurin, xanthin,* and others. (4) Salts, chiefly of potassium, calcium, and magnesium. (5) Water, the amount of which varies from 15 per cent. in dried bacon to 39 in pork, 51 to 53 in fat beef and mutton, to 72 per cent. in lean beef and mutton. (6) A certain amount of carbo-hydrate material is found in the flesh of young animals, in the form of *inosite, dextrin, grape sugar,* and (in young animals) *glycogen*.

Table of Per-centage Composition of Beef, Mutton, Pork, and Veal.—
(Letheby.)

	Water.	Albumen.	Fat.	Salts.
Beef.—Lean	72	19·3	3·6	5.1
" Fat	51	14·8	29·8	4·4
Mutton.—Lean . . .	72	18.3	4.9	4.8
" Fat . . .	53	12·4	31·1	3·5
Veal	63	16·5	15·8	4·7
Pork.—Fat	39	9.8	48·9	2·3

Together with the flesh of the above-mentioned animals, that of the *deer, hare, rabbit,* and *birds,* constituting *venison, game* and *poultry,* should be added as taking part in the supply of nitrogenous substances, and also *fish*—salmon, eels, etc., and *shell-fish, e.g.,* lobster, crab, mussels, oysters, shrimps, scollop, cockles, etc.

Table of Per-centage Composition of Poultry and Fish.—(Letheby.)

	Water.	Albumen.	Fats.	Salts.
Poultry	74	21	3·8	1·2

(Singularly devoid of fat, and so generally eaten with bacon or pork.)

	Water.	Albumen.	Fat.	Salts.
White Fish	78	18·1	2·9	1·
Salmon	77	16.1	5·5	1·4
Eels (very rich in fat) . .	75	9·9	13·8	1·3
Oysters . . .	75·74	11·72	2·42	2.73

Even now the list of fleshy foods is not complete, as nearly all animals have been occasionally eaten, and we may presume that the average composition of all is nearly the same.

II. *Milk*—Is intended as the entire food of young animals, and as such contains, when pure, all the elements of a typical diet. (1) Albuminous substances in the form of *casein* and, in small amount, of *serum-albumin.* (2) Fats in the cream. (3) Carbo-hydrates in the form of *lactose* or milk sugar. (4) Salts, chiefly *calcium phosphate;* and (5) Water. From it we obtain (α) *cheese,* which is the casein precipitated with more or less fat according as the cheese is made of skim milk (skim cheese), of fresh milk with its cream (Cheddar and Cheshire), or of fresh milk plus cream (Stilton and double Gloucester). The precipitated casein is allowed to ripen, by which process some of the albumen is split up with formation of fat. (β) *Cream,* which consists of the fatty globules incased in casein, and which being of low specific gravity float to the surface. (γ) *Butter,* or the fatty matter deprived of its casein envelope by the process of churning. (δ) *Buttermilk,* or the fluid obtained from cream after butter has been formed; very rich therefore in nitrogen. (ε) *Whey,* or the fluid which remains after the precipitation of casein; this contains sugar, salt, and a small quantity of albumen.

Table of Composition of Milk, Buttermilk, Cream, and Cheese.—(Letheby and Payen.)

	Nitrogenous matters.	Fats.	Lactose.	Salts.	Water.
Milk (Cow) . .	4·1	3·9	5.2	·8	86
Buttermilk . .	4·1	·7	6·4	·8	88
Cream . . .	2·7	26·7	2·8	1·8	66
Cheese.—Skim . .	44·8	6·3	—	4·9	44
" Cheddar .	28·4	31·1	—	4.5	36

			Non-nitrogenous matter and loss.		
" Neufchatel (fresh) 8·	40·71	36·58	·51	36.58

III. *Eggs.*—The yelk and albumen of eggs are in the same relation as food for the embryoes of oviparous animals that milk is to the young of mammalia, and afford another example of the natural admixture of the various alimentary principles.

Table of the Per-centage Composition of Fowls' Eggs.

	Nitrogenous substances.	Fats.	Salts.	Water.
White	20·4	—	1·6	78
Yelk	16	30·7	1·3	52

IV. *Leguminous fruits* are used by vegetarians, as the chief source of the nitrogen of the food. Those chiefly used are *peas, beans, lentils,* etc., they contain a nitrogenous substance called *legumin,* allied to albumen. They contain about 25·30 per cent. of this nitrogenous body, and twice as much nitrogen as wheat.

B.   Substances supplying principally Carbohydrate bodies.

α. *Bread,* made from the ground grain obtained from various so-called *cereals,* viz., wheat, rye, maize, barley, rice, oats, etc., is the direct form in which the carbohydrate is supplied in an ordinary diet. Flour, however, besides the starch, contains *gluten,* a nitrogenous body, and a small amount of fat.

Table of Per-centage Composition of Bread and Flour.

	Nitrogenous matters.	Carbohydrates.	Fats.	Salts.	Water.
Bread . . .	8·1	51·	1·6	2.3	37
Flour . .	10.8	70·85	2·	1.7	15

Various articles of course are made from flour, *e.g.*, macaroni, biscuits, etc., besides bread.

β. Vegetables, especially potatoes.

γ. Fruits contain sugar, and organic acids, tartaric, malic, citric, and others.

C. SUBSTANCES SUPPLYING PRINCIPALLY FATTY BODIES.

The chief are *butter, lard* (pig's fat), *suet* (beef and mutton fat).

D. SUBSTANCES SUPPLYING THE SALTS OF THE FOOD.

Nearly all the foregoing substances in A, B, and C, contain a greater or less amount of the salts required in food; but green vegetables and fruit supply certain salts, without which the normal health of the body is not maintained.

E. LIQUID FOODS.

Water is consumed alone, or together with certain other substances used to flavor it, *e.g.*, tea, coffee, etc. Tea in moderation is a stimulant, and contains an aromatic oil to which it owes its peculiar aroma, an astringent of the nature of tannin, and an alkaloid, *theine.* The composition of coffee is very nearly similar to that of tea. Cocoa, in addition to similar substances contained in tea and coffee, contains fat, albuminous matter, and starch, and must be looked upon more as a food.

Beer, in various forms, is an infusion of *malt* (barley which has sprouted, and in which the starch is converted in great part into sugar), boiled with hops and allowed to ferment. Beer contains from 1·2 to 8·8 per cent. of alcohol.

Cider and Perry, the fermented juice of the apple and pear.

Wine, the fermented juice of the grape, contains from 6 or 7 (Rhine wines, and white and red Bordeaux) to 24—25 (ports and sherries) per cent. of alcohol.

Spirits, obtained from the distillation of fermented liquors. They contain upward of 40—70 per cent. of absolute alcohol.

Effects cf cooking upon Food.—In general terms this may be said to make food more easily digestible, and this includes two other alterations, food is made more agreeable to the palate and also more pleasing to the eye. Cooking consists in exposing the food to various degrees of heat, either to the direct heat of the fire, as in roasting, or to the indirect heat of the fire, as in broiling, baking, or frying, or to hot water, as in boiling or stewing. The effect of heat upon flesh is to coagulate the albumen and coloring matter, to solidify fibrin, and to gelatinize tendons and fibrous connective tissue. Previous beating or bruising (as with steaks and chops, or keeping (as in the case of game), renders the meat more tender. Prolonged exposure to heat also develops on the surface certain empyreumatic bodies, which are agreeable both to the taste and smell. By placing meat into hot water, the external coating of albumen is coagulated, and very little, if any, of the constituents of the meat are lost afterward if boiling be prolonged, but if the constituents of the meat are to be extracted, it should be exposed to prolonged simmering at a much lower temperature, and the "*broth*" will then contain the gelatin and extractive matters of the meat, as well as a certain amount of albumen. The addition of salt will help to extract the myosin.

The effect of boiling upon an egg coagulates the albumen, and helps in rendering the article of food more suitable for adult dietary. Upon milk, the effect of heat is to produce a scum composed of serum-albumin and a little casein (the greater part of the casein being uncoagulated) with some fat. Upon vegetables, the cooking produces the necessary effect of rendering them softer, so that they can be more readily broken up in the mouth; it also causes the starch to swell up and burst, and so aids the digestive fluids to penetrate into their substance. The albuminous matters are coagulated, and the gummy, saccharine and saline matters are removed. The conversion of flour into bread is effected by mixing it with water, a little salt and a certain amount of yeast, which consists of the cells of an organized ferment (*Torula cerevisiæ*). By the growth of this plant, which lives upon the sugar produced from the starch of the flour, carbonic acid gas and a small amount of alcohol are formed. It is by means of the former that the *dough* rises. Another method consists in mixing the flour with water containing a large quantity of the gas in solution.

By the action of heat during baking the dough continues to expand, and the gluten being coagulated, the bread sets as a permanently vesiculated mass.

I.—Effects of an Insufficient Diet.

Hunger and Thirst.—The sensation of *hunger* is manifested in consequence of deficiency of food in the system. The mind refers the sensation to the stomach; yet since the sensation is relieved by the introduction of food either into the stomach itself, or into the blood through other channels than the stomach, it would appear not to depend on the state of the stomach alone. This view is confirmed by the fact, that the division of both pneumogastric nerves, which are the principal channels by which the brain is cognizant of the condition of the stomach, does not appear to allay the sensations of hunger. But that the stomach has some share in this sensation is proved by the relief afforded, though only temporarily, by the introduction of even non-alimentary substances into this organ. It may, therefore, be said that the sensation of hunger is caused both by a want in the system generally, and also by the condition of the stomach itself, by which condition, of course, its own nerves are more directly affected.

The sensation of *thirst*, indicating the want of fluid, is referred to the fauces, although, as in hunger, this is, in great part, only the local declaration of a general condition. For thirst is relieved for only a very short time by moistening the dry fauces; but may be relieved completely by the introduction of liquids into the blood, either through the stomach, or by injections into the blood-vessels, or by absorption from the surface of the skin or the intestines. The sensation of thirst is perceived most naturally whenever there is a disproportionately small quantity of water in the blood: as well, therefore, when water has been abstracted from the blood, as when saline or any solid matters have been abundantly added to it. And the cases of hunger and thirst are not the only ones in which the mind derives, from certain organs, a peculiar predominant sensation of some condition affecting the whole body. Thus, the sensation of the "necessity of breathing," is referred especially to the air-passages; but, as Volkmann's experiments show, it depends on the condition of the blood which circulates everywhere, and is felt even after the lungs of animals are removed; for they continue, even then, to gasp and manifest the sensation of want of breath.

Starvation.—The effects of total deprivation of food have been made the subject of experiments on the lower animals, and have been but too frequently illustrated in man. (1) One of the most notable effects of starvation, as might be expected, is *loss of weight;* the loss being greatest at first, as a rule, but afterward not varying very much, day by day, until death ensues. Chossat found that the ultimate proportional loss was, in different animals experimented on, almost exactly the same; death occurring when the body had lost two-fifths (forty per cent.) of its original weight. Different parts of the body lose weight in very different proportions. The following results are taken, in round numbers, from the table given by M. Chossat:—

Fat	loses 93 per cent.
Blood	. 75 "
Spleen	. 71 "
Pancreas	. 64
Liver	. 52
Heart	. 44 "
Intestines	. 42 "
Muscles of locomotion	. 42 "
Stomach	. 39 "
Pharynx, (Œsophagus)	. 34
Skin	. 33

Kidneys . . . . . . loses 31 per cent.
Respiratory apparatus 22   "
Bones , ... 16   "
Eyes 10   "
Nervous system 2   " (nearly).

(2.) The effect of starvation on the temperature of the various animals experimented on by Chossat was very marked. For some time the *variation* in the daily temperature was more marked than its absolute and continuous diminution, the daily fluctuation amounting to 5° or 6° F. (3° C.), instead of 1° or 2° F. (·5° to 1° C.), as in health. But a short time before death, the temperature fell very rapidly, and death ensued when the loss had amounted to about 30° F. (16·5°C.). It has been often said, and with truth, although the statement requires some qualification, that death by starvation is really death by cold; for not only has it been found that differences of time with regard to the period of the fatal result are attended by the same ultimate loss of heat, but the effect of the application of external warmth to animals cold and dying from starvation, is more effectual in reviving them than the administration of food. In other words, an animal exhausted by deprivation of nourishment is unable so to digest food as to use it as fuel, and therefore is dependent for heat on its supply from without. Similar facts are often observed in the treatment of exhaustive diseases in man.

(3.) The symptoms produced by starvation in the human subject are hunger, accompanied, or it may be replaced by pain, referred to the region of the stomach; insatiable thirst; sleeplessness; general weakness and emaciation. The exhalations both from the lungs and skin are fetid, indicating the tendency to decomposition which belongs to badly-nourished tissues; and death occurs, sometimes after the additional exhaustion caused by diarrhœa, often with symptoms of nervous disorder, delirium or convulsions.

(4.) In the human subject death commonly occurs within six to ten days after total deprivation of food. But this period may be considerably prolonged by taking a very small quantity of food, or even water only. The cases so frequently related of survival after many days, or even some weeks, of abstinence, have been due either to the last-mentioned circumstances, or to others no less effectual, which prevented the loss of heat and moisture. Cases in which life has continued after total abstinence from food and drink for many weeks, or months, exist only in the imagination of the vulgar.

(5.) The appearances presented after death from starvation are those of general wasting and bloodlessness, the latter condition being least noticeable in the brain. The stomach and intestines are empty and contracted, and the walls of the latter appear remarkably thinned and almost transparent. The various secretions are scanty or absent, with the exception of the bile, which, somewhat concentrated, usually fills the gall-bladder. All parts of the body readily decompose.

II.—Effects of improper Diet.

Experiments on Feeding.—Experiments illustrating the ill effects produced by feeding animals upon one or two alimentary substances only have been often performed.

Dogs were fed exclusively on *sugar and distilled water*. During the first seven or eight days they were brisk and active, and took their food and drink as usual; but in the course of the second week, they began to get thin, although their appetite continued good, and they took daily between six and eight ounces of sugar. The emaciation increased during the third week, and they became feeble, and lost their activity and appetite. At the same time an ulcer formed on each cornea, followed by an escape of the humors of the eye: this took place in repeated experiments. The animals still continued to eat three or four ounces of sugar daily; but became at length so feeble as to be incapable of motion, and died on a day varying from the thirty-first to the thirty-fourth. On dissection, their bodies presented all the appearances produced by death from starvation; indeed, dogs will live almost the same length of time without any food at all.

When dogs were fed exclusively on *gum*, results almost similar to the above ensued. When they were kept on *olive-oil and water*, all the phenomena produced were the same, except that no ulceration of the cornea took place; the effects were also the same with butter. The experiments of Chossat and Letellier prove the same; and in men, the same is shown by the various diseases to which those who consume but little nitrogenous food are liable, and especially by the affection of the cornea which is observed in Hindus feeding almost exclusively on rice. But it is not only the non-nitrogenous substances, which, taken alone, are insufficient for the maintenance of health. The experiments of the Academies of France and Amsterdam were equally conclusive that *gelatin* alone soon ceases to be nutritive.

Savory's observations on food confirm and extend the results obtained by Magendie, Chossat, and others. They show that animals fed exclusively on non-nitrogenous diet speedily emaciate and die, as if from starvation; that life is much more prolonged in those fed with nitrogenous than by those with non-nitrogenous food; and that animal heat is maintained as well by the former as by the latter—a fact which proves, if proof were wanting—that nitrogenous elements of food, as well as non-nitrogenous, may be regarded as calorifacient.

III.—EFFECT OF TOO MUCH FOOD.

Sometimes the excess of food is so great that it passes through the alimentary canal, and is at once got rid of by increased peristaltic action of the intestines. In other cases, the unabsorbed portions undergo putrefactive changes in the intestines, which are accompanied by the production of gases, such as carbonic acid, carburetted and sulphuretted hydrogen; a distended condition of the bowels, accompanied by symptoms of indigestion, is the result. An excess of the substances required as food may, however, undergo absorption. It is a well-known fact that numbers of people habitually eat too much; especially of nitrogenous food. Dogs can digest an immense amount of meat if fed often, and the amount of meat taken by some men would supply not only the nitrogen, but the carbon which is requisite for an ordinary natural diet. A method of getting rid of an excess of nitrogen is provided by the digestive processes in the duodenum, to be presently described, whereby the excess of the albuminous food is capable of being changed before absorption into nitrogenous crystalline matters, easily converted by the liver into urea, and so easily excreted by the kidneys, affording one variety of what is called *luxus consumption;* but after a time the organs, especially the liver, will yield to the strain of the over-work, and will not reduce the excess of nitrogenous material into urea, but into other less oxidized products, such as uric acid; and general plethora and gout may be the result. This state of things, however, is delayed for a long time, if not altogether obviated, when large meat-eaters take a considerable amount of exercise.

Excess of carbohydrate food produces an accumulation of fat, which may not only be an inconvenience by causing obesity, but may interfere with the proper nutrition of muscles, causing a feebleness of the action of the heart, and other troubles. The accumulation of fat is due to the excess of carbohydrate being stored up by the protoplasm in the form of fat. Starches when taken in great excess are almost certain to give rise in addition to dyspepsia, with acidity and flatulence. There is a limit to the absorption of starch and of fat, as, if taken beyond a certain amount, they appear unchanged in the fæces.

Requisites of a Normal Diet.—It will have been understood that it is necessary that a normal diet should be made up of various articles, that they should be well cooked, and should contain about the same amount of the carbon and nitrogen that are got rid of by the excreta. Without doubt these desiderata may be satisfied in numerous ways, and it would be simply absurd to believe that the diet of every adult should be exactly similar. The age, sex, strength, and circumstances of each individual should ultimately determine his diet. A dinner of bread and hard cheese with an onion contain all the requisites for a meal; but such diet would be suitable only for those possessing strong digestive powers. It is a well-known fact that the diet of the continental nations differs from that of our own country, and that of cold from that of hot climates; but the same principle underlies them all, viz., replacement of the loss of the excreta in the most convenient and economical way possible. Without going into detail in the matter, it may be said that any one in active work requires more nitrogenous matter than one at rest, and that children and women require less than adult men.

The quantity of food for a healthy adult man of average height and weight may be stated in the following table:—

Table of Water and Food required for a Healthy Adult.—(Parkes.)

	In laborious occupation.	At rest.
Nitrogenous substances, *e.g.*, flesh	6 to 7 oz. av.	2·5 oz.
Fats	3·5 to 4·5 oz.	1 oz.
Carbo-hydrates	16 to 18 oz.	12 oz.
Salts	1.2 to 1.5 oz.	.5 oz.
	26·7 to 31 oz.	16 oz.

The above is the dry food; but as this is nearly always combined with 50 to 60 per cent. of water, these numbers should be doubled, and they would then be 52 to 60 oz., and 32 oz. of so called solid food, and to this should be added 50 to 80 oz. of fluid.

Full diet scale for an adult male in hospital (*St. Bartholomew's Hospital*).

Breakfast.—1 pint of tea (with milk and sugar), bread and butter.
Dinner.—½ lb. of cooked meat, ½ lb. potatoes, bread and beer.
Tea.—1 pint of tea, bread and butter.
Supper.—Bread and butter, beer.
Daily allowance to each patient.—2 pints of tea, with milk and sugar; 14 oz. bread; ½ lb. of cooked meat: ½ lb. potatoes: 2 pints of beer, 1 oz. butter. 31 oz. solid, and 4 pints (80 oz.), liquid.

CHAPTER VIII.

DIGESTION.

THE object of digestion is to prepare the food to supply the waste of the tissues, which we have seen is its proper function in the economy. Few of the articles of diet are taken in the exact condition in which it is possible for them to be absorbed into the system by the blood-vessels and lymphatics, without which absorption they would be useless for the purposes they have to fulfil; almost the whole of the food undergoes various changes before it is fit for absorption. Having been received into the mouth, it is subjected to the action of the teeth and tongue, and is mixed with the first of the digestive juices—the *saliva*. It is then swallowed, and, passing through the pharynx and œsophagus into the stomach, is subjected to the action of the *gastric juice*. Thence it passes into the intestines, where it meets with the bile, the pancreatic juice and the intestinal juices, all of which exercise an influence upon that portion of the food not absorbed from the stomach. By this time most of the food is capable of absorption, and the residue of undigested matter leaves the body in the form of fæces by the anus.

The course of the food through the alimentary canal of man will be readily seen from the accompanying diagram (Fig. 165).

The Mouth is the cavity contained between the jaws and inclosed by the cheeks laterally, and by the lips in front; behind it opens into the pharynx by the *fauces*, and is separated from the nasal cavity by the hard palate in front, and the soft palate behind, which form its roof. The tongue forms the lower part or floor. In the jaws are contained the teeth; and when the mouth is shut these form its anterior and lateral boundaries. The whole of the mouth is lined with mucous membrane, covered by stratified squamous epithelium, which is continuous in front along the lips with the epithelium of the skin, and posteriorly with that of the pharynx. The mucous membrane is provided with numerous glands (small tubular), called mucous glands, and into it open the ducts of the salivary glands, three chief glands on each side. The tongue is not only a prehensile organ, but is also the chief seat of the sense of taste.

We shall now consider, in detail, the process of digestion, as it takes place in each stage of this journey of the food through the alimentary canal.

Mastication.—The act of chewing or mastication is performed by the biting and grinding movement of the lower range of teeth against the upper. The simultaneous movements of the tongue and cheeks assist partly by crushing the softer portions of the food against the hard palate, gums, etc., and thus supplementing the action of the teeth, and partly by returning the morsels of food to the action of the teeth, again and again,

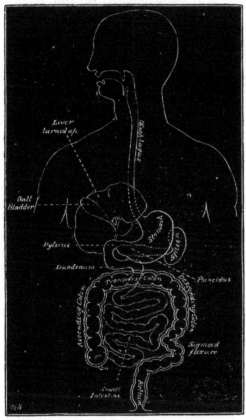

Fig. 165.—Diagram of the Alimentary Canal. The small intestine of man is from about 3 to 4 times as long as the large intestine.

as they are squeezed out from between them, until they have been sufficiently chewed.

The simple up and down, or *biting* movements of the lower jaw, are performed by the *temporal, masseter,* and *internal pterygoid* muscles, the action of which in closing the jaws alternates with that of the *digastric* and other muscles passing from the os hyoides to the lower jaw, which open them. The *grinding* or side to side movements of the lower jaw are performed mainly by the *external pterygoid* muscles, the muscle of one side acting alternately with the other. When both external pterygoids act together, the lower jaw is pulled directly forward, so that the lower incisor teeth are brought in front of the level of the upper.

Temporo-maxillary Fibro-cartilage.—The function of the inter-articular fibro-cartilage of the temporo-maxillary joint in mastication may be here mentioned. (1) As an elastic pad, it serves well to distrib-ute the pressure caused by the exceedingly powerful action of the masti-catory muscles. (2) It also serves as a joint-surface or socket for the condyle of the lower jaw, when the latter has been partially drawn for-ward out of the glenoid cavity of the temporal bone by the external ptery-goid muscle, some of the fibres of the latter being attached to its front surface, and consequently drawing it forward with the condyle which moves on it.

Nerve-mechanism of Mastication.—As in the case of so many other actions, that of mastication is partly voluntary and partly reflex and involuntary. The consideration of such *sensori-motor* actions will come hereafter (see Chapter on the Nervous System). It will suffice here to state that the nerves chiefly concerned are the *sensory* branches of the fifth and the glosso-pharyngeal, and the *motor* branches of the fifth and the ninth (hypoglossal) cerebral nerves. The nerve-centre through which the reflex action occurs, and by which the movements of the various muscles are harmonized, is situate in the medulla oblongata. In so far as mastication is voluntary or mentally perceived, it becomes so under the influence, in addition to the medulla oblongata, of the cerebral hemi-spheres.

Insalivation.—The act of mastication is much assisted by the saliva which is secreted by the salivary glands in largely increased amount during the process, and the intimate incorporation of which with the food, as it is being chewed, is termed *insalivation*.

THE SALIVARY GLANDS.

The salivary glands are the *parotid*, the *sub-maxillary*, and the *sub-lingual*, and numerous smaller bodies of similar structure, and with sep-arate ducts, which are scattered thickly beneath the mucous membrane of the lips, cheeks, soft palate, and root of the tongue.

Structure.—The salivary glands are usually described as compound tubular glands. They are made up of lobules. Each lobule consists of the branchings of a subdivision of the main duct of the gland, which are generally more or less convoluted toward their extremities, and some-times, according to some observers, sacculated or pouched. The convo-luted or pouched portions form the *alveoli*, or proper secreting parts of the gland. The alveoli are composed of a basement membrane of flattened cells joined together by processes to produce a fenestrated membrane, the spaces of which are occupied by a homogeneous ground-substance. With-in, upon this membrane, which forms the tube, the nucleated salivary secreting cells, of cubical or columnar form, are arranged parallel to one another surrounding a middle central canal. The granular appearance which is frequently seen in the salivary cells is due to the very dense network of fibrils..which they contain. When isolated, the cells not unfrequently are found to be branched. Connecting the alveoli into lobules is a considerable amount of fibrous connective tissue, which contains both flattened and granular protoplasmic cells, lymph corpuscles, and in some cases fat cells. The lobules are connected to form larger lobules (lobes), in a similar manner. The alveoli pass into the intralobular ducts by a narrowed portion (intercalary); lined with flattened epithelium with elongated nuclei. The intercalary ducts pass into the intralobular ducts by a narrowed neck, lined with cubical cells with small nuclei. The intralobular duct is larger in size, and is lined with large columnar nucleated

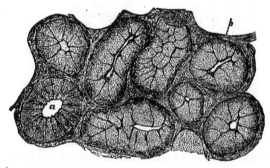

Fig. 166.—Section of submaxillary gland of dog. Showing gland-cells, *b*, and a duct, *a*, in section. (Kölliker.)

cells, the parts of which, toward the lumen of the tube, presents a fine longitudinal striation, due to the arrangement of the cell network. It is most marked in the submaxillary gland. The intralobular ducts pass into the larger ducts, and these into the main duct of the gland. As these ducts become larger they acquire an outside coating of connective tissue, and later on some unstriped muscular fibres. The lining of the larger ducts consists of one or more layers of columnar epithelium, containing an intracellular network of fibres arranged longitudinally.

Varieties.—Certain differences in the structure of salivary glands may be observed according as the glands secrete pure saliva, or saliva mixed with mucus, or pure mucus, and therefore the glands have been classified as: (1) *True salivary glands* (called most unfortunately by some *serous* glands), *e.g.*, the parotid of man and other animals, and the submaxillary of the rabbit and guinea-pig (Fig. 167). In this kind the alveolar lumen is small, and the cells lining the tubule are short, granular columnar cells, with nuclei presenting the intranuclear network. During rest the cells become larger, highly granular, with obscured nuclei, and the lumen becomes smaller. During activity, and after stimulation of the sympathetic, the cells become smaller and their contents more opaque; the granules first of all disappearing from the outer part of the cells, and then being found only at the extreme inner part and contiguous border of the cell. The nuclei reappear, as does also the lumen. (2) In *the true mucus-secreting glands*, as the sublingual of man and other animals, and

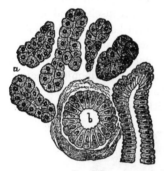
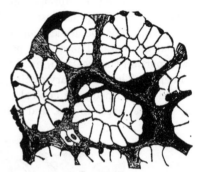

Fig. 167.   Fig. 168.

FIG. 167.—From a section through a true salivary gland. *a*, the gland alveoli, lined with albuminous "salivary cells;" *b*, intralobular duct cut transversely. (Klein and Noble Smith.)

FIG. 168.—From a section through a mucous gland in a quiescent state. The alveoli are lined with transparent mucous cells, and outside these are the demilunes of Heidenhain. The cells should have been represented as more or less granular. (Heidenhain.)

in the submaxillary of the dog, the tubes are larger, contain a larger lumen, and also have larger cells lining them. The cells are of two kinds, (*a*) *mucous or central cells,* which are transparent columnar cells with nuclei near the basement membrane. The cell substance is made up of a

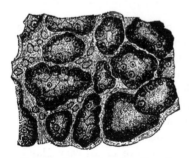

FIG. 169.—A part of a section through a mucous gland after prolonged electrical stimulation. The alveoli are lined with small granular cells. (Lavdovski.)

fine network, which in the resting state contains a transparent substance called *mucigen,* during which the cell does not stain well with logwood (Fig. 168). When the gland is secreting, *mucigen* is converted into *mucin,* and the cells swell up, appear more transparent, and stain deeply in logwood (Fig. 169). During rest, the cells become smaller and more granular from having discharged their contents, and the nuclei appear more distinct. (*b*) *Semilunes of Heidenhain* (Fig. 168), which are crescentic masses of granular parietal cells found here and there between the basement membrane and the central cells. These cells are small, and have a very dense reticulum, the nuclei are spherical, and increase in size during secretion. In the mucous gland there are some large tubes, lined with large transparent central cells, and have besides a few granular parietal cells; other small tubes are lined with small granular parietal cells alone; and a third variety are lined equally with each kind of cell. (3) In the muco-salivary or mixed glands, as the human submaxillary gland, part of the gland presents the structure of the mucous gland, whilst the remainder has that of the salivary glands proper.

Nerves and blood-vessels.—Nerves of large size are found in the salivary glands, they are contained in the connective tissue of the alveoli principally, and in certain glands, especially in the dog, are provided with ganglia. Some nerves have special endings in Pacinian corpuscles, some supply the blood-vessels, and others, according to Pflüger, penetrate the basement membrane of the alveoli and enter the salivary cells.

The blood-vessels form a dense capillary network around the ducts of the alveoli, being carried in by the fibrous trabeculæ between the alveoli, in which also begin the lymphatics by lacunar spaces.

Saliva.—Saliva, as it commonly flows from the mouth, is mixed with the secretion of the *mucous glands,* and often with air bubbles, which, being retained by its viscidity, make it frothy. When obtained from the parotid ducts, and free from mucus, saliva is a transparent watery fluid, the specific gravity of which varies from 1004 to 1008, and in which, when examined with the microscope, are found floating a number of minute particles, derived from the secreting ducts and vesicles of the glands. In the impure or mixed saliva are found, besides these particles, numerous epithelial scales separated from the surface of the mucous membrane of the mouth and tongue, and the so-called salivary corpuscles, discharged probably from the mucous glands of the mouth and the tonsils, which, when the saliva is collected in a deep vessel, and left at rest, subside in the form of a white opaque matter, leaving the supernatant salivary fluid transparent and colorless, or with a pale bluish-grey tint. In *reaction,* the saliva, when first secreted, appears to be always alkaline. During fasting, the saliva, although secreted alkaline, shortly becomes neutral; and it does so especially when secreted slowly and allowed to mix with the acid mucus of the mouth, by which its alkaline reaction is neutralized.

Chemical Composition of Mixed Saliva (Frerichs).

Water	994·10
Solids	5·90

Ptyalin	1·41
Fat	0.07
Epithelium and Proteids (including Serum-Albumin, Globulin, Mucin, &c.) . . .	2.13
Salts—Potassium Sulpho-Cyanate . . .	
Sodium Phosphate	
Calcium Phosphate	
Magnesium Phosphate	2·29
Sodium Chloride	
Potassium Chloride	
	5·90

The presence of potassium sulphocyanate (or *thiocyanate*) (C N K S) in saliva, may be shown by the blood-red coloration which the fluid gives with a solution of ferric chloride (Fe_2Cl_6), and which is bleached on the addition of a solution of mercuric chloride ($HgCl_2$).

Rate of Secretion and Quantity.—The rate at which saliva is secreted is subject to considerable variation. When the tongue and muscles concerned in mastication are at rest, and the nerves of the mouth are subject to no unusual stimulus, the quantity secreted is not more than sufficient, with the mucus, to keep the mouth moist. During actual secretion the flow is much accelerated.

The *quantity* secreted in twenty-four hours varies; its average amount is probably from 1 to 3 pints (1 to 2 litres).   ●

Uses of Saliva.—The purposes served by saliva are (1) mechanical and (2) chemical. I. *Mechanical.*—(1) It keeps the mouth in a due condition of moisture, facilitating the movements of the tongue in speaking, and the mastication of food. (2) It serves also in dissolving sapid substances, and rendering them capable of exciting the nerves of taste. But the principal mechanical purpose of the saliva is, (3) that by mixing with the food during mastication, it makes it a soft pulpy mass, such as may be easily swallowed. To this purpose the saliva is adapted both by quantity and quality. For, speaking generally, the quantity secreted during feeding is in direct proportion to the dryness and hardness of the food. The quality of saliva is equally adapted to this end. It is easy to see how much more readily it mixes with most kinds of food than water alone does; and the saliva from the parotid, labial, and other small glands, being more aqueous than the rest, is that which is chiefly *braided* and mixed with the food in mastication; while the more viscid mucous secretion of the submaxillary, palatine, and tonsillitic glands is spread over the surface of the softened mass, to enable it to slide more easily through the fauces and œsophagus. II. *Chemical.*—Saliva has the power of converting starch into glucose or grape-sugar. When saliva, or a portion of a salivary gland, is added to starch paste in a test-tube, and the mixture kept at a temperature of 100° F. (37·8° C.), the starch is very rapidly transformed into grape-sugar. There is an intermediate stage in which a part or the whole of the starch becomes dextrin.

Test for Glucose.—In such an experiment the presence of sugar is at once discovered by the application of Trommer's test, which consists in the addition of a drop or two of a solution of copper sulphate, followed by a larger quantity of caustic potash. When the liquid is boiled, an orange-red precipitate of copper suboxide indicates the presence of sugar; and when common raw starch is masticated and mingled with saliva, and kept with it at a temperature of 90° or 100° F. (30°—37.8° C.), the starch-grains are cracked or eroded, and their contents are transformed in the same manner as the starch-paste.

Saliva from the parotid is less viscid, less alkaline, clearer, and more watery than that from the submaxillary. It has, moreover, a less powerful action on starch. Sublingual saliva is the most viscid, and contains more solids than either of the other two, but does not appear to be so powerful in its action.

The salivary glands of children do not become functionally active till the age of 4 to 6 months, and hence the bad effect of feeding them before this age on starchy food, corn-flour, etc., which they are unable to render soluble and capable of absorption.

Action of Saliva on Starch.—This action is due to the presence in the saliva of the body called *ptyalin*. It is a nitrogenous body, and belongs to the order of *ferments*, which are bodies whose exact chemical composition is unknown, and which are capable of producing by their presence changes in other bodies, without themselves undergoing change. Ptyalin is called a *hydrolytic ferment*, that is to say, it acts by adding a molecule of water to the body changed. The reaction is supposed to be as follows:

$$3\ C_6H_{10}O_5 + 3\ H_2O = C_6H_{12}O_6 + 2\ (C_6H_{10}O_5) + 2\ H_2O = 3\ C_6H_{12}O_6.$$

Starch + Water. Glucose Dextrin Glucose

But it is not unlikely that the action is by no means so simple. In the first place, recent observers believe that a molecule of starch must be represented by a much more complex formula; next, that the stages in the reaction are more numerous and extensive; and thirdly, that the product of the reaction is not true glucose, but *maltose*. Maltose is a sugar more akin to cane than grape sugar, of very little sweetening power, and with less reducing power over copper salts. Its formula is $C_{12}H_{22}O_{11}$.

The action of saliva on starch is facilitated by: (*a*) Moderate heat, about 100° F. (37·8° C.). (*b*) A slightly alkaline medium. (*c*) Removal of the changed material from time to time. Its action is retarded by: (*a*) Cold; a temperature of 32° F. (0° C.) stops it for a time, but does not destroy it, whereas a high temperature above 140° F. (60° C.) destroys it. (*b*) Acids or strong alkalies either delay or stop the action altogether. (*c*) Presence of too much of the changed material. Ptyalin, in that it converts starch into sugar, is an *amylolytic* ferment.

Starch appears to be the only principle of food upon which saliva acts chemically: it has no apparent influence on any of the other ternary principles, such as sugar, gum, cellulose, or on fat, and seems to be equally destitute of power over albuminous and gelatinous substances.

Influence of the Nervous System.—The secretion of saliva is under the control of the nervous system. It is a reflex action, and in ordinary conditions is excited by the stimulation of the peripheral branches of two nerves, viz., the *gustatory or lingual* branch of the in-

ferior maxillary division of the fifth nerve, and the *glosso-pharyngeal* part
of the eighth pair of nerves, which are distributed to the mucous membrane of the tongue and pharynx.   The stimulation occurs on the introduction of sapid substances into the mouth, and the secretion is brought
about in the following way.   From the terminations of these sensory
nerves in the mucous membrane an impression is conveyed upward (afferent) to the special nerve centre situated in the medulla, which controls
the process, and by it is reflected to certain nerves supplied to the salivary
glands, which will be presently indicated.   In other words, the centre,
stimulated to action by the sensory impressions carried to it, sends out
impulses along efferent or secretory nerves supplied to the salivary glands,
which cause the saliva to be secreted by and discharged from the gland
cells.   Other stimuli, however, besides that of the food, and other sensory
nerves besides those mentioned, may produce reflexly the same effects.
Saliva may be caused to flow by irritation of the mucous membrane of the
mouth with mechanical, chemical, electrical, or thermal stimuli, also by
the irritation of the mucous membrane of the stomach in some way, as in
nausea, which precedes vomiting, when some of the peripheral fibres of
the *vagi* are irritated.   Stimulation of the *olfactory* nerves by smell of
food, of the *optic* nerves by the sight of it, and of the *auditory* nerves
by the sounds which are known by experience to accompany the preparation of a meal, may also, in the hungry, stimulate the nerve centre to
action.   In addition to these, as a secretion of saliva follows the movement of the muscles of mastication, it may be assumed that this movement stimulates the secreting nerve fibres of the gland, directly or reflexly.   From the fact that the flow of saliva may be increased or diminished by mental emotions, it is evident that impressions from the cerebrum also are capable of stimulating the centre to action or of inhibiting
its action.

Secretion may be excited by direct stimulation of the centre in the
medulla.

A.   On the Submaxillary Gland.—The submaxillary gland has been
the gland chiefly employed for the purpose of experimentally demonstrating the influence of the nervous system upon the secretion of saliva, because of the comparative facility with which, with its blood-vessels and
nerves, it may be exposed to view in the dog, rabbit, and other animals.
The chief nerves supplied to the gland are: (1) the *chorda tympani* (a
branch given off from the *facial* portio dura of the seventh pair of nerves),
in the canal through which it passes in the temporal bone, in its passage
from the interior of the skull to the face; and (2) branches of the *sympathetic* nerve from the plexus around the facial artery and its branches
to the gland.   The chorda (Fig. 170, *ch. t.*), after quitting the temporal
bone, passes downward and forward, under cover of the external pterygoid
muscle, and joins at an acute angle the lingual or gustatory nerve, pro-

ceeds with it for a short distance, and then passes along the submaxillary gland duct (Fig. 170, *sm. d.*), to which it is distributed, giving branches to the submaxillary ganglion (Fig. 170, *sm. gl.*), and sending others to terminate in the superficial muscle of the tongue. If this nerve be exposed and divided anywhere in its course from its exit from the skull to the gland, the secretion, if the gland be in action, is arrested, and no stimulation either of the lingual or of the glosso-pharyngeal will produce a flow of saliva. But if the peripheral end of the divided nerve be stimulated, an abundant secretion of saliva ensues, and the blood supply is enormously

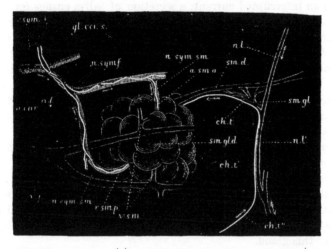

Fig. 170.—Diagrammatic representation of the submaxillary gland of the dog with its nerves and blood-vessels. (This is not intended to illustrate the exact anatomical relations of the several structures.) *sm. gld.*, the submaxillary gland into the duct (*sm. d.*), of which a cannula has been tied. The sublingual gland and duct are not shown. *n.l.*, *n.l'.*, the lingual or gustatory nerve; *ch. t.*, *ch. t'.*, the chorda tympani proceeding from the facial nerve, becoming conjoined with the lingual at *n. l'.*, and afterward diverging and passing to the gland along the duct; *sm. gl.*, submaxillary ganglion with its roots; *n. l.*, the lingual nerve proceeding to the tongue; *a. car.*, the carotid artery, two branches of which, *a. sm*, *a.* and *r. sm. p.*, pass to the anterior and posterior parts of the gland; *v. sm.*, the anterior and posterior veins from the gland ending in *v. j.*, the jugular vein; *v. sym.*, the conjoined vagus and sympathetic trunks; *gl. cer. s.*, the superior-cervical ganglion, two branches of which forming a plexus, *a. f.*, over the facial artery are distributed (*n. sym. sm.*) along the two glandular arteries to the anterior and posterior portion of the gland. The arrows indicate the direction taken by the nervous impulses; during reflex stimulations of the gland they ascend to the brain by the lingual and descend by the chorda tympani. (M. Foster.)

increased, the arteries being dilated. The veins even pulsate, and the blood contained within them is more arterial than venous in character.

When, on the other hand, the stimulus is applied to the *sympathetic* filaments (mere division producing no apparent effect), the arteries contract, and the blood stream is in consequence much diminished; and from the veins, when opened, there escapes only a sluggish stream of dark blood. The saliva, instead of being abundant and watery, becomes scanty and tenacious. If both chorda tympani and sympathetic branches be divided, the gland, released from nervous control, secretes continuously and abundantly (*paralytic*) secretion.

The abundant secretion of saliva, which follows stimulation of the chorda tympani, is not merely the result of a filtration of fluid from the blood-vessels, in consequence of the largely increased circulation through them. This is proved by the fact that, when the main duct is obstructed, the pressure within may considerably exceed the blood-pressure in the arteries, and also that when into the veins of the animal experimented upon some *atropin* has been previously injected, stimulation of the peripheral end of the divided chorda produces all the vascular effects as before, without any secretion of saliva accompanying them. Again, if an animal's head be cut off, and the chorda be rapidly exposed and stimulated with an interrupted current, a secretion of saliva ensues for a short time, although the blood supply is necessarily absent. These experiments serve to prove that the chorda contains two sets of nerve fibres, one set (*vaso-dilator*) which, when stimulated, act upon a local vaso-motor centre for regulating the blood supply, inhibiting its action, and causing the vessels to dilate, and so producing an increased supply of blood to the gland; while another set, which are paralyzed by injection of atropin, directly stimulate the cells themselves to activity, whereby they secrete and discharge the constituents of the saliva which they produce. These latter fibres very possibly terminate in the salivary cells themselves. If, on the other hand, the sympathetic fibres be divided, stimulation of the tongue by sapid substances, or of the trunk of the lingual, or of the glosso-pharyngeal, continues to produce a flow of saliva. From these experiments it is evident that the chorda tympani nerve is the principal nerve through which efferent impulses proceed from the centre to excite the secretion of this gland.

The sympathetic fibres appear to act principally as a vaso-constrictor nerve, and to exalt the action of the local vaso-motor centres. The sympathetic is more powerful in this direction than the chorda. There is not sufficient evidence in favor of the belief that the submaxillary gan-glion is ever the nerve centre which controls the secretion of the sub-maxillary gland.

B. On the Parotid Gland.—The nerves which influence secretion in the parotid gland are branches of the facial (lesser superficial petrosal) and of the sympathetic. The former nerve, after passing through the otic ganglion, joins the auriculo-temporal branch of the fifth cerebral nerve, and, with it, is distributed to the gland. The nerves by which the stimu-lus ordinarily exciting secretion is conveyed to the medulla oblongata, are, as in the case of the submaxillary gland, the fifth, and the glossopharyn-geal. The pneumogastric nerves convey a further stimulus to the secre-tion of saliva, when food has entered the stomach; the nerve centre is the same as in the case of the submaxillary gland.

Changes in the Gland Cells.—The method by which the salivary cells produce the secretion of saliva appears to be divided into two stages, which differ somewhat according to the class to which the gland belongs, viz., (1) the true salivary, or (2) the mucous type. In the former case, it has been noticed, as has been already described (p. 228), that during the rest which follows an active secretion the lumen of the alveoli becomes smaller, the gland cells larger, and very granular. During secretion the alveoli and their cells become smaller, and the granular appearance in the latter to a considerable extent disappears, and at the end of secretion, the granules are confined to the inner part of the cell nearest to the lumen, which is now quite distinct (Fig. 171).

It is supposed from these appearances that the first stage in the act of secretion consists in the protoplasm of the salivary cell taking up from the lymph certain materials from which it manufactures the elements of its own secretion, and which are stored up in the form of granules in the cell during rest, the second stage consisting of the actual discharge of

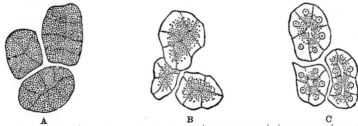

Fig. 171.—Alveoli of true salivary gland. A, at rest; B, in the first stage of secretion; C, after prolonged secretion. (Langley.)

these granules, with or without previous change. The granules are taken to represent the chief substance of the salivary secretion, i.e., the ferment ptyalin. In the case of the submaxillary gland of the dog, at any rate, the sympathetic nerve-fibres appear to have to do with the first stage of the process, and when stimulated the protoplasm is extremely active in manufacturing the granules, whereas the chorda tympani is concerned in the production of the second act, the actual discharge of the materials of secretion, together with a considerable amount of fluid, the latter being an actual secretion by the protoplasm, as it ceases to occur when atropin has been subcutaneously injected.

In the mucous-secreting gland, the changes in the cells during secretion have been already spoken of (p. 228). They consist in the gradual secretion by the protoplasm of the cell of a substance called *mucigen*, which is converted into *mucin*, and discharged on secretion into the canal of the alveoli. The mucigen is, for the most part. collected into the inner part of the cells during rest, pressing the nucleus and the small portion of the protoplasm which remains, against the limiting membrane of the alveoli.

The process of secretion in the salivary glands is identical with that of glands in general; the cells which line the ultimate branches of the ducts being the agents by which the special constituents of the saliva are formed.

The materials which they have incorporated with themselves are almost at once given up again, in the form of a fluid (secretion), which escapes from the ducts of the gland; and the cells, themselves, undergo disintegration,—again to be renewed, in the intervals of the active exercise of their functions. The source whence the cells obtain the materials of their secretion, is the blood, or, to speak more accurately, the plasma, which is filtered off from the circulating blood into the interstices of the glands as of all living textures.

THE PHARYNX.

That portion of the alimentary canal which intervenes between the mouth and the œsophagus is termed the *Pharynx* (Fig. 165). It will

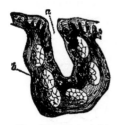

suffice here to mention that it is constructed of a series of three muscles with striated fibres (*constrictors*), which are covered by a thin fascia externally, and are lined internally by a strong fascia (pharyngeal aponeurosis), on the inner aspect of which is areolar (submucous) tissue and mucous membrane, continuous with that of the mouth, and, as regards the part concerned in swallowing, is identical with it in general structure. The epithelium of this part of the pharynx, like that of the mouth, is stratified and squamous.

Fig. 172.—Lingual follicle or crypt. *a*, involution of mucous membrane with its papillæ; *b*, lymphoid tissue, with several lymphoid sacs. (Frey.)

The pharynx is well supplied with mucous glands (Fig. 174).

The Tonsils.—Between the anterior and posterior arches of the soft palate are situated the *Tonsils*, one on each side. A tonsil consists of an elevation of the mucous membrane presenting 12 to 15 orifices, which lead into crypts or recesses, in the walls of which are placed nodules of adenoid or lymphoid tissue (Fig. 173). These nodules are enveloped in a less dense adenoid tissue which reaches the mucous surface. The surface is covered with stratified squamous epithelium, and the subepithelial or mucous membrane proper may present rudimentary papillæ formed of adenoid tissue. The tonsil is bounded by a fibrous capsule (Fig. 173, *e*). Into the crypts open a number of ducts of mucous glands.

The viscid secretion which exudes from the tonsils serves to lubricate the bolus of food as it passes them in the second part of the act of deglutition.

THE ŒSOPHAGUS OR GULLET.

The *Œsophagus* or Gullet (Fig. 165), the narrowest portion of the alimentary canal, is a muscular and mucous tube, nine or ten inches in length, which extends from the lower end of the pharynx to the cardiac orifice of the stomach.

Structure.—The œsophagus is made up of three coats—viz., the outer, *muscular;* the middle, *submucous;* and the inner, *mucous.* The *muscular* coat (Fig. 175, *g* and *i*) is covered externally by a varying amount of loose fibrous tissue. It is composed of two layers of fibres, the outer being arranged longitudinally, and the inner circularly. At the upper part of the œsophagus this coat is made up principally of striated muscle fibres, as they are continuous with the constrictor muscles of the pharynx; but lower down the unstriated fibres become more and more numerous, and toward the end of the tube form the entire coat. The muscular coat is connected with the mucous coat by a more or less developed layer of

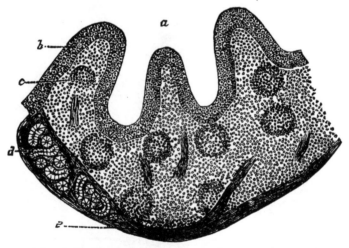

FIG. 173.—Vertical section through a crypt of the human tonsil. *a*, entrance to the crypt, which is divided below by the elevation which does not quite reach the surface; *b*, stratified epithelium; *c*, masses of adenoid tissue; *d*, mucous glands cut across; *e*, fibrous capsule. (V. D. Harris.)

areolar tissue, which forms the *submucous* coat (Fig 175, *f*), in which is contained in the lower half or third of the tube many mucous glands, the ducts of which, passing through the mucous membrane (Fig. 175, *c*) open on its surface. Separating this coat from the mucous membrane proper is a well-developed layer of longitudinal, unstriated muscle (*d*), called the *muscularis mucosæ.* The mucous membrane is composed of a closely felted meshwork of fine connective tissue, which, toward the surface, is elevated into rudimentary papillæ. It is covered with a stratified epithelium, of which the most superficial layers are squamous. The epithelium is arranged upon a basement membrane.

In newly-born children the mucous membrane exhibits, in many parts, the structure of lymphoid tissue (Klein).

Blood and lymph vessels, and nerves, are distributed in the walls of the œsophagus. Between the outer and inner layers of the muscular coat, *nerve-ganglia* of Auerbach are also found.

DEGLUTITION OR SWALLOWING.

When properly masticated, the food is transmitted in successive portions to the stomach by the act of *deglutition* or *swallowing*. This, for the purpose of description, may be divided into *three* acts. In the first, particles of food collected to a morsel are made to glide between the surface of the tongue and the palatine arch, till they have passed the anterior arch of the fauces; in the second, the morsel is carried through the

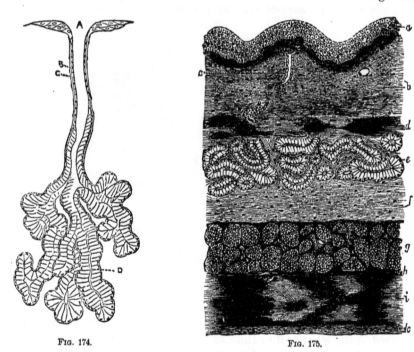

FIG. 174.　　　　　　　　　　　　FIG. 175.

FIG. 174.—Section of a mucous gland from the tongue. A, opening of the duct on the free surface; C, basement membrane with nuclei; B, flattened epithelial cells lining duct. The duct divides into several branches, which are convoluted and end blindly, being lined throughout by columnar epithelium. D, lumen of one of the tubuli of the gland. × 90. (Klein and Noble Smith.)

FIG. 175.—Longitudinal section of œsophagus of a dog toward the lower end. a, stratified epithelium of the mucous membrane; b, mucous membrane proper; c, duct of mucous gland; d, muscularis mucosæ; e, mucous glands; f, submucous coat; g, circular muscular layer; h, intermuscular layer, in which is contained the ganglion cells of Auerbach; i, longitudinal muscular layer; k, outside investment of fibrous tissue. × 100. (V. D. Harris.)

pharynx; and in the third, it reaches the stomach through the œsophagus. These three acts follow each other rapidly. (1.) The first act of deglutition may be voluntary, although it is usually performed unconsciously; the morsel of food, when sufficiently masticated, being pressed between the tongue and palate, by the agency of the muscles of the former, in such a manner as to force it back to the entrance of the pharynx. (2.) The second act is the most complicated, because the food must pass by the posterior orifice of the nose and the upper opening of the larynx without touching them. When it has been brought, by the first act, between the anterior arches of the palate, it is moved onward by the movement of the tongue backward, and by the muscles of the anterior arches contracting on it and then behind it. The root of the tongue being retracted, and the larynx being raised with the pharynx and carried forward under the base of the tongue, the epiglottis is pressed over the upper opening of the larynx, and the morsel glides past it; the closure of the glottis being additionally secured by the simultaneous contraction of its own muscles: so that, even when the epiglottis is destroyed, there is little danger of food or drink passing into the larynx so long as its muscles can act freely. At the same time, the raising of the soft palate, so that its posterior edge touches the back part of the pharynx, and the approximation of the sides of the posterior palatine arch, which move quickly inward like side curtains, close the passage into the upper part of the pharynx and the posterior nares, and form an inclined plane, along the under surface of which the morsel descends; then the pharynx, raised up to receive it, in its turn contracts, and forces it onward into the œsophagus. (3.) In the third act, in which the food passes through the œsophagus, every part of that tube, as it receives the morsel and is dilated by it, is stimulated to contract: hence an undulatory contraction of the œsophagus, which is easily observable in horses while drinking, proceeds rapidly along the tube. It is only when the morsels swallowed are large, or taken too quickly in succession, that the progressive contraction of the œsophagus is slow, and attended with pain. Division of both pneumogastric nerves paralyzes the contractile power of the œsophagus, and food accordingly accumulates in the tube. The second and third parts of the act of deglutition are involuntary.

Nerve Mechanism.—The nerves engaged in the reflex act of deglutition are:—*sensory*, branches of the fifth cerebral supplying the soft palate; glosso-pharyngeal, supplying the tongue and pharynx; the superior laryngeal branch of the vagus, supplying the epiglottis and the glottis; while the *motor* fibres concerned are:—branches of the fifth, supplying part of the digastric and mylo-hyoid muscles, and the muscles of mastication; the facial, supplying the levator palati; the glosso-pharyngeal, supplying the muscles of the pharynx; the vagus, supplying the muscles of the larynx through the inferior laryngeal branch, and the hypoglossal, the muscles of the tongue. The nerve-centre by which the muscles are harmonized in their action, is situate in the medulla oblongata. In the movements of the œsophagus, the ganglia contained in its walls, with the pneumogastrics, are the nerve-structures chiefly concerned.

It is important to note that the swallowing both of food and drink is a *muscular* act, and can, therefore, take place in opposition to the force of gravity. Thus, horses and many other animals habitually drink up-hill, and the same feat can be performed by jugglers.

The Stomach.

In man and those Mammalia which are provided with a single stomach, it consists of a dilatation of the alimentary canal placed between and continuous with the œsophagus, which enters its larger or cardiac end on the one hand, and the small intestine, which commences at its narrowed end or pylorus, on the other. It varies in shape and size according to its state of distension.

The *Ruminants* (ox, sheep, deer, etc.) possess very complex stomachs; in most of them four distinct cavities are to be distinguished (Fig 176). 1. The *Paunch* or *Rumen,*   large cavity which occupies the cardiac end, and into which large   tics of food are in the first instance swallowed with little or no mu   ication. 2. The *Reticulum,* or *Honeycomb* stomach, so called from the fact that its mucous membrane is disposed in a number of folds enclosing hexagonal cells. 3. The *Psalterium,*

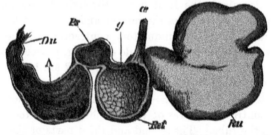

Fig. 176.— Stomach of sheep. *œ*, œsophagus; *Ru,* rumen; *Ret,* reticulum; *Ps,* psalterium, or manyplies; *A,* abomasum; *Du,* duodenum; *g,* groove from œsophagus to psalterium. (Huxley.)

or *Manyplies,* in which the mucous membrane is arranged in very prominent longitudinal folds. 4. *Abomasum, Reed,* or *Rennet,* narrow and elongated, its mucous membrane being much more highly vascular than that of the other divisions. In the process of rumination small portions of the contents of the rumen and reticulum are successively regurgitated into the mouth, and there thoroughly masticated and insalivated (chewing the cud): they are then again swallowed, being this time directed by a groove (which in the figure is seen running from the lower end of the œsophagus) into the manyplies, and thence into the abomasum. It will thus be seen that the first two stomachs (paunch and reticulum) have chiefly the mechanical functions of storing and moistening the fodder: the third (manyplies) probably acts as a strainer, only allowing the finely divided portions of food to pass on into the fourth stomach, where the gastric juice is secreted and the process of digestion carried on. The mucous membrane of the first three stomachs is lowly vascular, while that of the fourth is pulpy, glandular, and highly vascular.

In some other animals, as the pig, a similar distinction obtains between the mucous membrane in different parts of the stomach.

In the pig the glands' in the cardiac end are few and small, while toward the pylorus they are abundant and large.

A similar division of the stomach into a cardiac (receptive) and a pyloric (digestive) part, foreshadowing the complex stomach of ruminants, is seen in the common rat, in which those two divisions of the stomach are distinguished, not only by the characters of their lining membrane, but also by a well-marked constriction.

In birds the function of mastication is performed by the stomach (gizzard) which in granivorous orders, *e.g.* the common fowl, possesses very powerful muscular walls and a dense horny epithelium.

Structure.—The stomach is composed of four coats, called respectively—an external or (1) *peritoneal*, (2) *muscular*, (3) *submucous*, and (4) *mucous* coat; with blood-vessels, lymphatics, and nerves distributed in and between them.

(1) The *peritoneal* coat has the structure of serous membranes in general (p. 319). (1) The *muscular* coat consists of three separate layers or sets of fibres, which, according to their several directions, are named the longitudinal, circular, and oblique. The *longitudinal* set are the most superficial: they are continuous with the longitudinal fibres of the œsophagus, and spread out in a diverging manner over the cardiac end and sides of the stomach. They extend as far as the pylorus, being especially distinct at the lesser or upper curvature of the stomach, along which they pass in several strong bands. The next set are the *circular* or *transverse* fibres, which more or less completely encircle all parts of the stomach; they are most abundant at the middle and in the pyloric portion of the organ, and form the chief part of the thick projecting ring of the pylorus. These fibres are not simple circles, but form double or figure-of-8 loops, the fibres intersecting very obliquely. The next, and consequently deepest set of fibres, are the *oblique*, continuous with the circular muscular fibres of the œsophagus, and having the same double-looped arrangement that prevails in the preceding layer: they are comparatively few in number, and are placed only at the cardiac orifice and portion of the stomach, over both surfaces of which they are spread, some passing obliquely from left to right, others from right to left, around the cardiac orifice, to which, by their interlacing, they form a kind of sphincter, continuous with that around the lower end of the œsophagus. The muscular fibres of the stomach and of the intestinal canal are *unstriated*, being composed of elongated, spindle-shaped fibre-cells.

(3) and (4) The *mucous membrane* of the stomach, which rests upon a layer of loose cellular membrane, or *submucous tissue*, is smooth, level, soft, and velvety; of a pale pink color during life, and in the contracted state thrown into numerous, chiefly longitudinal, folds or rugæ, which disappear when the organ is distended.

The basis of the mucous membrane is a fine connective tissue, which approaches closely in structure to adenoid tissue; this tissue supports the tubular glands of which the superficial and chief part of the mucous membrane is composed, and passing up between them assists in binding them together. Here and there are to be found in this coat, immediately underneath the glands, masses of adenoid tissue sufficiently marked to be termed by some lymphoid follicles. The glands are separated from the rest of the mucous membrane by a very fine homogeneous basement membrane.

At the deepest part of the mucous membrane are two layers (circular and longitudinal) of unstriped muscular fibres, called the *muscularis mucosæ*, which separate the mucous membrane from the scanty submucous tissue.

When examined with a lens, the internal or free surface of the stomach presents a peculiar honeycomb appearance, produced by shallow polygonal depressions, the diameter of which varies generally from $\frac{1}{200}$ th to $\frac{1}{300}$th of an inch; but near the pylorus is as much as $\frac{1}{100}$th of an inch. They are separated by slightly elevated ridges, which sometimes, especially in certain morbid states of the stomach, bear minute, narrow vascular processes, which look like villi, and have given rise to the erroneous supposition that the stomach has absorbing villi, like those of the small intestines. In the bottom of these little pits, and to some extent between them, minute openings are visible, which are the orifices of the ducts of perpendicularly arranged tubular glands (Fig. 177), imbedded side by side in sets or bundles, on the surface of the mucous membrane, and composing nearly the whole structure.

Gastric Glands.—Of these there are two varieties, (*a*) Peptic, (*b*) Pyloric or Mucous.

(*a*) *Peptic* glands are found throughout the whole of the stomach except at the pylorus. They are arranged in groups of four or five, which are separated by a fine connective tissue. Two or three tubes often open into one duct, which forms about a third of the whole length of the tube and opens on the surface. The ducts are lined with columnar epithelium. Of the gland tube proper, *i.e.*, the part of the gland below the duct, the upper third is the *neck* and the rest the *body*. The neck is narrower than the body, and is lined with granular cubical cells which are continuous with the columnar cells of the duct. Between these cells and the membrana propria of the tubes, are large oval or spherical cells, opaque or granular in appearance, with clear oval nuclei, bulging out the membrana propria; these cells are called *peptic or parietal cells*. They do not form a continuous layer. The body, which is broader than the neck and terminates in a blind extremity or *fundus* near the muscularis mucosæ, is lined by cells continuous with the cubical or central cells of the neck, but longer, more columnar and more transparent. In this part are a few parietal cells of the same kind as in the neck (Fig. 177).

As the pylorus is approached the gland ducts become longer, and the tube proper becomes shorter, and occasionally branched at the fundus.

(*b*) *Pyloric Glands.*—These glands (Fig. 179) have much longer ducts than the peptic glands. Into each duct two or three tubes open by very short and narrow necks, and the body of each tube is branched, wavy, and convoluted. The lumen is very large. The ducts are lined with columnar epithelium, and the neck and body with shorter and more gran-

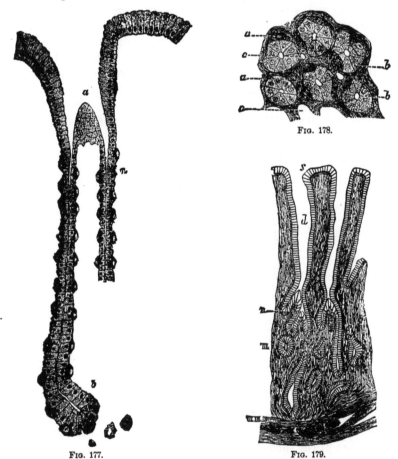

FIG. 178.

FIG. 177.   FIG. 179.

FIG. 177.—From a vertical section through the mucous membrane of the cardiac end of stomach. Two peptic glands are shown with a duct common to both, one gland only in part. *a*, duct with columnar epithelium becoming shorter as the cells are traced downward; *n*, neck of gland tubes, with central and parietal or so-called peptic cells; *b*, fundus with curved cæcal extremity—the parietal cells are not so numerous here. × 400. (Klein and Noble Smith.)

FIG. 178.—Transverse section through lower part of peptic glands of a cat. *a*, peptic cells; *b*, small spheroidal or cubical cells; *c*, transverse section of capillaries. (Frey.)

FIG. 179.—Section showing the pyloric glands. *s*, free surface; *d*, ducts of pyloric glands; *n*, neck of same; *m*, the gland alveoli; *m m*, muscularis mucosæ. (Klein and Noble Smith.)

ular cubical cells, which correspond with the central cells of the peptic glands. During secretion the cells become, as in the case of the peptic glands, larger and the granules restricted to the inner zone of the cell. As they approach the duodenum the pyloric glands become larger, more convoluted and more deeply situated. They are directly continuous with Brunner's glands in the duodenum. (Watney.)

Changes in the gland cells during secretion.—The chief or cubical cells of the peptic glands, and the corresponding cells of the pyloric glands during the early stage of digestion, if hardened in alcohol, appear swollen and granular, and stain readily. At a later stage the cells become smaller, but more granular and stain even more readily. The parietal cells swell up, but are otherwise not altered during digestion. The granules, however, in the alcohol-hardened specimen, are believed not to exist in the living cells, but to have been precipitated by the hardening reagent; for if examined during life they appear to be confined to the inner zone of the cells, and the outer zone is free from granules, whereas during rest the cell is granular throughout. These granules are thought to be

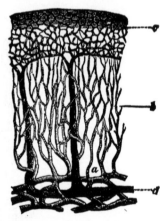

Fig. 180.—Plan of the blood-vessels of the stomach, as they would be seen in a vertical section. *a*, arteries, passing up from the vessels of submucous coat; *b*, capillaries branching between and around the tubes; *c*, superficial plexus of capillaries occupying the ridges of the mucous membrane; *d*, vein formed by the union of veins which, having collected the blood of the superficial capillary plexus, are seen passing down between the tubes. (Brinton.)

pepsin, or the substance from which pepsin is formed, *pepsinogen*, which is during rest stored chiefly in the inner zone of the cells and discharged into the lumen of the tube during secretion. (Langley.)

Lymphatics.—Lymphatic vessels surround the gland tubes to a greater or less extent. Toward the fundus of the peptic glands are found masses of lymphoid tissue, which may appear as distinct follicles, somewhat like the solitary glands of the small intestine.

Blood-vessels.—The blood-vessels of the stomach, which first break up in the submucous tissue, send branches upward between the closely packed glandular tubes, anastomosing around them by means of a fine capillary network, with oblong meshes. Continuous with this deeper plexus, or prolonged upward from it, so to speak, is a more superficial network of larger capillaries, which branch densely around the orifices of the tubes, and form the framework on which are moulded the small elevated ridges of mucous membrane bounding the minute, polygonal pits before referred to. From this *superficial* network the veins chiefly take their origin. Thence passing down between the tubes, with no very free connection with the deeper *inter-tubular* capillary plexus, they open finally into the venous network in the submucous tissue.

Nerves.—The nerves of the stomach are derived from the pneumo-gastric and sympathetic, and form a plexus in the submucous and mus-cnlar coats, containing many ganglia (Remak, Meissner).

DIGESTION IN THE STOMACH.

Gastric Juice.—The functions of the stomach are to secrete a diges-tive fluid (gastric juice), to the action of which the food is next subjected after it has entered the cavity of the stomach from the œsophagus; to thoroughly incorporate the fluid with the food by means of its muscular movements; and to absorb such substances as are capable of absorption. While the stomach contains no food, and is inactive, no gastric fluid is secreted; and mucus, which is either neutral or slightly alkaline, covers its surface. But immediately on the introduction of food or other sub-stance the mucous membrane, previously quite pale, becomes slightly turgid and reddened with the influx of a larger quantity of blood; the gastric glands commence secreting actively, and an acid fluid is poured out in minute drops, which gradually run together and flow down the walls of the stomach, or soak into the substances within it.

Chemical Composition of Gastric Juice.—The first accurate analysis of gastric juice was made by Prout: but it does not appear to have been collected in any large quantity, or pure and separate from food, until the time when Beaumont was enabled, by a fortunate circumstance, to obtain it from the stomach of a man named St. Martin, in whom there existed, as the result of a gunshot wound, an opening leading directly into the stomach, near the upper extremity of the great curvature, and three inches from the cardiac orifice. The introduction of any mechanical irritant, such as the bulb of a thermometer, into the stomach, excited at once the secretion of gastric fluid. This was drawn off, and was often obtained to the extent of nearly an ounce. The introduction of aliment-ary substances caused a much more rapid and abundant secretion than did other mechanical irritants. No increase of temperature could be detected during the most active secretion; the thermometer introduced into the stomach always stood at 100° F. (37·8° C.) except during muscn-lar exertion, when the temperature of the stomach, like that of other parts of the body, rose one or two degrees higher.

The chemical composition of human gastric juice has been also in-vestigated by Schmidt. The fluid in this case was obtained by means of an accidental gastric fistula, which existed for several years below the left mammary region of a patient between the cartilages of the ninth and tenth ribs. The mucous membrane was excited to action by the introduction of some hard matter, such as dry peas, and the secretion was removed by means of an elastic tube. The fluid thus obtained was found to be acid, limpid, odorless, with a mawkish taste—with a specific gravity of 1002, or a little more. It contained a few cells, seen with the microscope, and some fine granular matter. The analysis of the fluid obtained in this is given below. The gastric juice of dogs and other animals obtained by the introduction into the stomach of a clean sponge through an artificially made gastric fistula, shows a decided difference in composition, but possibly this is due, at least in part, to admixture with food.

CHEMICAL COMPOSITION OF GASTRIC JUICE.

	Dogs.	Human.
Water	971·17	994·4
Solids	28·82	5·39
Solids—		
Ferment—Pepsin	17·5	3·19
Hydrochloric acid (free)	2·7	·2
Salts—		
Calcium, sodium, and potassium, chlorides; and		
calcium, magnesium, and iron, phosphates .	8·57	2·19

The *quantity* of gastric juice secreted daily has been variously estimated; but the average for a healthy adult may be assumed to range from ten to twenty pints in the twenty-four hours. The acidity of the fluid is due to free *hydrochloric* acid, although other acids, *e.g.*, *lactic, acetic, butyric*, are not unfrequently to be found therein as products of gastric digestion. The amount of hydrochloric acid varies from 2 to ·2 per 1000 parts. In healthy gastric juice the amount of free acid may be as much as ·2 per cent.

As regards the formation of pepsin and acid, the former is produced by the central or chief cells of the peptic glands, and also most likely by the similar cells in the pyloric glands; the acid is chiefly found at the surface of the mucous membrane, but is in all probability formed by the secreting action of the parietal cells of the peptic glands, as no acid is formed by the pyloric glands in which this variety of cell is absent.

The ferment *Pepsin* (p. 246) can be procured by digesting portions of the mucous membrane of the stomach in cold water, after they have been macerated for some time in water at a temperature 80°—100° F. (27·°—37·8° C.). The warm water dissolves various substances as well as some of the pepsin, but the cold water takes up little else than pepsin, which is contained in a greyish-brown viscid fluid, on evaporating the cold solution. The addition of alcohol throws down the pepsin in greyish-white flocculi. Glycerine also has the property of dissolving out the ferment; and if the mucous membrane be finely minced and the moisture removed by 'absolute alcohol, a powerful extract may be obtained by throwing into glycerine.

Functions.—The digestive power of the gastric juice depends on the pepsin and acid contained in it, both of which are, under ordinary circumstances, necessary for the process.

The general effect of digestion in the stomach is the conversion of the food into *chyme,* a substance of various composition according to the nature of the food, yet always presenting a characteristic thick, pultaceous, grumous consistence, with the undigested portions of the food mixed in a more fluid substance, and a strong, disagreeable acid odor and taste.

The chief function of the gastric juice is to *convert proteids into peptones.* This action may be shown by adding a little gastric juice (natural or artificial) to some diluted egg-albumin, and keeping the mixture at a temperature of about 100° F. (37·8° C.); it is soon found that the albumin cannot be precipitated on boiling, but that if the solution be neutralized with an alkali, a precipitate of acid-albumin is thrown down. After a while the proportion of acid-albumin gradually diminishes, so that at last scarcely any precipitate results on neutralization, and finally it is found that all the albumin has been changed into another proteid substance which is not precipitated on boiling or on neutralization. This is called *peptone.*

Characteristics of Peptones.—Peptones have certain characteristics which distinguish them from other proteids. 1. They are *diffusible, i.e.,* they possess the property of passing through animal membranes. 2. They cannot be precipitated by heat, nitric, or acetic acid, or potassium ferrocyanide and acetic acid. They are, however, thrown down by tannic acid, by mercuric chloride and by picric acid. 3. They are very soluble in water and in neutral saline solutions.

In their diffusibility peptones differ remarkably from egg-albumin, and on this diffusibility depends one of their chief uses. Egg-albumin as such, even in a state of solution, would be of little service as food, inasmuch as its indiffusibility would effectually prevent its passing by absorption into the blood-vessels of the stomach and intestinal canal. Changed, however, by the action of the gastric juice into peptones, albuminous matters *diffuse* readily, and are thus quickly absorbed.

After entering the blood the peptones are very soon again modified, so as to re-assume the chemical characters of albumin, a change as necessary for preventing their diffusing out of the blood-vessels, as the previous change was for enabling them to pass in. This is effected, probably, in great part by the agency of the liver.

Products of Gastric Digestion.—The chief product of gastric digestion is undoubtedly peptone. We have seen, however, in the above experiment that there is a by-product, and this is almost identical with syntonin or acid albumin. This body is probably not exactly identical, however, with syntonin, and its old name of *parapeptone* had better be retained. The conversion of native albumin into acid albumin may be effected by the hydrochloric acid alone, but the further action is undoubtedly due to the ferment and the acid together, as although under high pressure any acid solution may, it is said, if strong enough, produce the entire conversion into peptone, under the condition of digestion in the stomach this would be quite impossible; and, on the other hand, pepsin will not act without the presence of acid. The production of two forms of peptone is usually recognized, called respectively *anti*-peptone and *hemi*-peptone. Their differences in chemical properties have not yet been made out, but they are distinguished by this remarkable fact, that the pancreatic juice, while possessing no action over the former, is able to convert the latter into leucin and tyrosin. Pepsin acts the part of a hydrolytic ferment (proteolytic), and appears to cause hydration of albumin, peptone being a highly hydrated form of albumin.

Circumstances favoring Gastric Digestion.—1. A temperature of about 100° F. (37·8° C.); at 32° F. (0° C.) it is delayed, and by boiling is altogether stopped. 2. An acid medium is necessary. Hydrochloric is the best acid for the purpose. Excess of acid or neutralization stops the process. 3. The removal of the products of digestion. Excess of peptone delays the action.

Action of the Gastric Juice on Bodies other than Proteids. —All proteids are converted by the gastric juice into peptones, and, therefore, whether they be taken into the body in meat, eggs, milk, bread, or other foods, the resultant still is peptone.

Milk is curdled, the casein being precipitated, and then dissolved. The curdling is due to a special ferment of the gastric juice (curdling ferment), and is not due to the action of the free acid only. The effect of rennet, which is a decoction of the fourth stomach of a calf in brine, has long been known, as it is used extensively to cause precipitation of casein in cheese manufacture.

The ferment which produces this curdling action is distinct from pepsin.

Gelatin is dissolved and changed into peptone, as are also *chondrin* and *elastin;* but *mucin*, and the *horny tissues*, keratin generally are unaffected.

On the *amylaceous* articles of food, and upon pure *oleaginous* principles the gastric juice has no action. In the case of adipose tissue, its effect is to dissolve the areolar tissue, albuminous cell-walls, etc., which enter into its composition, by which means the fat is able to mingle more uniformly with the other constituents of the *chyme.*

The gastric fluid acts as a general solvent for some of the *saline* constituents of the food, as, for example, particles of common salt, which may happen to have escaped solution in the saliva; while its acid may enable it to dissolve some other salts which are insoluble in the latter or in water. It also dissolves cane *sugar*, and by the aid of its mucus causes its conversion in part into grape sugar.

The action of the gastric juice in preventing and checking putrefaction has been often directly demonstrated. Indeed, that the secretions which the food meets with in the alimentary canal are *antiseptic* in their action, is what might be anticipated, not only from the proneness to decomposition of organic matters, such as those used as food, especially under the influence of warmth and moisture, but also from the well-known fact that decomposing flesh (*e.g.*, high game) may be eaten with impunity, while it would certainly cause disease were it allowed to enter the blood by any other route than that formed by the organs of digestion.

Time occupied in Gastric Digestion.—Under ordinary conditions, from three to four hours may be taken as the average time occupied by the digestion of a meal in the stomach. But many circumstances will modify the rate of gastric digestion. The chief are: the *nature* of the food taken and its *quantity* (the stomach should be fairly filled—not distended); the time that has elapsed since the last meal, which should be at least enough for the stomach to be quite clear of food; the amount of exercise previous and subsequent to a meal (gentle exercise being favorable, over-exertion injurious to digestion); the state of mind (tranquillity of temper being essential, in most cases, to a quick and due digestion); the bodily health; and some others.

Movements of the Stomach.—The gastric fluid is assisted in accomplishing its share in digestion by the movements of the stomach. In granivorous birds, for example, the contraction of the strong muscular gizzard affords a necessary aid to digestion, by grinding and triturating the hard seeds which constitute part of the food. But in the stomachs of man and other Mammalia the motions of the muscular coat are too feeble to exercise any such mechanical force on the food; neither are they needed, for mastication has already done the mechanical work of a gizzard; and experiments have demonstrated that substances enclosed in perforated tubes, and consequently protected from mechanical influence, are yet digested.

The normal actions of the muscular fibres of the human stomach appear to have a threefold purpose; (1) to adapt the stomach to the quantity of food in it, so that its walls may be in contact with the food on all sides, and, at the same time, may exercise a certain amount of compression upon it; (2) to keep the orifices of the stomach closed until the food is digested; and (3) to perform certain peristaltic movements, whereby the food, as it becomes chymified, is gradually propelled toward, and ultimately through, the pylorus. In accomplishing this latter end, the movements without doubt materially contribute toward effecting a thorough intermingling of the food and the gastric fluid.

When digestion is not going on, the stomach is uniformly contracted, its orifices not more firmly than the rest of its walls; but, if examined shortly after the introduction of food, it is found closely encircling its contents, and its orifices are firmly closed like sphincters. The cardiac orifice, every time food is swallowed, opens to admit its passage to the stomach, and immediately again closes. The pyloric orifice, during the first part of gastric digestion, is usually so completely closed, that even when the stomach is separated from the intestines, none of its contents escape. But toward the termination of the digestive process, the pylorus seems to offer less resistance to the passage of substances from the stomach; first it yields to allow the successively digested portions to go through it; and then it allows the transit of even undigested substances. It appears that food, so soon as it enters the stomach, is subjected to a kind of peristaltic action of the muscular coat, whereby the digested portions are gradually moved toward the pylorus. The movements were observed to increase in rapidity as the process of chymification advanced, and were continued until it was completed.

The contraction of the fibres situated toward the pyloric end of the stomach seems to be more energetic and more decidedly peristaltic than those of the cardiac portion. Thus, it was found in the case of St. Martin, that when the bulb of the thermometer was placed about three inches from the pylorus, through the gastric fistula, it was tightly embraced from time to time, and drawn toward the pyloric orifice for a distance of three or four inches. The object of this movement appears to be, as just said, to carry the food toward the pylorus as fast as it is formed into chyme, and to propel the chyme into the duodenum; the undigested portions of food being kept back until they are also reduced into chyme, or until all that is digestible has passed out. The action of these fibres is often seen in the contracted state of the pyloric portion of the stomach after death, when it alone is contracted and firm, while the cardiac portion forms a dilated sac. Sometimes, by a predominant action of strong circular fibres placed between the cardia and pylorus, the two portions, or ends as they are called, of the stomach, are partially separated from each other by a kind of hour-glass contraction. By means of the peristaltic action of the muscular coats of the stomach, not merely is chymified food gradually propelled through the pylorus, but a kind of double current is continually kept up among the contents of the stomach, the circumferential parts of the mass being gradually moved onward toward the pylorus by the contraction of the muscular fibres, while the central portions are propelled in the opposite direction, namely, toward the cardiac orifice; in this way is kept up a constant circulation of the contents of the viscus, highly conducive to their free mixture with the gastric fluid and to their ready digestion.

Vomiting.—The expulsion of the contents of the stomach in vomiting, like that of mucous or other matter from the lungs in *coughing*, is preceded by an inspiration; the glottis is then closed, and immediately afterward the abdominal muscles strongly act; but here occurs the difference in the two actions. Instead of the vocal cords yielding to the action of the abdominal muscles, they remain tightly closed. Thus the diaphragm being unable to go up, forms an unyielding surface against which the stomach can be pressed. In this way, as well as by its own contraction, it is *fixed*, to use a technical phrase. At the same time the *cardiac* sphincter-muscle being relaxed, and the orifice which it naturally guards being actively dilated, while the *pylorus* is closed, and the stomach itself also contracting, the action of the abdominal muscles, by these means assisted, expels the contents of the organ through the œsophagus, pharynx, and mouth. The reversed peristaltic action of the œsophagus probably increases the effect.

It has been frequently stated that the stomach itself is quite passive during vomiting, and that the expulsion of its contents is effected solely by the pressure exerted upon it when the capacity of the abdomen is diminished by the contraction of the diaphragm, and subsequently of the abdominal muscles. The experiments and observations, however, which are supposed to confirm this statement, only show that the contraction of the abdominal muscles alone is sufficient to expel matters from an unresisting bag through the œsophagus; and that, under very abnormal circumstances, the stomach, by itself, cannot expel its contents. They by no means show that in ordinary vomiting the stomach is passive; and, on the other hand, there are good reasons for believing the contrary.

It is true that facts are wanting to demonstrate with certainty this action of the stomach in vomiting; but some of the cases of fistulous opening into the organ appear to support the belief that it does take place; and the analogy of the case of the stomach with that of the other hollow viscera, as the rectum and bladder, may be also cited in confirmation.

The *muscles* concerned in the act of vomiting, are chiefly and primarily *those of the abdomen;* the *diaphragm* also acts, but usually not as the muscles of the abdominal walls do. They contract and compress the stomach more and more toward the diaphragm; and the diaphragm (which is usually drawn down in the deep inspiration that precedes each act of vomiting) is fixed, and presents an unyielding surface against which the stomach may be pressed. The diaphragm is, therefore, as a rule, passive during the actual expulsion of the contents of the stomach. But there are grounds for believing that sometimes this muscle actively contracts, so that the stomach is, so to speak, squeezed between the descending diaphragm and the retracting abdominal walls.

Some persons possess the power of *vomiting at will,* without applying any undue irritation to the stomach, but simply by a voluntary effort. It seems also, that this power may be acquired by those who do not naturally possess it, and by continual practice may become a habit. There are cases also of rare occurrence in which persons habitually swallow their food hastily, and nearly unmasticated, and then at their leisure regurgitate it, piece by piece, into their mouth, remasticate, and again swallow it, like members of the ruminant order of Mammalia.

The various *nerve-actions* concerned in vomiting are governed by a nerve-centre situate in the medulla oblongata.

The sensory nerves are the fifth, glosso-pharyngeal and vagus principally; but, as well, vomiting may occur from stimulation of sensory nerves from many organs, *e.g.*, kidney, testicle, etc. The centre may also be stimulated by impressions from the cerebrum and cerebellum, so called *central* vomiting occurring in disease of those parts. The efferent impulses are carried by the phrenics and the spinal nerves.

Influence of the Nervous System on Gastric Digestion.—The normal movements of the stomach during gastric digestion are directly connected with the plexus of nerves and ganglia contained in its walls, the presence of food acting as a stimulus which is conveyed to the ganglia and reflected to the muscular fibres. The stomach is, however, also directly connected with the higher nerve-centres by means of branches of the vagus and solar plexus of the sympathetic. The vaso-motor fibres of the latter are derived, probably, from the splanchnic nerves.

The exact function of the vagi in connection with the movements of the stomach is not certainly known. Irritation of the vagi produces contraction of the stomach, if digestion is proceeding; while, on the other hand, peristaltic action is retarded or stopped, when these nerves are divided.

Bernard, watching the act of gastric digestion in dogs which had fistulous openings into their stomachs, saw that on the instant of dividing their vagic nerves, the process of digestion was stopped, and the mucous membrane of the stomach, previously turgid with blood, became pale, and ceased to secrete. These facts may be explained by the theory that the vagi are the media by which, during digestion, an *inhibitory* impulse is conducted to the vaso-motor centre in the medulla; such impulse being reflected along the splanchnic nerves to the blood-vessels of the stomach, and causing their dilatation (Rutherford). From other experiments it may be gathered, that although division of both vagi always temporarily suspends the secretion of gastric fluid, and so arrests the process of digestion, being occasionally followed by death from inanition; yet the digestive powers of the stomach may be completely restored after the operation, and the formation of chyme and the nutrition of the animal may be carried on almost as perfectly as in health. This would indicate the existence of a special local nervous mechanism which controls the secretion.

Bernard found that galvanic stimulus of these nerves excited an active secretion of the fluid, while a like stimulus applied to the sympathetic nerves issuing from the semilunar ganglia, caused a diminution and even complete arrest of the secretion.

The influence of the higher nerve-centres on gastric digestion, as in the case of mental emotion, is too well known to need more than a reference.

Digestion of the Stomach after Death.—If an animal die during the process of gastric digestion, and when, therefore, a quantity of gastric juice is present in the interior of the stomach, the walls of this organ itself are frequently themselves acted on by their own secretion, and to such an extent, that a perforation of considerable size may be produced, and the contents of the stomach may in part escape into the cavity of the abdomen. This phenomenon is not unfrequently observed in *post-mortem* examinations of the human body. If a rabbit be killed during a period of digestion, and afterward exposed to artificial warmth to prevent its temperature from falling, not only the stomach, but many of the surrounding parts, will be found to have been dissolved (Pavy).

From these facts, it becomes an interesting question why, during life, the stomach is free from liability to injury from a secretion which, after death, is capable of such destructive effects?

It is only necessary to refer to the idea of Bernard, that the living stomach finds protection from its secretion in the presence of epithelium and mucus, which are constantly renewed in the same degree that they are constantly dissolved, in order to remark that, although the gastric mucus is probably protective, this theory, so far as the *epithelium* is concerned, has been disproved by experiments of Pavy's, in which the mucous membrane of the stomachs of dogs was dissected off for a small space, and, on killing the animals some days afterward, no sign of digestion of the stomach was visible. "Upon one occasion, after removing the mucous membrane, and exposing the muscular fibres over a space of about an inch and a half in diameter, the animal was allowed to live for ten days. It ate food every day, and seemed scarcely affected by the operation. Life was destroyed whilst digestion was being carried on, and the lesion in the stomach was found very nearly repaired: new matter had been deposited in the place of what had been removed, and the denuded spot had contracted to much less than its original dimensions."

Pavy believes that the natural alkalinity of the blood, which circulates so freely during life in the walls of the stomach, is sufficient to neutralize the acidity of the gastric juice; and as may be gathered from what has been previously said, the neutralization of the acidity of the gastric secretion is quite sufficient to destroy its digestive powers; but the experiments adduced in favor of this theory are open to many objections, and afford only a negative support to the conclusions they are intended to prove. Again, the pancreatic secretion acts best on proteids in an *alka-*

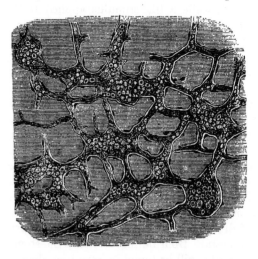

Fig. 181.—Auerbach's nerve-plexus in small intestine. The plexus consists of fibrillated substance, and is made up of trabeculæ of various thicknesses. Nucleus-like elements and ganglion-cells are imbedded in the plexus, the whole of which is enclosed in a nucleated sheath. (Klein.)

line medium; but it has no digestive action on the living intestine. It must be confessed that no entirely satisfactory theory has been yet stated.

THE INTESTINES.

The Intestinal Canal is divided into two chief portions, named from their differences in diameter, the (I.) *small* and (II.) *large* intestine (Fig. 165). These are continuous with each other, and communicate by means of an opening guarded by a valve, the *ileo-cœcal* valve, which allows the passage of the products of digestion from the small into the large bowel, but not, under ordinary circumstances, in the opposite direction.

I. The Small Intestine.—The Small Intestine, the average length of which in an adult is about twenty feet, has been divided, for convenience of description, into three portions, viz., the *duodenum*, which extends for eight or ten inches beyond the pylorus; the *jejunum*, which forms two-fifths, and the *ileum*, which forms three-fifths of the rest of the canal.

Structure.—The small intestine, like the stomach, is constructed of four principal coats, viz., the serous, muscular, submucous, and mucous.

(1) The *serous* coat, formed by the visceral layer of the peritoneum, and has the structure of serous membranes in general.

(2) The *muscular* coats consist of an internal circular and an external longitudinal layer: the former is usually considerably the thicker. Both ·

alike consist of bundles of unstriped muscular tissue supported by connective tissue. They are well provided with lymphatic vessels, which form a set distinct from those of the mucous membrane.

. Between the two muscular coats is a nerve-plexus (Auerbach's plexus, plexos myentericus) (Fig. 181) similar in structure to Meissner's (in the submucous tissue), but with more numerous ganglia. This plexus regulates the peristaltic movements of the muscular coats of the intestines.

(3) Between the mucous and muscular coats, is the *submucous* coat, which consists of connective tissue, in which numerous blood-vessels and lymphatics ramify. A fine plexus, consisting mainly of non-medullated nerve-fibres, "Meissner's plexus," with ganglion cells at its nodes, occurs

FIG. 182.—Horizontal section of a small fragment of the mucous membrane, including one entire crypt of Lieberkühn and parts of several others: *a*, cavity of the tubular glands or crypts; *b*, one of the lining epithelial cells; *c*, the lymphoid or retiform spaces, of which some are empty, and others occupied by lymph cells, as at *d*.

in the submucous tissue from the stomach to the anus. From the position of this plexus and the distribution of its branches, it seems highly probable that it is the local centre for regulating the calibre of the blood-vessels supplying the intestinal mucous membrane, and presiding over the processes of secretion and absorption.

(4) The *mucous membrane* is the most important coat in relation to the function of digestion. The following structures, which enter into its composition, may now be successively described;—the *valvulæ conniventes;* the *villi;* and the *glands.* The general structure of the mucous membrane of the intestines resembles that of the stomach (p. 241), and, like it, is lined on its inner surface by columnar epithelium. Adenoid tissue (Fig. 182, *c* and *d*) enters largely into its construction; and on its deep surface is the *muscularis mucosæ* (*m m*, Fig. 183), the fibres of which are arranged in two layers: the outer longitudinal and the inner circular.

Valvulæ Conniventes.—The *valvulæ conniventes* (Fig. 184) commence in the duodenum, about one or two inches beyond the pylorus, and becoming larger and more numerous immediately beyond the entrance of the bile duct, continue thickly arranged and well developed throughout the jejunum; then, gradually diminishing in size and number, they cease near the middle of the ileum. They are formed by a doubling inward of the mucous membrane; the crescentic, nearly circular, folds thus formed being arranged transversely to the axis of the intestine, and each individual fold seldom extending around more than ½ or ⅔ of the bowel's circumference. Unlike the rugæ in the œsophagus and stomach, they do not disappear on distension of the canal. Only an imperfect notion of their natural position and function can be obtained by looking at them after the intestine has been laid open in the usual manner. To under-

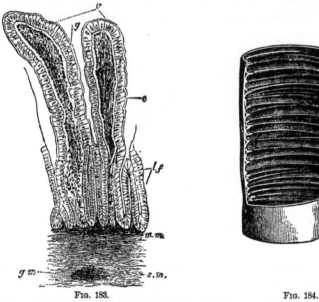

FIG. 183.　　　　　　　　　FIG. 184.

FIG. 183.—Vertical section through portion of small intestine of dog, v, two villi showing e, epithelium; g, goblet cells. The free surface is seen to be formed by the "striated basilar border," while inside the villus the adenoid tissue and unstriped muscle-cells are seen; lf, Lieberkühn's follicles: m, muscularis mucosæ, sending up fibres between the follicles into the villi; sm, submucous tissue; containing (gm), ganglion cells of Meissner's plexus. (Schofield.)
FIG. 184.—Piece of small intestine (previously distended and hardened by alcohol) laid open to show the normal position of the valvulæ conniventes.

stand them aright, a piece of gut should be distended either with air or alcohol, and not opened until the tissues have become hardened. On then making a section it will be seen that, instead of disappearing, they stand out at right angles to the general surface of the mucous membrane (Fig. 184). Their functions are probably less—Besides (1) offering a largely increased surface for secretion and absorption, they probably (2) prevent the too rapid passage of the very liquid products of gastric digestion, immediately after their escape from the stomach, and (3), by their projection, and consequent interference with a uniform and untroubled current of the intestinal contents, probably assist in the more perfect mingling of the latter with the secretions poured out to act on them.

Glands of the Small Intestine.—The glands are of three principal kinds:—viz., those of (1) Lieberkühn, (2) Brunner, and (3) Peyer.

(1.) The *glands* or *crypts of Lieberkühn* are simple tubular depressions
of the intestinal mucous membrane, thickly distributed over the whole surface both of the large and small intestines. In the small intestine they
are visible only with the aid of a lens; and their orifices appear as minute
dots scattered between the villi. They are larger in the large intestine,
and increase in size the nearer they approach the anal end of the intestinal tube; and in the rectum their orifices may be visible to the naked
eye. In length they vary from $\frac{1}{30}$ to $\frac{1}{10}$ of a line. Each tubule (Fig.
186) is constructed of the same essential parts as the intestinal mucous
membrane, viz., a fine *membrana propria*, or basement membrane, a

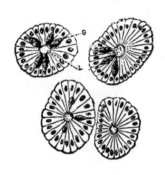

FIG. 185.   FIG. 186.

FIG. 185.—Transverse section through four crypts of Lieberkühn from the large intestine of the
pig. They are lined by columnar epithelial cells, the nuclei being placed in the outer part of the
cells. The divisions between the cells are seen as lines radiating from L, the lumen of the crypt; G,
epithelial cells, which have become transformed into goblet cells. × 350. (Klein and Noble Smith.)
FIG. 186.—A gland of Lieberkühn in longitudinal section. (Brinton.)

layer of cylindrical epithelium lining it, and capillary blood-vessels covering its exterior, the free surface of the columnar cells presenting an
appearance precisely similar to the "striated basilar border" which covers
the villi. Their contents appear to vary, even in health; the varieties
being dependent, probably, on the period of time in relation to digestion
at which they are examined.

Among the columnar cells of Lieberkühn's follicles, goblet-cells frequently occur (Fig. 185).

(2.) *Brunner's glands* (Fig. 188) are confined to the duodenum; they
are most abundant and thickly set at the commencement of this portion
of the intestine, diminishing gradually as the duodenum advances. They
are situated beneath the mucous membrane, and imbedded in the submucous tissue, each gland is a branched and convoluted tube, lined with
columnar epithelium. As before said, in structure they are very similar
to the pyloric glands of the stomach, and their epithelium undergoes a
VOL. I.—17.

similar change during secretion; but they are more branched and convoluted and their ducts are longer. (Watney.) The duct of each gland passes through the muscularis mucosæ, and opens on the surface of the mucous membrane.

(3.) The *glands of Peyer* occur chiefly but not exclusively in the *small* intestine. They are found in greatest abundance in the lower part of the

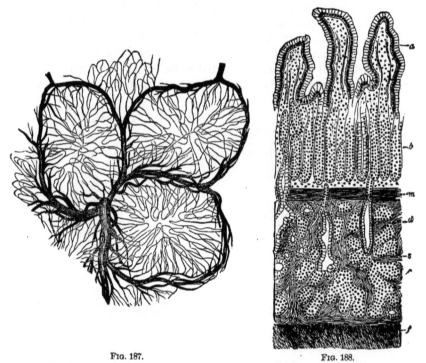

FIG. 187.   FIG. 188.

FIG. 187.—Transverse section of injected Peyer's glands (from Kölliker). The drawing was taken from a preparation made by Frey: it represents the fine capillary-looped network spreading from the surrounding blood-vessels into the interior of three of Peyser's capsules from the intestine of the rabbit.

FIG. 188.—Vertical section of duodenum, showing *a*, villi; *b*, crypts of Lieberkühn, and *c*, Brunner's glands in the submucosa *s*, with ducts, *d*: muscularis mucosæ, *m*; and circular muscular coat *f*. (Schofield.)

ileum near to the ileo-cæcal valve. They are met with in two conditions, viz., either scattered singly, in which case they are termed *glandulæ solitariæ*, or aggregated in groups varying from one to three inches in length and about half-an-inch in width, chiefly of an oval form, their long axis parallel with that of the intestine. In this state, they are named *glandulæ agminatæ*, the groups being commonly called *Peyer's patches* (Fig. 189), and almost always placed opposite the attachment of the mesentery. In structure, and in function, there is no essential difference between the solitary glands and the individual bodies of which each group or patch is made up. They are really single or aggregated masses of adenoid tissue forming lymph-follicles. In the condition in which they have been most commonly examined, each gland appears as a circular opaque-white rounded body, from $\frac{1}{24}$ to $\frac{1}{12}$ inch in diameter, according to the degree in which it is developed. They are principally contained in the submucous coat, but sometimes project through the *muscularis mucosæ* into the mucous membrane. In the agminate glands, each follicle reaches the free surface of the intestine, and is covered with columnar epithelium. Each gland is surrounded by the openings of Lieberkühn's follicles.

The adjacent glands of a Peyer's patch are connected together by adenoid tissue. Sometimes the lymphoid tissue reaches the free surface, replacing the epithelium, as is also the case with some of the lymphoid follicles of the tonsil (p. 236).

Peyer's glands are surrounded by lymphatic sinuses which do not penetrate into their interior; the interior is, however, traversed by a very rich blood capillary plexus. If the vermiform appendix of a rabbit, which consists largely of Peyer's glands, be injected with blue, by pressing the

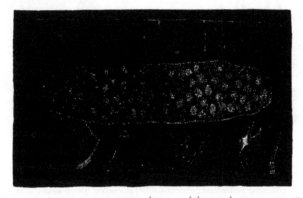

Fig. 189.—Agminate follicles, or Peyer's patch, in a state of distension. × 5. (Boehm.)

point of a fine syringe into one of the lymphatic sinuses, the Peyer's glands will appear as greyish white spaces surrounded by blue; if now the arteries of the same be injected with red, the greyish patches will change to red, thus proving that they are *surrounded* by lymphatic spaces, but *penetrated* by blood-vessels. The lacteals passing out of the villi communicate with the lymph sinuses round Peyer's glands.

It is to be noted that they are largest and most prominent in children and young persons.

Villi.—The *Villi* (Figs. 183, 188, 190, and 191), are confined exclusively to the mucous membrane of the *small* intestine. They are minute vascular processes, from a quarter of a line to a line and two-thirds in length, covering the surface of the mucous membrane, and giving it a peculiar velvety, fleecy appearance. Krause estimates them at fifty to ninety in number in a square line, at the upper part of the small intestine, and at forty to seventy in the same area at the lower part. They vary in form even in the same animal, and differ according as the lymphatic vessels they contain are empty or full of chyle; being usually, in the former case, flat and pointed at their summits, in the latter cylindrical or cleavate.

Each villus consists of a small projection of mucous membrane, and its interior is therefore supported throughout by fine adenoid tissue, which forms the framework or stroma in which the other constituents are contained.

The surface of the villus is clothed by columnar epithelium, which rests on a fine basement membrane; while within this are found, reckoning from without inward, blood-vessels, fibres of the *muscularis mucosæ*, and a single lymphatic or lacteal vessel rarely looped or branched (Fig. 192); besides granular matter, fat-globules, etc.

FIG. 190.   FIG. 191.

FIG. 190.—Section of small intestine showing villi, Lieberkühn's glands and a Peyer's solitary gland. *m, m,* muscularis mucosæ. (Klein and Noble Smith.)

FIG. 191.—Vertical section of a villus of the small intestine of a cat. *a,* striated basilar border of the epithelium; *b,* columnar epithelium; *c,* goblet cells; *d,* central lymph-vessel; *e,* smooth muscular fibres; *f,* adenoid stroma of the villus in which lymph corpuscles lie. (Klein.)

The *epithelium* is of the columnar kind, and continuous with that lining the other parts of the mucous membrane. The cells are arranged with their long axis radiating from the surface of the villus (Fig. 191), and their smaller ends resting on the basement membrane. The free surface of the epithelial cells of the villi, like that of the cells which cover the general surface of the mucous membrane, is covered by a fine border which exhibits very delicate striations, whence it derives its name, "striated basilar border."

Beneath the basement or limiting membrane there is a rich supply of *blood-vessels*. Two or more minute arteries are distributed within each villus; and from their capillaries, which form a dense network, proceed one or two small veins, which pass out at the base of the villus.

The layer of the *muscularis mucosæ* in the villus forms a kind of thin hollow cone immediately around the central lacteal, and is, therefore, situate beneath the blood-vessels. It is without doubt instrumental in the propulsion of chyle along the lacteal.

The *lacteal vessel* enters the base of each villus, and passing up in the middle of it, extends nearly to the tip, where it ends commonly by a closed and somewhat dilated extremity. In the larger villi there may be two small lacteal vessels which end by a loop (Fig. 192), or the lacteals may form a kind of network in the villus. The last method of ending, however, is rarely or never seen in the human subject, although common in some of the lower animals (A, Fig. 192).

FIG. 192.—A. Villus of sheep. B. Villi of man. (Slightly altered from Teichmann.)

The office of the villi is the absorption of chyle and other liquids from the intestine. The mode in which they affect this will be considered in the Chapter on ABSORPTION.

II. The Large Intestine.—The Large Intestine, which in an adult is from about 4 to 6 feet long, is subdivided for descriptive purposes into three portions (Fig. 165), viz.:—the *cæcum*, a short wide pouch, communicating with the lower end of the small intestine through an opening, guarded by the *ileo-cæcal* valve; the *colon*, continuous with the cæcum, which forms the principal part of the large intestine, and is divided into an ascending, transverse and descending portion; and the *rectum*, which, after dilating at its lower part, again contracts, and immediately afterward opens externally through the *anus.* Attached to the cæcum is the small *appendix vermiformis.*

Structure.—Like the *small* intestine, the *large* is constructed of four principal coats, viz., the serous, muscular, submucous, and mucous. The *serous* coat need not be here particularly described. Connected with it are the small processes of peritoneum, containing fat, called *appendices epiploicæ.* The fibres of the *muscular* coat, like those of the small intestine, are arranged in two layers—the outer longitudinal, the inner circular. In the cæcum and colon, the longitudinal fibres, besides being, as in the small intestine, thinly disposed in all parts of the wall of the bowel,

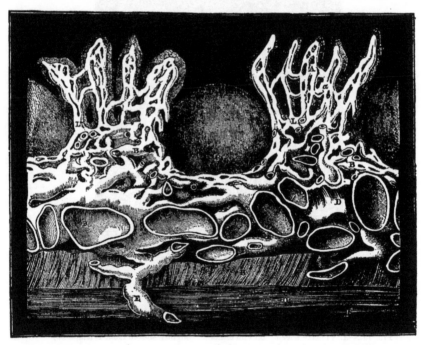

FIG. 193.—Diagram of lacteal vessels in small intestine. A, lacteals in villi; P, Peyer's glands; B and D, superficial and deep network of lacteals in submucous tissue; L, Lieberkühn's glands; E, small branch of lacteal vessel on its way to mesenteric gland; H and O, muscular fibres of intestine; s, peritoneum. (Teichmann.)

are collected, for the most part, into three strong bands, which being shorter, from end to end, than the other coats of the intestine, hold the canal in folds, bounding intermediate sacculi. On the division of these bands, the intestine can be drawn out to its full length, and it then assumes, of course, a uniformly cylindrical form. In the rectum, the fasciculi of these longitudinal bands spread out and mingle with the other longitudinal fibres, forming with them a thicker layer of fibres than exists on any other part of the intestinal canal. The circular muscular fibres are spread over the whole surface of the bowel, but are somewhat more marked in the intervals between the sacculi. Toward the lower end of the rectum they become more numerous, and at the anus they form a strong band called the *internal sphincter* muscle.

The *mucous membrane* of the large, like that of the small intestine, is lined throughout by columnar epithelium, but, unlike it, is quite smooth and destitute of villi, and is not projected in the form of *valvulæ conniventes*. Its general microscopic structure resembles that of the small intestine: and it is bounded below by the *muscularis mucosæ*.

The general arrangement of ganglia and nerve-fibres in the large intestine resembles that in the small (p. 255).

Glands of the Large Intestine.—The glands with which the large intestine is provided are of two kinds, (1) the *tubular* and (2) the *lymphoid*.

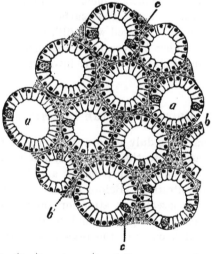

Fig. 194.—Horizontal section through a portion of the mucous membrane of the large intestine, showing Lieberkühn's glands in transverse section. *a*, lumen of gland—lining of columnar cells with *c*, goblet cells, *b*, supporting connective tissue. Highly magnified. (V. D. Harris.)

(1.) The *tubular* glands, or glands of Lieberkühn, resemble those of the small intestine, but are somewhat larger and more numerous. They are also more uniformly distributed.

(2.) Follicles of *adenoid* or *lymphoid* tissue are most numerous in the cæcum and vermiform appendix. They resemble in shape and structure, almost exactly, the solitary glands of the small intestine.

Peyer's patches are not found in the large intestine.

Ileo-Cæcal Valve.—The ileo-cæcal valve is situate at the place of junction of the small with the large intestine, and guards against any reflex of the contents of the latter into the ileum. It is composed of two semilunar folds of mucous membrane. Each fold is formed by a doubling inward of the mucous membrane, and is strengthened on the outside by some of the circular muscular fibres of the intestine, which are contained between the outer surfaces of the two layers of which each fold is composed. While the circular muscular fibres, however, of the bowel at the junction of the ileum with the cæcum are contained between the outer opposed surfaces of the folds of mucous membrane which form the valve, the longitudinal muscular fibres and the peritoneum of the small and large intestine respectively are continuous with each other, without dipping in to follow the circular fibres and the mucous membrane. In this manner, therefore, the folding inward of these two last-named structures is preserved, while, on the other hand, by dividing the longitudinal muscular fibres and the peritoneum, the valve can be made to disappear, just as the constrictions between the sacculi of the large intestine can be made to disappear by performing a similar operation. The inner surface of the folds is smooth; the mucous membrane of the ileum being continuous with that of the cæcum. That surface of each fold which looks toward the small intestine is covered with villi, while that which looks to the cæcum has none. When the cæcum is distended, the margin of the folds are stretched, and thus are brought into firm apposition one with the other.

DIGESTION IN THE INTESTINES.

After the food has been duly acted upon by the stomach, such as has not been absorbed passes into the duodenum, and is there subjected to the action of the secretions of the pancreas and liver, which enter that portion of the small intestine. Before considering the changes which the food undergoes in consequence, attention should be directed to the structure and secretion of these glands, and to the secretion (succus entericus) which is poured out into the intestines from the glands lining them.

THE PANCREAS, AND ITS SECRETION.

· The Pancreas is situated within the curve formed by the duodenum; and its main duct opens into that part of the small intestine, through a small opening, or through a duct common to it and to the liver, about two and a half inches from the pylorus.

Structure.—In structure the pancreas bears some resemblance to the salivary glands. Its capsule and septa, as well as the blood-vessels and lymphatics, are similarly distributed. It is, however, looser and softer, the lobes and lobules being less compactly arranged. The main duct divides into branches (lobar ducts), one for each lobe, and these branches subdivide into intralobular ducts, and these again by their division and branching form the gland tissue proper. The intralobular ducts correspond to a lobule, while between them and the secreting tubes or *alveoli* are longer or shorter *intermediary* ducts. The larger ducts possess a very distinct lumen and a membrana propria lined with columnar epithelium, the cells of which are longitudinally striated, but are shorter than those found in the ducts of the salivary glands. In the intralobular ducts the epithelium is short and the lumen is smaller. The intermediary ducts opening into the alveoli possess a distinct lumen, with a membrana propria lined with a single layer of flattened elongated cells. The alveoli are branched and convoluted tubes, with a membrana propria lined with a single layer of columnar cells. They have no distinct lumen, its place being taken by fusiform or branched cells. Heidenhain has observed that the alveoli cells in the pancreas of a fasting dog consist of two zones, an inner or central zone, which is finely granular, and which stains feebly,

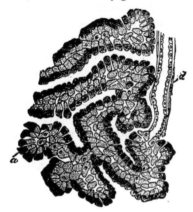

Fig. 195.—Section of the pancreas of a dog during digestion. *a*, alveoli lined with cells, the outer zone of which is well stained with hæmatoxylin; *d*, intermediary duct lined with squamous epithelium. × 350. (Klein and Noble Smith.)

and a smaller parietal zone of finely striated protoplasm, which stains easily. The nucleus is partly in one, partly in the other zone. During digestion, it is found that the outer zone increases in size, and the central zone diminishes; the cell itself becoming smaller from the discharge of the secretion. At the end of digestion the first condition again appears, the inner zone enlarging at the expense of the outer. It appears that the granules are formed by the protoplasm of the cells, from material supplied to it by the blood. The granules are thought to be not the ferment itself, but material from which, under certain conditions, the ferments of the gland are made, and therefore called *Zymogen*.

Pancreatic Secretion.—The secretion of the pancreas has been obtained for purposes of experiment from the lower animals, especially the dog, by opening the abdomen and exposing the duct of the gland, which is then made to communicate with the exterior. A pancreatic fistula is thus established.

An extract of pancreas made from the gland, which has been removed from an animal killed during digestion, possesses the active properties of pancreatic secretion. It is made by first dehydrating the gland, which has been cut up into small pieces, by keeping it for some days in absolute alcohol, and then, after the entire removal of the alcohol, placing it in strong glycerin: A glycerin extract is thus obtained. It is a remarkable fact, however, that the amount of the ferment *trypsin* greatly increases if the gland be exposed to the air for twenty-four hours before placing in alcohol; indeed, a glycerin extract made from the gland immediately upon removal from the body often appears to contain none of that ferment. This seems to indicate that the conversion of zymogen in the gland into the ferment only takes place during the act of secretion, and that the gland, although it always contains in its cells the materials (trypsinogen) out of which trypsin is formed, yet the conversion of the one into the other only takes place by degrees. Dilute acid appears to assist and accelerate the conversion, and if a recent pancreas be rubbed up with dilute acid before dehydration, a glycerin extract made afterward, even though the gland may have been only recently removed from the body, is very active.

Properties.—Pancreatic juice is colorless, transparent, and slightly viscid, alkaline in reaction. It varies in specific gravity from 1010 to 1015, according to whether it is obtained from a permanent fistula—then more watery—or from a newly-opened duct. The solids vary in a temporary fistula from 80 to 100 parts per thousand, and in a permanent one from 16 to 50 per thousand.

CHEMICAL COMPOSITION OF THE PANCREATIC SECRETION.

From a permanent fistula. (Bernstein.)

Water	975
Solids—Ferments:	
Proteids, including Serum—Albumin, Casein, Leucin and Tyrosin, Fats and Soaps . }	17
Inorganic residue, especially Sodium Carbonate .	8
	— 25
	1000

Functions.—(1.) It converts *proteids into peptones*, the intermediate product being not akin to syntonin or acid-albumin, as in gastric digestion, but to alkali-albumin. Kühne believes that the intermediate products, both in the peptic and pancreatic digestion of proteids, are two, viz., antialbumose and hemialbumose, and that the peptones formed correspond to these, viz., antipeptone and hemipeptone. The hemipeptone is capable of being converted by the action of the pancreatic ferment—

trypsin—into leucin and tyrosin, but is not so changed by pepsin; the antipeptone cannot be further split up. The products of pancreatic digestion are sometimes further complicated by the appearance of certain fæcal substances, of which indol and naphthilamine are the most important. (Kühne.)

When the digestion goes on for a long time the indol is formed in considerable quantities, and emits a most disagreeable fæcal odor, which was attributed to putrefaction till Kühne showed its true nature. All the albuminous or proteid substances which have not been converted into peptone, and absorbed in the stomach, and the partially changed substances, *i.e.*, the parapeptones, are converted into peptone by the pancreatic juice, and then in part into leucin and tyrosin.

(2.) *Nitrogenous bodies other than proteids, are not to any extent altered.* Mucin can, however, be dissolved, but not gelatin or horny tissues.

(3.) *Starch is converted into glucose* in an exactly similar manner to that which happens with the saliva. As mentioned before, it seems not unlikely that glucose is not formed at once from starch, but that certain dextrines are intermediate products. If the sugar which is at first formed, as is stated by some chemists, be not glucose but maltose, at any rate the pancreatic juice after a time completes the whole change of starch into glucose. There is a distinct amylolytic ferment (Amylopsin) in the pancreatic juice which cannot be distinguished from ptyalin.

(4.) *Oils and fats are both emulsified and split up into their fatty acids and glycerin by pancreatic secretion.* Even if part of this action is due to the alkalinity of the medium, it is probable that there is a third distinct ferment (Steapsin) which facilitates the change.

Several cases have been recorded in which the pancreatic duct being obstructed, so that its secretion could not be discharged, fatty or oily matter was abundantly discharged from the intestines. In nearly all these cases, indeed, the liver was coincidently diseased, and the change or absence of the bile might appear to contribute to the result; yet the frequency of extensive disease of the liver, unaccompanied by fatty discharges from the intestines, favors the view that, in these cases, it is to the absence of the pancreatic fluid from the intestines that the excretion or non-absorption of fatty matter should be ascribed.

(5.) *It possesses the property of curdling milk,* containing a special (rennet) ferment for that purpose. The ferment is distinct from trypsin, and will act in the presence of an acid (W. Roberts).

Conditions favorable to the Action of the Pancreatic Juice.— These are similar to those which are favorable to the action of the saliva, and the reverse (p. 231).

THE LIVER.

The Liver, the largest gland in the body, situated in the abdomen, chiefly on the right side, is an extremely vascular organ, and receives its supply of blood from two distinct vessels, the *portal vein* and *hepatic artery*, while the blood is returned from it into the vena cava inferior by the *hepatic veins*. Its secretion, the *bile*, is conveyed from it by the *hepatic duct*, either directly into the intestine, or, when digestion is not going on, into the *cystic* duct, and thence into the gall-bladder, where it

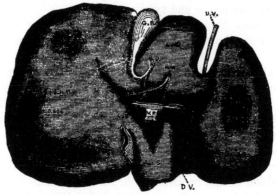

Fig. 196.—The under surface of the liver. G. B., gall-bladder; H. D., common bile-duct; H. A., hepatic artery; Y. P., portal vein; L. Q., lobulus quadratus; L. S., lobulus spigelii; L. C., lobulus caudatus; D. V., ductus venosus; U. V., umbilical vein. (Noble Smith.)

accumulates until required. The portal vein, hepatic artery, and hepatic duct branch together throughout the liver, while the hepatic veins and their tributaries run by themselves.

On the outside the liver has an incomplete covering of peritoneum, and beneath this is a very fine coat of areolar tissue, continuous over the whole surface of the organ. It is thickest where the peritoneum is absent, and is continuous on the general surface of the liver with the fine and, in the human subject, almost imperceptible, areolar tissue investing the lobules. At the transverse fissure it is merged in the areolar investment called Glisson's capsule, which, surrounding the portal vein, hepatic artery, and hepatic duct, as they enter at this part, accompanies them in their branchings through the substance of the liver.

Structure.—The liver is made up of small roundish or oval portions called *lobules*, each of which is about $\frac{1}{20}$ of an inch in diameter, and composed of the minute branches of the portal vein, hepatic artery, hepatic duct, and hepatic vein; while the interstices of these vessels are filled by the liver cells. The hepatic cells (Fig. 197), which form the glandular or secreting part of the liver, are of a spheroidal form, somewhat polygonal from mutual pressure about $\frac{1}{800}$ to $\frac{1}{1000}$ inch in diameter, possessing one, sometimes two nuclei. The cell-substance contains numerous fatty molecules, and some yellowish-brown granules of bile-pigment. The cells sometimes exhibit slow amœboid movements. They are held together by a very delicate sustentacular tissue, continuous with the interlobular connective tissue.

To understand the distribution of the blood-vessels in the liver, it will be well to trace, first, the two blood-vessels and the duct which enter the organ on the under surface at the transverse fissure, viz., the portal vein, hepatic artery, and hepatic duct. As before remarked, all three run in company, and their appearance on longitudinal section is shown in

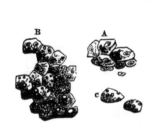

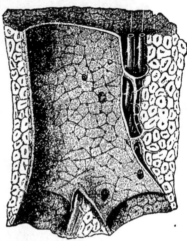

Fig. 197.   Fig. 198.

Fig. 197.—A. Liver-cells. B, Ditto, containing various sized particles of fat.
Fig. 198.—Longitudinal section of a portal canal, containing a portal vein, hepatic artery, and hepatic duct, from the pig. P, branch of vena portæ, situate in a portal canal formed amongst the lobules of the liver, *l l*, and giving off vaginal branches; there are also seen within the large portal vein numerous orifices of the smallest interlobular veins arising directly from it; *a*, hepatic artery; *d*, hepatic duct. × 5. (Kiernan.)

Fig. 198. Running together through the substance of the liver, they are contained in small channels called *portal canals*, their immediate investment being a sheath of areolar tissue (Glisson's capsule).

To take the distribution of the portal vein first:—In its course through the liver this vessel gives off small branches which divide and subdivide between the lobules surrounding them and limiting them, and from this circumstance called *inter*-lobular veins. From these small vessels a dense capillary network is prolonged into the substance of the lobule, and this network, gradually gathering itself up, so to speak, into larger vessels, converges finally to a single small vein, occupying the centre of the lobule, and hence called *intra*-lobular. This arrangement is well seen in Fig. 199, which represents a transverse section of a lobule.

The small *intra*-lobular veins discharge their contents into veins called *sub*-lobular (*h h h*, Fig. 200); while these again, by their union, form

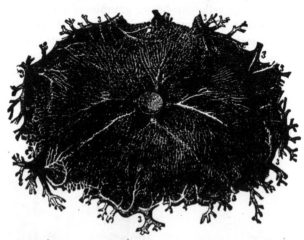

FIG. 199.—Cross-section of a lobule of the human liver, in which the capillary network between the portal and hepatic veins has been fully injected. 1, section of the *intra*-lobular vein; 2, its smaller branches collecting blood from the capillary network; 3, *inter*-lobular branches of the vena portæ with their smaller ramifications passing inward toward the capillary network in the substance of the lobule. × 60. (Sappey.)

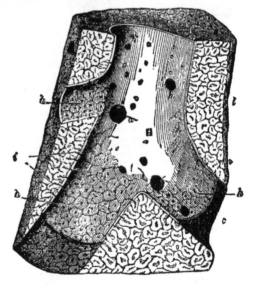

FIG. 200.—Section of a portion of liver passing longitudinally through a considerable hepatic vein, from the pig. H, hepatic venous trunk, against which the sides of the lobules (*l*) are applied; *h, h, h,* sublobular hepatic veins, on which the bases of the lobules rest, and through the coats of which they are seen as polygonal figures; *i*, mouth of the intralobular veins, opening into the sublobular veins; *i'*, intralobular veins shown passing up the centre of some divided lobules; *l, l,* cut surface of the liver; *c, c,* walls of the hepatic venous canal, formed by the polygonal bases of the lobules. × 5. (Kiernan.)

the main branches of the *hepatic* veins, which leave the posterior border of the liver to end by two or three principal trunks in the interior vena cava, just before its passage through the diaphragm. The *sub*-lobular and *hepatic veins,* unlike the *portal* vein and its companions, have little or no areolar tissue around them, and their coats being very thin, they form little more than mere channels in the liver substance which closely surrounds them.

The manner in which the lobules are connected with the *sub-lobular* veins by means of the small *intra-lobular* veins is well seen in the diagram (Fig. 200 and in Fig. 201), which represent the parts as seen in a longitudinal section. The appearance has been likened to a twig having leaves without footstalks—the lobules representing the leaves, and the *sub-lobular* vein the small branch from which it springs. On a transverse section, the appearance of the *intra-lobular* veins is that of 1, Fig. 199, while both a transverse and longitudinal section are exhibited in Fig. 176.

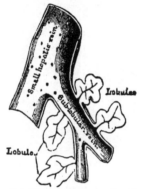

Fig. 201.—Diagram showing the manner in which the lobules of the liver rest on the sublobular veins. (After Kiernan.)

The hepatic artery, the function of which is to distribute blood for nutrition to Glisson's capsule, the walls of the ducts and blood-vessels, and other parts of the liver, is distributed in a very similar manner to the portal vein, its blood being returned by small branches either into the ramifications of the portal vein, or into the capillary plexus of the lobules which connects the *inter* and *intra* lobular veins.

Fig. 202.—Capillary network of the lobules of the rabbit's liver. The figure is taken from a very successful injection of the hepatic veins, made by Harting: it shows nearly the whole of two lobules, and parts of three others; *p,* portal branches running in the interlobular spaces; *h,* hepatic veins penetrating and radiating from the centre of the lobules. × 45. (Kölliker.)

The hepatic duct divides and subdivides in a manner very like that of the portal vein and hepatic artery, the larger branches being lined by *cylindrical,* and the smaller by small *polygonal* epithelium.

The bile-capillaries commence between the hepatic cells, and are bounded by a delicate membranous wall of their own. They appear to be always bounded by hepatic cells on all sides, and are thus separated from the nearest blood-capillary by at least the breadth of one cell (Figs. 203 and 204).

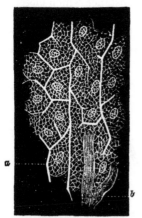

The Gall-Bladder.—The Gall-bladder (G, B, Fig. 196) is a pyriform bag, attached to the under surface of the liver, and supported also by the peritoneum, which passes below it. The larger end or *fundus*, projects beyond the front margin of the liver; while the smaller end contracts into the cystic duct.

Structure.—The walls of the gall-bladder are constructed of three principal coats. (1) Externally (excepting that part which is in contact with the liver), is the *serous* coat, which has the same structure as the peritoneum with which it is continuous. Within this is (2) the *fibrous* or areolar coat, constructed of tough fibrous and elastic tissue, with which is mingled a considerable number of plain muscular fibres, both longitudinal and circular. (3) Internally the gall-bladder is lined by mucous membrane, and a layer of columnar epithelium. The surface of the mucous membrane presents to the naked eye a minutely honeycombed appearance from a number of tiny polygonal depressions with intervening ridges, by which its surface is mapped out. In the cystic

Fig. 203.—Portion of a lobule of liver. *a*, bile capillaries between liver-cells, the network in which is well seen; *b*, blood capillaries. × 350. (Klein and Noble Smith.)

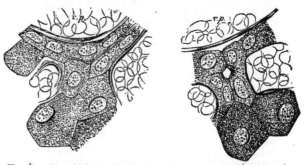

Fig. 204.—Hepatic cells and bile capillaries, from the liver of a child three months old. Both figures represent fragments of a section carried through the periphery of a lobule. The red corpuscles of the blood are recognized by their circular contour; *vp*, corresponds to an interlobular vein in immediate proximity with which are the epithelial cells of the biliary ducts, to which, at the lower part of the figures, the much larger hepatic cells suddenly succeed. (E. Hering.)

duct the mucous membrane is raised up in the form of crescentic folds, which together appear like a spiral valve, and which minister to the function of the gall-bladder in retaining the bile during the intervals of digestion.

The gall-bladder and all the main biliary ducts are provided with mucous glands, which open on their internal surface.

Functions of the Liver.—The functions of the Liver may be classified under the following heads:—1. The Secretion of Bile. 2. The Elaboration of Blood; under this head may be included the Glycogenic Function.

I. THE SECRETION OF BILE.

Properties of the Bile.—The bile is a somewhat viscid fluid, of a yellow or reddish-yellow color, a strongly bitter taste, and, when fresh, with a scarcely perceptible odor: it has a neutral or slightly alkaline reaction, and its specific gravity is about 1020. Its color and degree of consistence vary much, apparently independent of disease; but, as a rule, it becomes gradually more deeply colored and thicker as it advances along its ducts, or when it remains long in the gall-bladder, wherein, at the same time, it becomes more viscid and ropy, of a darker color, and more bitter taste, mainly from its greater degree of concentration, on account of partial absorption of its water, but partly also from being mixed with mucus.

Chemical Composition of Human Bile. (Frerichs.)

Water	859·2
Solids	140·8
	1000·0
Bile salts or Bilin	91·5
Fat	9·2
Cholesterin	2.6
Mucus and coloring matters	29.8
Salts	7·7
	140·8

Bile salts, or *Bilin*, can be obtained as colorless, exceedingly deliquescent crystals, soluble in water, alcohol, and alkaline solutions, giving to the watery solution the taste and general characters of bile. They consist of sodium salts of glycocholic and taurocholic acids. The former salt is composed of cholic acid conjugated with glycin (see Appendix), the latter of the same acid conjugated with taurin. The proportion of these two salts in the bile of different animals varies, *e.g.*, in ox bile the glycocholate is in great excess, whereas the bile of the dog, cat, bear, and other carnivora contains taurocholate alone; in human bile both are present in about the same amount (glycocholate in excess?).

Preparation of Bile Salt.—Bile salts may be prepared in the fol-
VOL. I.—18.

lowing manner: mix bile which has been evaporated to a quarter of its bulk with animal charcoal, and evaporate to perfect dryness in a water bath. Next extract the mass whilst still warm with absolute alcohol. Separate the alcoholic extract by filtration, and to it add perfectly anhydrous ether as long as a precipitate is thrown down. The solution and precipitate should be set aside in a closely stoppered bottle for some days, when crystals of the bile salts or bilin will have separated out. The gly-cocholate may be separated from the taurocholate by dissolving bilin in water, and adding to it a solution of neutral lead acetate, and then a little basic lead acetate, when lead glycocholate separates out. Filter and add to the filtrate lead acetate and ammonia, a precipitate of lead taurocholate will be formed, which may be filtered off. In both cases, the lead may be got rid of by suspending or dissolving in hot alcohol, adding hydrogen sulphate, filtering and allowing the acids to separate out by the addition of water.

The test for bile salts is known as Pettenkofer's. If to an aqueous solution of the salts strong sulphuric acid be added, the bile acids are first of all precipitated, but on the further addition of the acid are re-dissolved. If to the solution a drop of solution of cane sugar be added, a fine purple color is developed.

The re-action will also occur on the addition of grape or fruit sugar instead of cane sugar, slowly with the first, quickly with the last; and a color similar to the above is produced by the action of sulphuric acid and sugar on albumen, the crystalline lens, nerve tissue, oleic acid, pure ether, cholesterin, morphia, codeia and amylic alcohol.

The spectrum of Pettenkofer's reaction, when the fluid is moderately diluted, shows four bands—the most marked and largest at E, and a little to the left; another at F; a third between D and E, nearer to D; and the fourth near D.

The yellow *coloring matter* of the bile of man and the Carnivora is termed *Bilirubin* or *Bilifulvin* ($C_{16}H_{18}N_2O_3$) crystallizable and insoluble in water, soluble in chloroform or carbon disulphate; a green coloring matter, *Biliverdin* ($C_{16}H_{20}N_2O_6$), which always exists in large amount in the bile of Herbivora, being formed from bilirubin on exposure to the air, or by subjecting the bile to any other oxidizing agency, as by adding nitric acid. When the bile has been long in the gall-bladder, a third pigment, *Biliprasin*, may be also found in small amount.

In cases of biliary obstruction, the coloring matter of the bile is re-absorbed, and circulates with the blood, giving to the tissues the yellow tint characteristic of jaundice.

The coloring matters of human bile do not appear to give characteristic absorption spectra; but the bile of the guinea pig, rabbit, mouse, sheep, ox, and crow do so, the most constant of which appears to be a band at

F. The bile of the sheep and ox give three bands in a thick layer, and four or five bands with a thinner layer, one on each side of D, one near E, and a faint line at F. (McMunn.) .

. There seems to be a close relationship between the color-matter of the blood and of the bile, and it may be added, between these and that of the urine (urobilin), and of the fæces (stercobilin) also; it is probable they are, all of them, varieties of the same pigment, or derived from the same source. Indeed it is maintained that *Urobilin* is identical with *Hydro-bilirubin*, a substance which is obtained from bilirubin by the action of sodium amalgam, or by the action of sodium amalgam on alkaline hæmatin; both urobilin and hydrobilirubin giving a characteristic absorption band between b and F. They are also identical with stercobilin, which is formed in the alimentary canal from bile pigments.

A common *test* (Gmelin's) for the presence of *bile-pigment* consists of the addition of a small quantity of nitric acid, yellow with nitrous acid; if bile be present, a play of colors is produced, beginning with green and passing through blue and violet to red, and lastly to yellow. The spectrum of Gmelin's test gives a black band extending from near b to beyond F.

Fatty substances are found in variable proportions in the bile. Besides the ordinary saponifiable fats, there is a small quantity of *Cholesterin*, a so-called non-saponifiable fat, which, with the other free fats, is probably held in solution by the bile salts. It is a body belonging to the class of monatomic alcohols ($C_{26}H_{44}O$), and crystallizes in rhombic plates (Fig. 205). It is insoluble in water and cold alcohol, but dissolves easily in boiling alcohol or ether. It gives a red

FIG. 205.—Crystalline scales of cholesterin.

color with strong sulphuric acid, and with nitric acid and ammonia; also a play of colors beginning with blood red and ending with green on the addition of sulphuric acid and chloroform. *Lecithin* ($C_{44}H_{90}NPO_9$), a phosphorus-containing body and *Neurin* ($C_5H_{15}NO_2$), are also found in bile, the latter probably as a decomposition product of the former.

The *Mucus* in bile is derived from the mucous membrane and glands of the gall-bladder, and of the hepatic ducts. It constitutes the residue after bile is treated with alcohol. The epithelium with which it is mixed may be detected in the bile with the microscope in the form of cylindrical cells, either scattered or still held together in layers. To the presence of the mucus is probably to be ascribed the rapid decomposition undergone by the bilin; for, according to Berzelius, if the mucus be separated, bile will remain unchanged for many days.

The *Saline* or *inorganic constituents* of the bile are similar to those found in most other secreted fluids. It is possible that the carbonate and neutral phosphate of sodium and potassium, found in the ashes of bile, are formed in the incineration, and do not exist as such in the fluid. Oxide of iron is said to be a common constituent of the ashes of bile, and copper is generally found in healthy bile, and constantly in biliary calculi.

Gas—A certain small amount of carbonic acid, oxygen, and nitrogen, may be extracted from bile.

Mode of Secretion and Discharge.—The process of secreting bile is continually going on, but appears to be retarded during fasting, and accelerated on taking food. This has been shown by tying the common bile-duct of a dog, and establishing a fistulous opening between the skin and gall-bladder, whereby all the bile secreted was discharged at the surface. It was noticed that when the animal was fasting, sometimes not a drop of bile was discharged for several hours; but that, in about ten minutes after the introduction of food into the stomach, the bile began to flow abundantly, and continued to do so during the whole period of digestion. (Blondlot, Bidder and Schmidt.)

The bile is formed in the hepatic cells; then, being discharged into the minute hepatic ducts, it passes into the larger trunks, and from the main hepatic duct may be carried at once into the duodenum. But, probably, this happens only while digestion is going on; during fasting, it regurgitates from the common bile-duct through the cystic duct, into the gall-bladder, where it accumulates till, in the next period of digestion, it is discharged into the intestine. The gall-bladder thus fulfils what appears to be its chief or only office, that of a reservoir; for its presence enables bile to be constantly secreted, yet insures its employment in the service of digestion, although digestion is periodic, and the secretion of bile constant.

The mechanism by which the bile passes into the gall-bladder is simple. The orifice through which the common bile-duct communicates with the duodenum is narrower than the duct, and appears to be closed, except when there is sufficient pressure behind to force the bile through it. The pressure exercised upon the bile secreted during the intervals of digestion appears insufficient to overcome the force with which the orifice of the duct is closed; and the bile in the common duct, finding no exit in the intestine, traverses the cystic duct, and so passes into the gall-bladder, being probably aided in this retrograde course by the peristaltic action of the ducts. The bile is discharged from the gall-bladder and enters the duodenum on the introduction of food into the small intestine: being pressed on by the contraction of the coats of the gall-bladder, and of the common bile-duct also; for both these organs contain unstriped muscular fibre-cells. Their contraction is excited by the stimulus of the food in the duodenum acting so as to produce a reflex movement, the force of which is sufficient to open the orifice of the common bile-duct.

Bile, as such, is not pre-formed in the blood. As just observed, it is formed by the hepatic cells, although some of the material may be brought to them almost in the condition for immediate secretion. When it is, however, prevented by an obstruction of some kind, from escaping into the intestine (as by the passage of a *gall-stone* along the hepatic duct) it is absorbed in great excess into the blood, and, circulating with it, gives rise to the well-known phenomena of jaundice. This is explained by the fact that the pressure of secretion in the ducts is normally very low, and if it exceeds ⅗ inch of mercury (16 mm.) the secretion ceases to be poured out, and if the opposing force be increased, the bile finds its way into the blood.

Quantity.—Various estimates have been made of the *quantity* of bile discharged into the intestines in twenty-four hours: the quantity doubtless varying, like that of the gastric fluid, in proportion to the amount of food taken. A fair average of several computations would give 20 to 40 oz. (600—900 cc.) as the quantity daily secreted by man.

Uses.—(1) As an *excrementitious* substance, the bile may serve especially as a medium for the separation of excess of carbon and hydrogen from the blood; and its adaptation to this purpose is well illustrated by the peculiarities attending its secretion and disposal in the fœtus. During intra-uterine life, the lungs and the intestinal canal are almost inactive; there is no respiration of open air or digestion of food; these are unnecessary, on account of the supply of well elaborated nutriment received by the vessels of the fœtus at the placenta. The liver, during the same time, is proportionately larger than it is after birth, and the secretion of bile is active, although there is no food in the intestinal canal upon which it can exercise any digestive property. At birth, the intestinal canal is full of thick bile, mixed with intestinal secretion; the *meconium*, or fæces of the fœtus, containing all the essential principles of bile.

Composition of Meconium (Frerichs):

Biliary resin	15.6
Common fat and cholesterin . . .	15.4
Epithelium, mucus, pigment, and salts .	69.0
	100.0

In the fœtus, therefore, the main purpose of the secretion of bile must be the purification of blood by *direct* excretion, *i.e.*, by separation from the blood, and ejection from the body without further change. Probably all the bile secreted in fœtal life is incorporated in the meconium, and with it discharged, and thus the liver may be said to discharge a function in some sense vicarious of that of the lungs. For, in the fœtus, nearly all the blood coming from the placenta passes through the liver, previous to its distribution to the several organs of the body; and the abstraction of carbon, hydrogen, and other elements of bile will purify it, as in extra-uterine life it is purified by the separation of carbonic acid and water at the lungs.

The evident disposal of the fœtal bile by excretion, makes it highly probable that the bile in extra-uterine life is also, at least in part, destined to be discharged as excrementitious. The analysis of the fæces of both children and adults shows that (except when rapidly discharged in pur-gation) they contain very little of the bile secreted, probably not more than one-sixteenth part of its weight, and that this portion includes chiefly its coloring, and some of its fatty matters, and to only a very slight degree, its salts, almost all of which have been re-absorbed from the intestines into the blood.

The elementary composition of bile salts shows, however, such a pre-ponderance of carbon and hydrogen, that probably, after absorption, it combines with oxygen, and is excreted in the form of carbonic acid and water. The change after birth, from the direct to the indirect mode of excretion of the bile, may, with much probability, be connected with a purpose in relation to the development of heat. The temperature of the fœtus is maintained by that of the parent, and needs no source of heat within itself; but, in extra-uterine life, there is (as one may say) a waste of material for heat when any excretion is discharged unoxidized; the carbon and hydrogen of the bilin, therefore, instead of being ejected in the fæces, are re-absorbed, in order that they may be combined with oxygen, and that in the combination heat may be generated.

A substance, which has been discovered in the fæces, and named ster-corin is closely allied to cholesterin; and it has been suggested that while one great function of the liver is to excrete cholesterin from the blood, as the kidney excretes urea, the stercorin of fæces is the modified form in which cholesterin finally leaves the body. Ten grains and a half of ster-corin are excreted daily (A. Flint).

From the peculiar manner in which the liver is supplied with much of the blood that flows through it, it is probable that this organ is exore-tory, not only for such hydro-carbonaceous matters as may need expulsion from any portion of the blood, but that it serves for the direct purification of the stream which, arriving by the portal vein, has just gathered up various substances in its course through the digestive organs—substances which may need to be expelled, almost immediately after their absorption. For it is easily conceivable that many things may be taken up during digestion, which not only are unfit for purposes of nutrition, but which would be positively injurious if allowed to mingle with the general mass of the blood. The liver, therefore, may be supposed placed in the only road by which such matters can pass unchanged into the general current, jealously to guard against their further progress, and turn them back again into an excretory channel. The frequency with which metallic poisons are either excreted by the liver, or intercepted and retained, often for a considerable time, in its own substance, may be adduced as evidence for the probable truth of this supposition.

(2). *As a digestive fluid.*—Though one chief purpose of the secretion of bile may thus appear to be the purification of the blood by ultimate excretion, yet there are many reasons for believing that, while it is in the intestines, it performs an important part in the process of digestion. In nearly all animals, for example, the bile is discharged, not through an excretory duct communicating with the external surface or with a simple reservoir, as most excretions are, but is made to pass into the intestinal canal, so as to be mingled with the chyme directly after it leaves the stomach; an arrangement, the constancy of which clearly indicates that the bile has some important relations to the food with which it is thus mixed. A similar indication is furnished also by the fact that the secretion of bile is most active, and the quantity discharged into the intestines much greater, during digestion than at any other time; although, without doubt, this activity of secretion during digestion may, however, be in part ascribed to the fact that a greater quantity of blood is sent through the portal vein to the liver at this time, and that this blood contains some of the materials of the food absorbed from the stomach and intestines, which may need to be excreted, either temporarily (to be afterward reabsorbed) or permanently.

Respecting the functions discharged by the bile in digestion there is little doubt that it, (*a.*) assists in *emulsifying the fatty portions* of the food, and thus rendering them capable of being absorbed by the lacteals. For it has appeared in some experiments in which the common bile-duct was tied, that, although the process of digestion in the stomach was unaffected, chyle was no longer well formed; the contents of the lacteals consisting of clear, colorless fluid, instead of being opaque and white, as they ordinarily are, after feeding.

(*b.*) It is probable, also, that the *moistening of the mucous membrane* of the intestines by bile facilitates absorption of fatty matters through it.

(*c.*) The bile, like the gastric fluid, has a considerable *antiseptic* power, and may serve to prevent the decomposition of food during the time of its sojourn in the intestines. Experiments show that the contents of the intestines are much more fœtid after the common bile-duct has been tied than at other times; moreover, it is found that the mixture of bile with a fermenting fluid stops or spoils the process of fermentation.

(*d.*) The bile has also been considered to act as a *natural purgative,* by promoting an increased secretion of the intestinal glands, and by stimulating the intestines to the propulsion of their contents. This view receives support from the constipation which ordinarily exists in jaundice, from the diarrhœa which accompanies excessive secretion of bile, and from the purgative properties of ox-gall.

(*e.*) The bile appears to have the power of *precipitating the gastric parapeptones and peptones, together with the pepsin* which is mixed up with them, as soon as the contents of the stomach meet it in the duodenum. The purpose of this operation is probably both to delay any change in the parapeptones until the pancreatic juice can act upon them, and also to prevent the pepsin from exercising its solvent action on the ferments of the pancreatic juice.

Nothing is known with certainty respecting the changes which the reabsorbed portions of the bile undergo. That they are much changed appears from the impossibility of detecting them in the blood; and that part of this change is effected in the liver is probable from an experiment of Magendie, who found that when he injected bile into the portal vein, a dog was unharmed, but was killed when he injected the bile into one of the systemic vessels.

II. The Liver as a Blood-elaborating Gland.

The secretion of bile, as already observed, is only one of the purposes fulfilled by the liver. Another very important function appears to be that of so acting upon certain constituents of the blood passing through it, as to render some of them capable of assimilation with the blood generally, and to prepare others for being duly eliminated in the process of respiration. It appears that the peptones, conveyed from the alimentary canal by the blood of the portal vein, require to be submitted to the influence of the liver before they can be assimilated by the blood; for if such albuminous matter is injected into the jugular vein, it speedily appears in the urine; but if introduced into the portal vein, and thus allowed to traverse the liver, it is no longer ejected as a foreign substance, but is incorporated with the albuminous part of the blood. Albuminous matters are also subject to decomposition by the liver in another way to be immediately noticed (p. 281). The formation of urea by the liver will be again referred to (p. 371).

Glycogenic Function.—One of the chief uses of the liver in connection with elaboration of the blood is comprised in what is known as its *glycogenic function*. The important fact that the liver normally forms *glucose* or grape sugar, or a substance readily convertible into it, was discovered by Claude Bernard in the course of some experiments which he undertook for the purpose of finding out in what part of the circulatory system the saccharine matter disappeared, which was absorbed from the alimentary canal. With this purpose he fed a dog for seven days with food containing a large quantity of sugar and starch; and, as might be expected, found sugar in both the portal and hepatic veins. He then fed a dog with meat only, and, to his surprise, still found sugar in the hepatic veins. Repeated experiments gave invariably the same result; no sugar being found, under a meat diet, in the portal vein, if care were taken, by applying a ligature on it at the transverse fissure, to prevent reflux of blood from the hepatic venous system. Bernard found sugar also in the substance of the liver. It thus seemed certain that the liver formed sugar, even when, from the absence of saccharine and amyloid matters in the food, none could be brought directly to it from the stomach or intestines.

Excepting cases in which large quantities of starch and sugar were taken as food, no sugar was found in the blood after it had passed through the lungs; the sugar formed by the liver, having presumably disappeared by combustion, in the course of the pulmonary circulation.

Bernard found, subsequently to the before-mentioned experiments, that a liver, removed from the body, and from which all sugar had been completely washed away by injecting a stream of water through its blood-vessels, will be found, after the lapse of a few hours, to contain sugar in abundance. This *post-mortem* production of sugar was a fact which could only be explained in the supposition that the liver contained a substance, readily convertible into sugar in the course merely of post-mortem decomposition; and this theory was proved correct by the discovery of a substance in the liver allied to starch, and now generally termed *glycogen*. We may believe, therefore, that the liver does not form sugar directly from the materials brought to it by the blood, but that glycogen is first formed and stored in its substance; and that the sugar, when present, is the result of the transformation of the latter.

Quantity of Glycogen formed.—Although, as before mentioned, glycogen is produced by the liver when neither starch nor sugar is present in the food, its amount is much less under such a diet.

Average amount of Glycogen in the Liver of Dogs under various Diets.
(Pavy.)

Diet.	Amount of Glycogen in Liver.
Animal food	7·19 per cent.
Animal food with sugar (about ¼ lb. of sugar daily)	14·5 "
Vegetable diet (potatoes, with bread or barley-meal)	17·23 "

The dependence of the formation of glycogen on the food taken is also well shown by the following results, obtained by the same experimenter:

Average quantity of Glycogen found in the Liver of Rabbits after Fasting and after a diet of Starch and Sugar respectively.

	Average amount of Glycogen in Liver.
After fasting for three days	Practically absent.
" diet of starch and grape-sugar . . .	15·4 per cent.
" " cane-sugar	16·9 "

Regarding these facts there is no dispute. All are agreed that glycogen is formed, and laid up in store, temporarily, by the liver-cells; and that it is not formed exclusively from saccharine and amylaceous foods, but from albuminous substances also; the albumen, in the latter case, being probably split up into glycogen, which is temporarily stored in the liver, and urea, which is excreted by the kidneys.

Destination of Glycogen.—There are two chief theories on the subject of the destination of glycogen. (1.) That the conversion of glycogen into sugar takes place rapidly during life by the agency of a ferment also formed in the liver: and the sugar is conveyed away by the blood of the hepatic veins, and soon undergoes combustion. (2.) That the conversion into sugar only occurs after death, and that during life no sugar exists in healthy livers; glycogen not undergoing this transformation. The chief arguments advanced in support of this view are, (a) that scarcely a trace of sugar is found in blood drawn during life from the right ventricle, or in blood collected from the right side of the heart *immediately* after an animal has been killed; while if the examination be delayed for a very short time after death, sugar in abundance may be found in such blood; (b), that the liver, like the venous blood in the heart, is, at the moment of death, completely free from sugar, although afterward its tissue speedily becomes saccharine, unless the formation of sugar be prevented by freezing, boiling, or other means calculated to interfere with the action of a ferment on the amyloid substance of the organ. Instead of adopting Bernard's view, that normally, during life, glycogen passes as sugar into the hepatic venous blood, and thereby is conveyed to the lungs to be further disposed of, Pavy inclines to the belief that it may represent an intermediate stage in the formation of fat from materials absorbed from the alimentary canal.

Liver-sugar and Glycogen.—To demonstrate the presence of sugar in the liver, a portion of this organ, after being cut into small pieces, is bruised in a mortar to a pulp with a small quantity of water, and the pulp is boiled with sodium-sulphate in order to precipitate albuminous and coloring matters. The decoction is then filtered and may be tested for glucose (p. 230).

Glycogen ($C_6H_{10}O_5$) is an amorphous, starch-like substance, odorless and tasteless, soluble in water, insoluble in alcohol. It is converted into glucose by boiling with dilute acids, or by contact with any animal ferment. It may be obtained by taking a portion of liver from a recently killed rabbit, and, after cutting it into small pieces, placing it for a short time in boiling water. It is then bruised in a mortar, until it forms a pulpy mass, and subsequently boiled in distilled water for about a quarter of an hour. The glycogen is precipitated from the filtered decoction by the addition of alcohol. Glycogen has been found in many other structures than the liver. (See Appendix.)

Glycosuria.—The facility with which the glycogen of the liver is transformed into sugar would lead to the expectation that this chemical change, under many circumstances, would occur to such an extent that sugar would· be present not only in the hepatic veins, but in the blood generally. Such is frequently the case; the sugar when in excess in the blood being secreted by the kidneys, and thus appearing in variable quantities in the urine (Glycosuria).

Influence of the Nervous System in producing Glycosuria.—Glycosuria may be experimentally produced by puncture of the medulla oblongata in the region of the vaso-motor centre. The better fed the animal the larger is the amount of sugar found in the urine; whereas in the case of a starving animal no sugar appears. It is, therefore, highly probable that the sugar comes from the hepatic glycogen, since in the one case glycogen is in excess, and in the other it is almost absent. The nature of the influence is uncertain. It may be exercised in dilating the hepatic vessels, or possibly on the liver cells themselves. The whole course of the nervous stimulus cannot be traced to the liver, but at first it passes from the medulla down the spinal cord as far as—in rabbits—the fourth dorsal vertebra, and thence to the first thoracic ganglion.

Many other circumstances will cause glycosuria. It has been observed after the administration of various drugs, after the injection of urari, poisoning with carbonic oxide gas, the inhalation of ether, chloroform, etc., the injection of oxygenated blood into the portal venous system. It has been observed in man after injuries to the head, and in the course of various diseases.

The well-known disease, *diabetus mellitus*, in which a large quantity of sugar is persistently secreted daily with the urine, has, doubtless, some close relation to the normal glycogenic function of the liver; but the nature of the relationship is at present quite unknown.

The Intestinal Secretion, or Succus Entericus.—On account of the difficulty in isolating the secretion of the glands in the wall of the intestine (Brunner's and Lieberkühn's) from other secretions poured into the canal (gastric juice, bile, and pancreatic secretion), but little is known regarding the composition of the former fluid (intestinal juice, *succus entericus*).

It is said to be a yellowish alkaline fluid with a specific gravity of 1011, and to contain about 2·5 per cent. of solid matters (Thiry).

Functions.—The secretion of Brunner's glands is said to be able to convert proteids into peptones, and that of Lieberkühn's is believed to convert starch into sugar. To these functions of the *succus entericus* the powers of converting cane into grape sugar, and of turning cane sugar into lactic, and afterward into butyric acid, are added by some physiologists. It also probably contains a milk-curdling ferment (W. Roberts).

The reaction which represents the conversion of cane sugar into grape sugar may be represented thus:—

$$2\,C_{12}H_{22}O_{11} + 2\,H_2O = C_{12}H_{24}O_{12} + C_{12}H_{24}O_{12}$$

Saccharose   Water   Dextrose   Lævulose

The conversion is probably effected by means of a hydrolytic ferment. (Inversive ferment, Bernard.)

The *length and complexity* of the digestive tract seem to be closely connected with the character of the food on which an animal lives. Thus, in all carnivorous animals, such as the cat and dog, and pre-eminently in carnivorous birds, as hawks and herons, it is exceedingly short. The seals, which, though carnivorous, possess a very long intestine, appear to furnish an exception; but this is doubtless to be explained as an adaptation to their aquatic habits: their constant exposure to cold requiring that they should absorb as much as possible from their intestines.

Herbivorous animals, on the other hand, and the ruminants especially, have very long intestines (in the sheep 30 times the length of the body) which is no doubt to be connected with their lowly nutritious diet. In others, such as the rabbit, though the intestines are not excessively long, this is compensated by the great length and capacity of the cæcum. In man, the length of the intestines is intermediate between the extremes of the carnivora and herbivora, and his diet also is intermediate.

Summary of the Digestive Changes in the Small Intestine.

In order to understand the changes in the food which occur during its passage through the small intestine, it will be well to refer briefly to the state in which it leaves the stomach through the pylorus. It has been said before, that the chief office of the stomach is not only to mix into a uniform mass all the varieties of food that reach it through the œsophagus, but especially to dissolve the nitrogenous portion by means of the gastric juice. The fatty matters, during their sojourn in the stomach, become more thoroughly mingled with the other constituents of the food taken, but are not yet in a state fit for absorption. The conversion of starch into sugar, which began in the mouth, has been interfered with, if not altogether stopped. The soluble matters—both those which were so from the first, as sugar and saline matter, and the gastric peptones—have begun to disappear by absorption into the blood-vessels, and the same thing has befallen such fluids as may have been swallowed, —wine, water, etc.

The thin pultaceous chyme, therefore, which during the whole period of gastric digestion, is being constantly squeezed or strained through the pyloric orifice into the duodenum, consists of albuminous matter, broken down, dissolving and half dissolved; fatty matter broken down and melted, but not dissolved at all; starch very slowly in process of conversion into sugar, and as it becomes sugar, also dissolving in the fluids with which it is mixed; while, with these are mingled gastric fluid, and fluid that has been swallowed, together with such portions of the food as are not digestible, and will be finally expelled as part of the fæces.

On the entrance of the chyme into the duodenum, it is subjected to the influence of the bile and pancreatic juice, which are then poured out, and also to that of the succus entericus. All these secretions have a more or less alkaline reaction, and by their admixture with the gastric chyme its acidity becomes less and less until at length, at about the middle of the small intestine, the reaction becomes alkaline and continues so as far as the ileo-cæcal valve.

The special digestive functions of the small intestine may be taken in the following order:—

(1.) One important duty of the small intestine is the alteration of the *fat* in such a manner as to make it fit for absorption; and there is no doubt that this change is chiefly effected in the upper part of the small intestine. What is the exact share of the process, however, allotted respectively to the bile, to the pancreatic secretion, and to the intestinal juice, is still uncertain,—probably the pancreatic juice is the most important. The fat is changed in two ways. (*a*). To a slight extent it is chemically decomposed by the alkaline secretions with which it is mingled, and a soap is the result. (*b*). It is emulsionized, *i.e.*, its particles are minutely subdivided and diffused, so that the mixture assumes the condition of a milky fluid, or emulsion. As will be seen in the next Chapter, most of the fat is absorbed by the lacteals of the intestine, but a small part, which is saponified, is also absorbed by the blood-vessels.

(2.) The *albuminous* substances which have been partly dissolved in the stomach, and have not been absorbed, are subjected to the action of the pancreatic and intestinal secretions. The pepsin is rendered inert by being precipitated together with the gastric peptones and parapeptones, as soon as the chyme meets with bile. By these means the pancreatic ferment trypsin is enabled to proceed with the further conversion of the parapeptones into peptones, and of part of the peptones (hemipeptone, Kühne) into leucin and tyrosin. *Albuminous* substances, which are chemically altered in the process of digestion (peptones), and gelatinous matters similarly changed, are absorbed by both the blood-vessels and lymphatics of the intestinal mucous membrane. Albuminous matters, in a state of solution, which have not undergone the peptonic change, are probably, from the difficulty with which they *diffuse*, absorbed, if at all, almost solely by the lymphatics.

(3.) The *starchy*, or amyloid portions of the food, the conversion of which into dextrin and sugar was more or less interrupted during its stay in the stomach, is now acted on briskly by the pancreatic juice and the succus entericus; and the sugar, as it is formed, is dissolved in the intestinal fluids, and is absorbed chiefly by the blood-vessels.

(4.) *Saline* and *saccharine* matters, as common salt, or cane sugar, if not in a state of solution beforehand in the saliva or other fluids which may have been swallowed with them, are at once dissolved in the stomach, and if not here absorbed, are soon taken up in the small intestine; the blood-vessels, as in the last case, being chiefly concerned in the absorption. Cane sugar is in part or wholly converted into grape-sugar before its absorption. This is accomplished partially in the stomach, but also by a ferment in the succus entericus.

(5.) The *liquids,* including in this term the ordinary drinks, as water, wine, ale, tea, etc., which may have escaped absorption in the stomach, are absorbed probably very soon after their entrance into the intestine; the fluidity of the contents of the latter being preserved more by the constant secretion of fluid by the intestinal glands, pancreas, and liver, than by any given portion of fluid, whether swallowed or secreted, remaining long unabsorbed. From this fact, therefore, it may be gathered that there is a kind of circulation constantly proceeding from the intestines into the blood, and from the blood into the intestines again; for as all the fluid—a very large amount—secreted by the intestinal glands, must come from the blood, the latter would be too much drained, were it not that the same fluid after secretion is again re-absorbed into the current of blood —going into the blood charged with nutrient products of digestion—coming out again by secretion through the glands in a comparatively uncharged condition.

At the lower end of the small intestine, the chyme, still thin and pultaceous, is of a light yellow color, and has a distinctly fæcal odor. This odor depends upon the formation of indol. In this state it passes through the ileo-cæcal opening into the large intestine.

Summary of the Digestive Changes in the Large Intestine.

The changes which take place in the chyme in the *large* intestine are probably only the continuation of the same changes that occur in the course of the food's passage through the upper part of the intestinal canal. From the absence of villi, however, we may conclude that absorption, especially of fatty matter, is in great part completed in the small intestine; while, from the still half-liquid, pultaceous consistence of the chyme when it first enters the cæcum, there can be no doubt that the absorption of liquid is not by any means concluded. The peculiar odor, moreover, which is acquired after a short time by the contents of the large bowel, would seem to indicate a further chemical change in the alimentary matters or in the digestive fluids, or both. The acid reaction, which had disappeared in the small bowel, again becomes very manifest in the cæcum —probably from acid fermentation-processes in some of the materials of the food.

There seems no reason to conclude that any special "secondary digestive" process occurs in the cæcum or in any other part of the large intestine. Probably any constituent of the food which has escaped digestion and absorption in the small bowel may be digested in the large intestine; and the power of this part of the intestinal canal to digest fatty, albuminous, or other matters, may be gathered from the good effects of nutrient enemata, so frequently given when from any cause there is difficulty in introducing food into the stomach. In ordinary healthy digestion, however, the changes which ensue in the chyme after its passage into the large intestine, are mainly the absorption of the more liquid parts, and the completion of the changes which were proceeding in the small intestine,—the process being assisted by the secretion of the numerous tubular glands therein present.

Fæces.—By these means the contents of the large intestine, as they proceed toward the rectum, become more and more solid, and losing their more liquid and nutrient parts, gradually acquire the odor and consistence characteristic of *fæces*. After a sojourn of uncertain duration in the sigmoid flexure of the colon, or in the rectum, they are finally expelled by the act of defæcation.

The average quantity of solid fæcal matter evacuated by the human adult in twenty-four hours is about six or eight ounces.

COMPOSITION OF FÆCES.

Water . . . . 733·00
Solids . . . . 267·00
    Special excrementitious constituents:—Excretin, ⎫
      excretoleic acid (Marcet), and stercorin (Aus- |
      tin Flint). |
    Salts:—Chiefly phosphate of magnesium and phos- |
      phate of calcium, with small quantities of iron, |
      soda, lime, and silica. |
    Insoluble residue of the food (chiefly starch grains, |
      woody tissue, particles of cartilage and fibrous ⎬ 267·00
      tissue, undigested muscular fibres or fat, and |
      the like, with insoluble substances accidentally |
      introduced with the food). |
    Mucus, epithelium, altered coloring matter of bile, |
      fatty acids, etc. |
    Varying quantities of other constituents of bile, |
      and derivatives from them. ⎭

Length of Intestinal Digestive Period.—The time occupied by the journey of a given portion of food from the stomach to the anus, varies considerably even in health, and on this account, probably, it is that such different opinions have been expressed in regard to the subject. About twelve hours are occupied by the journey of an ordinary meal through the *small* intestine, and twenty-four to thirty-six hours by the passage through the *large* bowel. (Brinton.)

Defæcation.—Immediately before the act of voluntary expulsion of fæces (*defæcation*) there is usually, first an inspiration, as in the case of coughing, sneezing, and vomiting; the glottis is then closed, and the diaphragm fixed. The abdominal muscles are contracted as in expiration; but as the glottis is closed, the whole of their pressure is exercised on the abdominal contents. The sphincter of the rectum being relaxed, the evacuation of its contents takes place accordingly; the effect being, of course, increased by the peristaltic action of the intestine. As in the other actions just referred to, there is as much tendency to the escape of the contents of the lungs or stomach as of the rectum; but the pressure is relieved only at the orifice, the sphincter of which instinctively or involuntarily yields (see Fig. 144).

Nervous Mechanism of Defæcation.—The anal sphincter muscle is normally in a state of tonic contraction. The nervous centre which governs this contraction is probably situated in the lumbar region of the spinal cord, inasmuch as in cases of division of the cord above this region the sphincter regains, after a time, to some extent the tonicity which is lost immediately after the operation. By an effort of the will, acting through the centre, the contraction may be relaxed or increased. In ordinary cases the apparatus is set in action by the gradual accumulation of fæces in the sigmoid flexure and rectum pressing against the sphincter and causing its relaxation; this sensory impulse acting through the brain and reflexly through the spinal centre. Peristaltic action, especially of the sigmoid flexure in pressing onward the fæces against the sphincter, is a very important part of the act.

The Gases contained in the Stomach and Intestines.—Under ordinary circumstances, the alimentary canal contains a considerable quantity of gaseous matter. Any one who has had occasion, in a postmortem examination, either to lay open the intestines, or to let out the gas which they contain, must have been struck by the small space afterward occupied by the bowels, and by the large degree, therefore, in which the gas, which naturally distends them, contributes to fill the cavity of the abdomen. Indeed, the presence of air in the intestines is so constant, and, within certain limits, the amount in health so uniform, that there can be no doubt that its existence here is not a mere accident, but intended to serve a definite and important purpose, although, probably, a mechanical one.

Sources.—The sources of the gas contained in the stomach and bowels may be thus enumerated:—

1. Air introduced in the act of swallowing either food or saliva; 2. Gases developed by the decomposition of alimentary matter or of the

secretions and excretions mingled with it in the stomach and intestines; 3. It is probable that a certain mutual interchange occurs between the gases contained in the alimentary canal, and those present in the blood of these gastric and intestinal blood-vessels; but the conditions of the exchange are not known, and it is very doubtful whether anything like a true and definite secretion of gas from the blood into the intestines or stomach ever takes place. There can be no doubt, however, that the intestines may be the proper excretory organs for many odorous and other substances, either absorbed from the air taken into the lungs in inspiration, or absorbed in the upper part of the alimentary canal, again to be excreted at a portion of the same tract lower down—in either case assuming rapidly a gaseous form after their excretion, and in this way, perhaps, obtaining a more ready egress from the body. It is probable that, under ordinary circumstances, the gases of the stomach and intestines are derived chiefly from the second of the sources which have been enumerated (Brinton).

COMPOSITION OF GASES CONTAINED IN THE ALIMENTARY CANAL.

(TABULATED FROM VARIOUS AUTHORITIES BY BRINTON.)

Whence obtained.	Composition by Volume.					
	Oxygen.	Nitrog.	Carbon. Acid.	Hydrog.	Carburet. Hydrogen.	Sulphuret. Hydrogen.
Stomach	11	71	14	4	—	—
Small Intestines . . .	—	32	30	38	—	⎫
Cæcum	—	66	12	8	13	⎬ trace
Colon	—	35	57	6	8	⎭
Rectum	—	46	43	—	11	
Expelled *per anum* . .	—	22	41	19	19	½

Movements of the Intestines.—It remains only to consider the manner in which the food and the several secretions mingled with it are moved through the intestinal canal, so as to be slowly subjected to the influence of fresh portions of intestinal secretion, and as slowly exposed to the absorbent power of all the villi and blood-vessels of the mucous membrane. The movement of the intestines is *peristaltic* or *vermicular*, and is effected by the alternate contractions and dilatations of successive portions of the intestinal coats. The contractions, which may commence at any point of the intestine, extend in a wave-like manner along the tube. In any given portion, the longitudinal muscular fibres contract first, or more than the circular; they draw a portion of the intestine upward, or, as it were, backward, over the substance to be propelled, and then the circular fibres of the same portion contracting in succession from above downward, or, as it were, from behind forward, press on the substance into the portion next below, in which at once the same succession of action next ensues. These movements take place slowly, and, in health, are commonly unperceived by the mind; but they are perceptible when they are accelerated under the influence of any irritant.

The movements of the intestines are sometimes retrograde; and there is no hindrance to the backward movement of the contents of the small intestine. But almost complete security is afforded against the passage of the contents of the large into the small intestine by the ileo-cæcal valve. Besides,—the orifice of communication between the ileum and cæcum (at the borders of which orifice are the folds of mucous membrane which form the valve) is encircled with muscular fibres, the contraction of which prevents the undue dilatation of the orifice.

Proceeding from above downward, the muscular fibres of the large intestine become, on the whole, stronger in direct proportion to the greater strength required for the onward moving of the fæces, which are gradually becoming firmer. The greatest strength is in the rectum, at the termination of which the circular unstriped muscular fibres form a strong band called the *internal* sphincter; while an *external* sphincter muscle with striped fibres is placed rather lower down, and more externally, and as we have seen above, holds the orifice close by a constant slight tonic contraction.

Experimental irritation of the brain or cord produces no evident or constant effect on the movements of the intestines during life; yet in consequence of certain conditions of the mind the movements are accelerated or retarded; and in paraplegia the intestines appear after a time much weakened in their power, and costiveness, with a tympanitic condition, ensues. Immediately after death, irritation of both the sympathetic and pneumogastric nerves, if not too strong, induces genuine peristaltic movements of the intestines. Violent irritation stops the movements. These stimuli act, no doubt, not directly on the muscular tissue of the intestine, but on the ganglionic plexus before referred to.

Influence of the Nervous System on Intestinal Digestion.— As in the case of the œsophagus and stomach, the peristaltic movements of the intestines are directly due to reflex action through the ganglia and nerve fibres distributed so abundantly in their walls (p. 255); the presence of chyme acting as the stimulus, and few or no movements occurring when the intestines are empty. The intestines are, moreover, connected with the higher nerve-centres by the splanchnic nerves, as well as other branches of the sympathetic which come to them from the cœliac and other abdominal plexuses.

The splanchnic nerves are in relation to the intestinal movements, *inhibitory*—these movements being retarded or stopped when the splanchnics are irritated. As the vaso-motor nerves of the intestines, the splanchnics are also much concerned in intestinal digestion.

CHAPTER IX.

ABSORPTION.

The process of Absorption has, for one of its objects, the introduction into the blood of fresh materials from the food and air, and of whatever comes into contact with the external or internal surfaces of the body; and, for another, the gradual removal of parts of the body itself, when they need to be renewed. In both these offices, *i.e.*, in both absorption from without and absorption from within, the process manifests some variety, and a very wide range of action; and in both two sets of vessels are, or may be, concerned, namely, the *Blood-vessels*, and the Lymph-vessels or *Lymphatics* to which the term Absorbents has been also applied.

The Lymphatic Vessels and Glands.

Distribution.—The principal vessels of the lymphatic system are, in structure and general appearance, like very small and thin-walled veins, and like them are provided with valves. By one extremity they commence by fine microscopic branches, the *lymphatic capillaries* or *lymph-capillaries*, in the organs and tissues of the body, and by their other extremities they end directly or indirectly in two trunks which open into the large veins near the heart (Fig. 206). Their contents, the *lymph* and *chyle*, unlike the blood, pass only in one direction, namely, from the fine branches to the trunk and so to the large veins, on entering which they are mingled with the stream of blood, and form part of its constituents. Remembering the course of the fluid in the lymphatic vessels, viz., its passage in the direction only *toward* the large veins in the neighborhood of the heart, it will readily be seen from Fig. 206 that the greater part of the contents of the lymphatic system of vessels passes through a comparatively large trunk called the *thoracic duct*, which finally empties its contents into the blood-stream, at the junction of the internal jugular and subclavian veins of the left side. There is a smaller duct on the right side. The lymphatic vessels of the intestinal canal are called *lacteals*, because, during digestion, the fluid contained in them resembles milk in appearance; and the *lymph* in the lacteals during the period of digestion is called *chyle*. There is no essential distinction, however, between lacteals and lymphatics. In some parts of their course all lymphatic vessels pass through certain bodies called *lymphatic glands.*

Lymphatic vessels are distributed in nearly all parts of the body. Their existence, however, has not yet been determined in the placenta, the umbilical cord, the membranes of the ovum, or in any of the non-vascular parts, as the nails, cuticle, hair and the like.

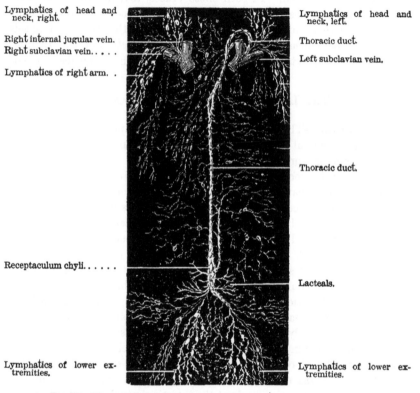

Lymphatics of head and neck, right.

Right internal jugular vein.
Right subclavian vein.. . . .

Lymphatics of right arm. .

Receptaculum chyli.. . . . .

Lymphatics of lower ex-
tremities.

Lymphatics of head and neck, left.

Thoracic duct.

Left subclavian vein.

Thoracic duct.

Lacteals.

Lymphatics of lower ex-
tremities.

Fig. 206.—Diagram of the principal groups of lymphatic vessels (from Quain).

Origin of Lymph Capillaries.—The lymphatic *capillaries* commence most commonly either in closely-meshed networks, or in irregular lacunar spaces between the various structures of which the different organs are composed. Such irregular spaces, forming what is now termed the *lymph-canalicular system,* have been shown to exist in many tissues. In serous membranes, such as the omentum and mesentery, they occur as a connected system of very irregular branched spaces partly occupied by connective-tissue corpuscles, and both in these and in many other tissues are found to communicate freely with regular lymphatic vessels. In many cases, though they are formed mostly by the chinks and crannies between the blood-vessels, secreting ducts, and other parts which may happen to form the framework of the organ in which they exist, they are lined by a distinct layer of endothelium.

The lacteals offer an illustration of another mode of origin, namely, in blind dilated extremities (Figs. 192 and 193); but there is no essential difference in structure between these and the lymphatic capillaries of other parts.

Structure of Lymph Capillaries.—The structure of lymphatic capillaries is very similar to that of blood-capillaries: their walls consist of a single layer of endothelial cells of an elongated form and sinuous outline, which cohere along their edges to form a delicate membrane.

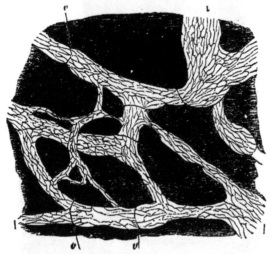

Fig. 207.—Lymphatics of central tendon of rabbit's diaphragm, stained with silver nitrate. The ground substance has been shaded diagrammatically to bring out the lymphatics clearly. *l.* Lymphatics lined by long narrow endothelial cells, and showing *v.* valves at frequent intervals (Schofield).

They differ from blood capillaries mainly in their larger and very variable calibre, and in their numerous communications with the spaces of the lymph-canalicular system.

Communications of the Lymphatics.—The fluid part of the blood constantly exudes or is strained through the walls of the blood-capillaries, so as to moisten all the surrounding tissues, and occupies the interspaces which exist among their different elements. These same interspaces have been shown, as just stated, to form the beginnings of the lymph-capillaries; and the latter, therefore, are the means of collecting the exuded blood-plasma, and returning that part which is not directly absorbed by the tissues into the blood-stream. For many years, the notion of the existence of any such channels between the blood-vessels and lymph-vessels as would admit blood-corpuscles, has been given up; observations having proved that, for the passage of such corpuscles, it is not necessary to assume the presence of any special channels at all, inasmuch as blood-corpuscles can pass bodily, without much difficulty, through the walls of the blood-capillaries and small veins (p. 159), and could pass with still less trouble, probably, through the comparatively ill-defined walls of the capillaries which contain lymph.

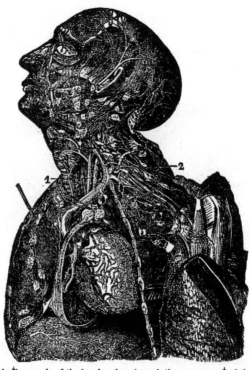

Fig. 208.—Lymphatic vessels of the head and neck and the upper part of the trunk (Mascagni). 1-6.— The chest and pericardium have been opened on the left side, and the left mamma detached and thrown outward over the left arm, so as to expose a great part of its deep surface. The principal lymphatic vessels and glands are shown on the side of the head and face, and in the neck, axilla, and mediastinum. Between the left internal jugular vein and the common carotid artery, the upper ascending part of the thoracic duct marked 1, and above this, and descending to 2, the arch and last part of the duct. The termination of the upper lymphatics of the diaphragm in the mediastinal glands, as well as the cardiac and the deep mammary lymphatics, is also shown.

It is worthy of note that, in many animals, both arteries and veins, especially the latter, are often found to be more or less completely ensheathed in large lymphatic channels. In turtles, crocodiles, and many other animals, the abdominal aorta is enclosed in a large lymphatic vessel.

Stomata.—In certain parts of the body openings exist by which lymphatic capillaries directly communicate with parts hitherto supposed to be closed cavities. If the peritoneal cavity be injected with milk, an injection is obtained of the plexus of lymphatic vessels of the central tendon of the diaphragm (Fig. 207); and on removing a small portion of the central tendon, with its peritoneal surface uninjured, and examining the process of absorption under the microscope, the milk-globules run toward small natural openings or *stomata* between the epithelial cells, and disappear by passing vortex-like through them. The *stomata*, which have a roundish outline, are only wide enough to admit two or three milk-globules abreast, and never exceed the size of an epithelial cell.

Fig. 209.   Fig. 210.

Fig. 209.—Superficial lymphatics of the forearm and palm of the hand, 1-5. 5. Two small glands at the bend of the arm. 6. Radial lymphatic vessels. 7. Ulnar lymphatic vessels. 8, 8. Palmar arch of lymphatics. 9, 9. Outer and inner sets of vessels. b. Cephalic vein. d. Radial vein. e. Median vein. f. Ulnar vein. The lymphatics are represented as lying on the deep fascia. (Mascagni.)

Fig. 210.—Superficial lymphatics of right groin and upper part of thigh, 1-6. 1, upper inguinal glands. 2, 2', Lower inguinal or femoral glands. 3, 3'. Plexus of lymphatics in the course of the long saphenous vein. (Mascagni.)

Pseudostomata.—When absorption into the lymphatic system takes place in membranes covered by epithelium or endothelium through the interstitial or intercellular cement-substance, it is said to take place through *pseudostomata.*

Demonstration of Lymphatics of Diaphragm.—The stomata on the peritoneal surface of the diaphragm are the openings of short vertical canals which lead up into the lymphatics, and are lined by cells like those of germinating endothelium (p. 23). By introducing a solution of Berlin blue into the peritoneal cavity of an animal shortly after death, and suspending it, head downward, an injection of the lymphatic vessels of the diaphragm, through the stomata on its peritoneal surface, may readily be obtained, if artificial respiration be carried on for about half an hour. In this way it has been found that in the rabbit the lymphatics are arranged between the tendon bundles of the centrum tendineum; and they are hence termed *interfascicular.* The centrum tendineum is coated by endothelium on its pleural and peritoneal surfaces, and its substance consists of tendon bundles arranged in concentric rings toward the pleural side and in radiating bundles toward the peritoneal side.

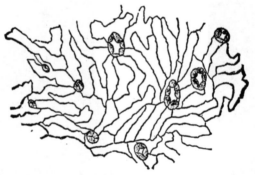

FIG. 211.—Peritoneal surface of septum cisternæ lymphaticæ magnæ of frog. The stomata, some of which are open, some collapsed, are surrounded by germinating endothelium. × 160. (Klein.)

The lymphatics of the anterior half of the diaphragm open into those of the anterior mediastinum, while those of the posterior half pass into a lymphatic vessel in the posterior mediastinum, which soon enters the thoracic duct. Both these sets of vessels, and the glands into which they pass, are readily injected by the method above described; and there can be little doubt that during life the flow of lymph along these channels is chiefly caused by the action of the diaphragm during respiration. As it descends in inspiration, the spaces between the *radiating* tendon bundles dilate, and lymph is sucked from the peritoneal cavity, through the widely open stomata, into the interfascicular lymphatics. During expiration, the spaces between the *concentric* tendon bundles dilate, and the lymph is squeezed into the lymphatics toward the pleural surface. (Klein.) It thus appears probable that during health there is a continued sucking in of lymph from the .peritoneum into the lymphatics by the "pumping" action of the diaphragm; and there is doubtless an equally continuous exudation of fluid from the general serous surface of the peritoneum. When this balance of transudation and absorption is disturbed, either by increased transudation or some impediment to absorption, an accumulation of fluid necessarily takes place (ascites).

Stomata have been found in the pleura; and as they may be presumed to exist in other serous membranes, it would seem as if the serous cavities, hitherto supposed closed, form but a large lymph-sinus, or widening out, so to speak, of the lymph-capillary system with which they directly communicate.

. **Structure of Lymphatic Vessels.**—The larger vessels are very like veins, having an external coat of fibro-cellular tissue, with elastic filaments; within this, a thin layer of fibro-cellular tissue, with plain muscular fibres, which have, principally, a circular direction, and are much more abundant in the small than in the larger vessels; and again, within this, an inner elastic layer of longitudinal fibres, and a lining of epithelium; and numerous valves. The valves, constructed like those of veins, and with the free edges turned toward the heart, are usually arranged in pairs, and, in the small vessels, are so closely placed, that when the vessels are full, the valves constricting them where their edges are attached, give them a peculiar beaded or knotted appearance.

Current of the Lymph.—With the help of the valvular mechanism (1) all occasional pressure on the exterior of the lymphatic and lacteal vessels propels the lymph toward the heart: thus muscular and other external pressure accelerates the flow of the lymph as it does that of the blood in the veins. The actions of (2) the muscular fibres of the small intestine, and probably the layer of organic muscle present in each intestinal villus, seem to assist in propelling the chyle: for, in the small intestine of a mouse, the chyle has been seen moving with intermittent propulsions that appeared to correspond with the peristaltic movements of the intestine. But for the general propulsion of the lymph and chyle, it is probable that, together with (3) the *vis a tergo* resulting from absorption (as in the ascent of sap in a tree), and from external pressure, some of the force may be derived (4) from the contractility of the vessel's own walls. The respiratory movements, also, (5) favor the current of lymph through the thoracic duct as they do the current of blood in the thoracic veins (p. 206).

Lymphatic Glands are small round or oval compact bodies varying in size from a hempseed to a bean, interposed in the course of the lymphatic vessels, and through which the chief part of the lymph passes in its course to be discharged into the blood-vessels. They are found in great numbers in the mesentery, and along the great vessels of the abdomen, thorax, and neck; in the axilla and groin; a few in the popliteal space, but not further down the leg, and in the arm as far as the elbow. Some lymphatics do not, however, pass through glands before entering · the thoracic duct.

Structure.—A lymphatic gland is covered externally by a capsule of connective tissue, generally containing some unstriped muscle. At the inner side of the gland, which is somewhat concave (*hilus*) (Fig. 212, *a*), the capsule sends processes inward in which the blood-vessels are contained, and these join with other processes called *trabeculæ* (Fig. 215, *t.r.*)

prolonged from the inner surface of the part of the capsule covering the convex or outer part of the gland; they have a structure similar to that of the capsule, and entering the gland from all sides, and freely communicating, form a fibrous supporting *stroma*. The interior of the gland is seen on section, even when examined with the naked eye, to be made up of two parts, an outer or *cortical* (Fig. 212, *c, c*), which is light-colored, and an inner of redder appearance, the *medullary* portion (Fig. 212). In the outer or cortical part of the gland (Fig. 215, *c*) the intervals between the trabeculæ are comparatively large and more or less trian-

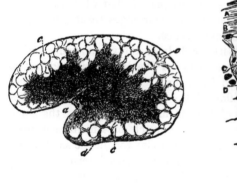

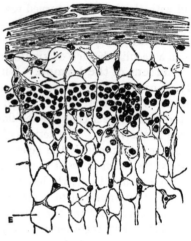

FIG. 212.

FIG. 213.

FIG. 212.—Section of a mesenteric gland from the ox, slightly magnified. *a*, Hilus; *b* (in the central part of the figure), medullary substance; *c*, cortical substance with indistinct alveoli; *d*, capsule (Kölliker.)

FIG. 213.—From a vertical section through the capsule, cortical sinus and peripheral portion of follicle of a human compound lymphatic gland. The section had been shaken, so as to get rid of most of the lymph corpuscles. A. Outer stratum of capsule, consisting of bundles of fibrous tissue cut at various angles. B. Inner stratum, showing fibres of connective tissue with nuclei of flattened connective-tissue-corpuscles. Beneath this (between B and C) is the lymph-sinus or lymph-path, containing a reticulum coated by flat nucleated endothelial cells. C. Fine nucleated endothelial membrane, marking boundary of the lymph-follicle. The rest of the section from C to E is the adenoid tissue of the lymph-follicle, which consists of a fine reticulum, E. with numerous lymph-corpuscles, D. They are so closely packed that the adenoid reticulum is invisible till the section has been shaken so as to dislodge a number of the lymph-corpuscles. × 350. (Klein and Noble Smith.)

gular, the intercommunicating spaces being termed *alveoli;* whilst in the more central or medullary part a finer meshwork is formed by the more free anastomosis of the trabecular processes. In the alveoli of the cortex and in the meshwork formed by the trabeculæ in the medulla, is contained the proper gland structure. In the former it is arranged as follows (Fig. 215): occupying the central and chief part of each alveolus, is a more or less wedge-shaped mass (*l.h.*) of adenoid tissue, densely packed with lymph corpuscles; but at the periphery surrounding the central portion and immediately next the capsule and trabeculæ, is a more open meshwork of adenoid tissue constituting the *lymph sinus* or *channel* (*l.s.*), and containing fewer lymph corpuscles. The central mass is enclosed in endothelium, the cells of which join by their processes, the processes of the adenoid framework of the lymph sinus. The trabeculæ are also covered with endothelium. The lining of the central mass does not prevent the passage

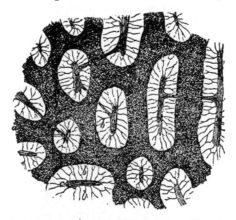

Fig. 214.—Section of medullary substance of an inguinal gland of an ox: *a, a*, glandular substance or pulp forming rounded cords joining in a continuous net (dark in the figure); *c, c*, trabeculæ; the space, *b, b*, between these and the glandular substance is the lymph-sinus, washed clear of corpuscles and traversed by filaments of retiform connective-tissue. × 90. (Kölliker.)

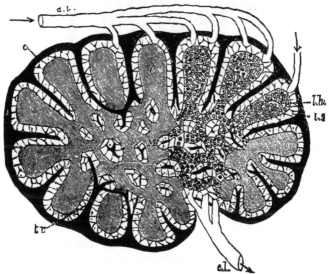

Fig. 215.—Diagrammatic section of Lymphatic gland. *a. l.*, Afferent; *e. l.*, efferent lymphatics; *C*, cortical substance; *l. h.*, reticulating cords of medullary substance; *l. s.*, lymph-sinus; *c.*, fibrous coat sending in trabeculæ; *t. r.*, into the substance of the gland. (Sharpey.)

of fluids and even of corpuscles into the lymph sinus. The framework of the adenoid tissue of the lymph sinus is nucleated, that of the central mass is non-nucleated. At the inner part of the alveolus, the wedge-shaped central mass bifurcates (Fig. 215) or divides into two or more smaller rounded or cord-like masses, and here joining with those from the other alveoli, form a much closer arrangement of the gland tissue (Fig. 214, *a*) than in the cortex; spaces (Fig. 214, *b*) are left within those anastomosing cords, in which are found portions of the trabecular mesh-work and the continuation of the lymph sinus (*b, c*).

The essential structure of lymphatic-gland substance resembles that which was described as existing, in a simple form, in the interior of the solitary and agminated intestinal follicles.

The lymph enters the gland by several afferent vessels (Fig. 215, *a.l.*) which open beneath the capsule into the lymph-channel or lymph-path;

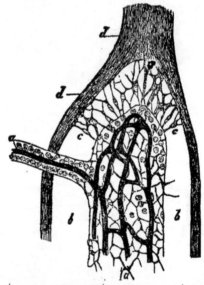

Fig. 216.—A small portion of medullary substance from a mesenteric gland of the ox. *d, d,* tra-beculæ; *a,* part of a cord of glandular substances from which all but a few of the lymph-corpuscles have been washed out to show its supporting meshwork of retiform tissue and its capillary blood-ves-sels (which have been injected, and are dark in the figure); *b, b,* lymph-sinus, of which the retiform tissue is represented only at *c, c.* × 300. (Kölliker.)

at the same time they lay aside all their coats except the endothelial lining, which is continuous with the lining of the lymph-path. The *efferent* vessels (Fig. 215, *e.l.*) begin in the medullary part of the gland, and are continuous with the lymph-path here as the afferent vessels were with the cortical portion; the endothelium of one is continuous with that of the other.

The efferent vessels leave the gland at the *hilus,* the more or less con-cave inner side of the gland, and generally either at once or very soon after join together to form a single vessel.

Blood-vessels which enter and leave the gland at the hilus are freely distributed to the trabecular tissue and to the gland-pulp (Fig. 216).

The tonsils, in part, and Peyer's glands of ·the intestine, are really
lymphatic glands, and doubtless discharge similar functions.

THE LYMPH AND CHYLE.

The *lymph*, contained in the lymphatic vessels, is, under ordinary cir--
cumstances, a clear, transparent, and yellowish fluid. It is devoid of
smell, is slightly alkaline, and has a saline taste. As seen with the
microscope in the small transparent vessels of the tail of the tadpole, it.
usually contains no corpuscles or particles of any kind; and it is only in
the larger trunks in which any corpuscles are to be found. These corpus·
cles are similar to colorless blood-corpuscles. The fluid in which the corpuscles float is albuminous, and contains no fatty particles or molecular base;
but is liable to variations according to the general state of the blood, and
to that of the organ from which the lymph is derived. As it advances.
toward the thoracic duct, and after passing through the lymphatic glands,.
it becomes spontaneously coagulable and the number of corpuscles is much
increased. The fluid contained in the lacteals is clear and transparent
during fasting, and differs in no respect from ordinary lymph; but, during
digestion, it becomes milky, and is termed *chyle*.

Chyle is an opaque, whitish, milky fluid, neutral or slightly alkaline
in reaction. Its whiteness and opacity are due to the presence of innumerable particles of oily or fatty matter, of exceedingly minute though
nearly uniform size, measuring on the average about $\frac{1}{30000}$ of an inch.
These constitute what is termed the *molecular base* of chyle. Their
number, and consequently the opacity of the chyle, are dependent upon
the quantity of fatty matter contained in the food. The fatty nature of
the molecules is made manifest by their solubility in ether, and, when
the ether evaporates, by their being deposited in various-sized drops of
oil. Each molecule probably consists of oil coated over with albumen, in
the manner in which oil always becomes covered when set free in minute
drops in an albuminous solution. This is proved when water or dilute
acetic acid is added to chyle, many of the molecules are lost sight of, and
oil-drops appear in their place, as the investments of the molecules have
been dissolved, and their oily contents have run together.

Except these molecules, the chyle taken from the villi or from lacteals.
near them, contains no other solid or organized bodies. The fluid in
which the molecules float is albuminous, and does not spontaneously
coagulate. But as the chyle passes on toward the thoracic duct, and
especially, while it traverses one or more of the mesenteric glands, it is
elaborated. The quantity of molecules and oily particles gradually diminishes; cells, to which the name of *chyle-corpuscles* is given, are developed in it; and it acquires the property of coagulating spontaneously.
The higher in the thoracic duct the chyle advances, the more is it, in all

these respects, developed; the greater is the number of chyle-corpuscles, and the larger and firmer is the clot which forms in it when withdrawn and left at rest. Such a clot is like one of blood without the red corpuscles, having the chyle corpuscles entangled in it, and the fatty matter forming a white creamy film on the surface of the serum. But the clot of chyle is softer and moister than that of blood. Like blood, also, the chyle often remains for a long time in its vessels without coagulating, but coagulates rapidly on being removed from them. The existence of the materials which, by their union, form fibrin, is, therefore, certain; and their increase appears to be commensurate with that of the corpuscles.

The structure of the chyle-corpuscles was described when speaking of the white corpuscles of the blood, with which they are identical.

Chemical Composition of Lymph and Chyle.—From what has been said, it will appear that perfect chyle and lymph are, in essential characters, nearly similar, and scarcely differ, except in the preponderance of fatty and proteid matter in the chyle.

CHEMICAL COMPOSITION OF LYMPH AND CHYLE.   (Owen Rees.)

	I. Lymph (Donkey).	II. Chyle (Donkey).	III. Mixed Lymph & Chyle (Human).
Water	96·536	90·237	90·48
Solids	3·454	9·763	9·52
Solids—			
Proteids, including Serum- Albumin, Fibrin, and Globulin . . .	1·320	3·886	7·08
Extractives, including in (I and I) Sugar, Urea, Leu- cin and Cholesterin .	1.559	1·565	·108
Fatty matter . . .	a trace	3·601	·92
Salts	·585	·711	·44

From the above analyses of lymph and chyle, it appears that they contain essentially the same constituents that are found in the blood. Their composition, indeed, differs from that of the blood in degree rather than in kind. They do not, however, unless by accident, contain colored corpuscles.

Quantity.—The quantity which would pass into a cat's blood in twenty-four hours has been estimated to be equal to about one-sixth of the weight of the whole body. And, since the estimated weight of the blood in cats is to the weight of their bodies as 1·7, the quantity of lymph daily traversing the thoracic duct would appear to be about equal to the quantity of blood at any time contained in the animals. By another series of experiments, the quantity of lymph traversing the thoracic duct of a dog in twenty-four hours was found to be about equal to two-thirds of the blood in the body. (Bidder and Schmidt.)

Absorption by the Lacteals.—During the passage of the chyme along the whole tract of the intestinal canal, its completely digested parts are absorbed by the blood-vessels and lacteals distributed in the mucous membrane. The blood-vessels appear to absorb chiefly the dissolved portions of the food, and these, including especially the albuminous and saccharine, they imbibe without choice; whatever can mix with the blood passes into the vessels, as will be presently described. But the lacteals appear to absorb only certain constituents of the food, including particularly the fatty portions. The absorption by both sets of vessels is carried on most actively but not exclusively, in the villi of the small intestine; for in these minute processes, both the capillary blood-vessels and the lacteals are brought almost into contact with the intestinal contents. There seems to be no doubt that absorption of fatty matters during digestion, from the contents of the intestines, is effected chiefly between the epithelial cells which line the intestinal tract (Watney), and especially those which clothe the surface of the villi. Thence, the fatty particles are passed on into the interior of the lacteal vessels (Fig. 216, *a*), but how they pass, and what laws govern their so doing, are not at present exactly known.

The process of absorption is assisted by the pressure exercised on the contents of the intestines by their contractile walls; and the absorption of fatty particles is also facilitated by the presence of the bile, and the pancreatic and intestinal secretions, which moisten the absorbing surface. For it has been found by experiment, that the passage of oil through an animal membrane is made much easier when the latter is impregnated with an alkaline fluid.

Absorption by the Lymphatics.—The real source of the lymph, and the mode in which its absorption is effected by the lymphatic vessels, were long matters of discussion. But the problem has been much simplified by more accurate knowledge of the anatomical relations of the lymphatic capillaries. The lymph is, without doubt, identical in great part with the *liquor sanguinis*, which, as before remarked, is always exuding from the blood-capillaries into the interstices of the tissues in which they lie; and as these interstices form in most parts of the body the beginnings of the lymphatics, the source of the lymph is sufficiently obvious. In connection with this may be mentioned the fact that changes in the character of the lymph correspond very closely with changes in the character of either the whole mass of blood, or of that in the vessels of the part from which the lymph is exuded. Thus it appears that the coagulability of the lymph is directly proportionate to that of the blood; and that when fluids are injected into the blood-vessels in sufficient quantity to distend them, the injected substance may be almost directly afterward found in the lymphatics.

Some other matters than those originally contained in the exuded liquor sanguinis may, however, find their way with it into the lymphatic vessels. Parts which having entered into the composition of a tissue, and, having fulfilled their purpose, require to be removed, may not be altogether excrementitious, but may admit of being reorganized and adapted again for nutrition; and these may be absorbed by the lymphatics, and elaborated with the other contents of the lymph in passing through the glands.

Lymph-Hearts.—In reptiles and some birds, an important auxiliary to the movement of the lymph and chyle is supplied in certain muscular sacs, named *lymph-hearts* (Fig. 217), and it has been shown that the caudal heart of the eel is a lymph-heart also. The number and position of these organs vary. In frogs and toads there are usually four, two anterior and two posterior; in the frog, the posterior lymph-heart on each side is situated in the ischiatic region, just beneath the skin; the anterior lies deeper, just over the transverse process of the third vertebra. Into each of these cavities several lymphatics open, the orifices of the vessels being guarded by valves, which prevent the retrograde passage of the

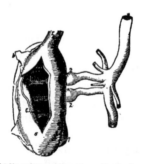

FIG. 217.—Lymphatic heart (9 lines long, 4 lines broad) of a large species of serpent, the Python bivittatus. 4. The external cellular coat. 5. The thick muscular coat. Four muscular columns run across its cavity, which communicates with three lymphatics (1—only one is seen here), and with two veins (2, 2). 6. The smooth lining membrane of the cavity. 7. A small appendage, or auricle, the cavity of which is continuous with that of the rest of the organ (after E. Weber).

lymph. From each heart a single vein proceeds and conveys the lymph directly into the venous system. In the frog, the inferior lymphatic heart, on each side, pours its lymph into a branch of the ischiatic vein; by the superior, the lymph is forced into a branch of the jugular vein, which issues from its anterior surface, and which becomes turgid each time that the sac contracts. Blood is prevented from passing from the vein into the lymphatic heart by a valve at its orifice.

The muscular coat of these hearts is of variable thickness; in some cases it can only be discovered by means of the microscope; but in every case it is composed of striped fibres. The contractions of the heart are rhythmical, occurring about sixty times in a minute, slowly, and, in comparison with those of the blood-hearts, feebly. The pulsations of the cervical pair are not always synchronous with those of the pair in the ischiatic region, and even the corresponding sacs of opposite sides are not always synchronous in their action.

Unlike the contractions of the blood-heart, those of the lymph-heart appear to be directly dependent upon a certain limited portion of the spinal cord. For Volkmann found that so long as the portion of spinal cord corresponding to the third vertebra of the frog was uninjured, the cervical pair of lymphatic hearts continued pulsating after all the rest of the spinal cord and the brain were destroyed; while destruction of this portion, even though all other parts of the nervous centres were uninjured, instantly arrested the heart's movements. The posterior, or ischiatic, pair of lymph-hearts were found to be governed, in like manner, by the portion of spinal cord corresponding to the eighth vertebra. Division of the posterior spinal roots did not arrest the movements; but division of the anterior roots caused them to cease at once.

Absorption by Blood-vessels.—In the absorption by the lymphatic or lacteal vessels just described, there appears something like the exercise of choice in the materials admitted into them. But the absorption by blood-vessels presents no such appearance of selection of materials; rather, it appears, that every substance, whether gaseous, liquid, or a soluble, or minutely divided solid, may be absorbed by the blood-vessels, provided it is capable of permeating their walls, and of mixing with the blood; and that of all such substances, the mode and measure of absorption are determined solely by their physical or chemical properties and conditions, and by those of the blood and the walls of the blood-vessels.

Osmosis.—The phenomena are, indeed, to a great extent, comparable to that passage of fluids through membrane, which occurs quite independently of vital conditions, and the earliest and best scientific investigation of which was made by Dutrochet. The instrument which he employed in his experiments was named an *endosmometer*. It may consist of a graduated tube expanded into an open-mouthed bell at one end, over which a portion of membrane is tied (Fig. 218). If now the bell be filled with a solution of a salt— say sodium chloride, and be immersed in water, the water will pass into the solution, and part of the salt will pass out into the water; the water, however, will pass into the solution much more rapidly than the salt will pass out into the water, and the diluted solution will rise in the tube. To this passage of fluids through membrane the term *Osmosis* is applied.

Fig. 218. Endosmometer.

The nature of the membrane used as a septum, and its affinity for the fluids subjected to experiment, have an important influence, as might be anticipated, on the rapidity and duration of the osmotic current. Thus, if a piece of ordinary bladder be used as the septum between water and alcohol, the current is almost solely from the water to the alcohol, on account of the much greater affinity of water for this kind of membrane; while, on the other hand, in the case of a membrane of caoutchouc, the alcohol, from its greater affinity for this substance, would pass freely into the water.

Osmosis by Blood-vessels.—Absorption by blood-vessels is the consequence of their walls being, like the membranous septum of the endosmometer, porous and capable of imbibing fluids, and of the blood being so composed that most fluids will mingle with it. The process of absorption, in an instructive, though very imperfect degree, may be observed in any portion of vascular tissue removed from the body. If such a one be placed in a vessel of water, it will shortly swell, and become heavier and moister, through the quantity of water imbibed or soaked into it; and if now, the blood contained in any of its vessels be let out, it will be found diluted with water, which has been absorbed by the blood-vessels and mingled with the blood. The water round the piece of tissue also will become blood-stained; and if all be kept at perfect rest, the stain derived from the solution of the coloring matter of the blood (together with which chemistry would detect some of the albumen and other parts of the liquor sanguinis) will spread more widely every day. The same will happen if the piece of tissue be placed in a saline solution instead of water, or in a solution of coloring or odorous matter, either of which will give their tinge or smell to the blood, and receive, in exchange, the color of the blood.

Colloids and Crystalloids.—Various substances have been classified according to the degree in which they possess the property of passing, when in a state of solution in water, through membrane; those which pass freely, inasmuch as they are usually capable of crystallization, being termed *crystalloids,* and those which pass with difficulty, on account of their, physically, glue-like characters, *colloids.* (Graham.)

This distinction, however, between colloids and crystalloids, which is made the basis of their classification, is by no means the only difference between them. The colloids, besides the absence of power to assume a crystalline form, are characterized by their inertness as acids or bases, and feebleness in all ordinary chemical relations. Examples of them are found in albumin, gelatin, starch, hydrated alumina, hydrated silicic acid, etc.; while the crystalloids are characterized by qualities the reverse of those just mentioned as belonging to colloids. Alcohol, sugar, and ordinary saline substances are examples of crystalloids.

Rapidity of Absorption.—The rapidity with which matters may be absorbed from the stomach, probably by the blood-vessels chiefly, and diffused through the textures of the body, may be gathered from the history of some experiments. From these it appears that even in a quarter of an hour after being given on an empty stomach, lithium chloride may be diffused into all the vascular textures of the body, and into some of the non-vascular, as the cartilage of the hip-joint, as well as into the aqueous humor of the eye. Into the outer part of the crystalline lens it may pass after a time, varying from half an hour to an hour and a half. Lithium carbonate, when taken in five or ten-grain doses on an empty stomach, may be detected in the urine in 5 or 10 minutes; or, if the stomach be full at the time of taking the dose, in 20 minutes. It may sometimes be detected in the urine, moreover, for six, seven, or eight days. (Bence Jones.)

Some experiments on the absorption of various mineral and vegetable poisons, have brought to light the singular fact, that, in some cases, absorption takes place more rapidly from the rectum than from the stomach. Strychnia, for example, when in solution, produces its poisonous effects much more speedily when introduced into the rectum than into the stomach. When introduced in the solid form, however, it is absorbed more rapidly from the stomach than from the rectum, doubtless because of the greater solvent property of the secretion of the former than of that of the latter. (Savory.)

With regard to the degree of absorption by living blood-vessels, much depends on the facility with which the substance to be absorbed can penetrate the membrane or tissue which lies between it and the blood-vessels. Thus, absorption will hardly take place through the epidermis, but is quick when the epidermis is removed, and the same vessels are covered with only the surface of the cutis, or with granulations. In general, the absorption through membranes is in an inverse proportion to the thickness of their epithelia; so that the urinary bladder of a frog is traversed in less than a second; and the absorption of poisons by the stomach or lungs appears sometimes accomplished in an immeasurably small time.

Conditions for Absorption.—1. The substance to be absorbed must, as a general rule, be in the liquid or gaseous state, or, if a solid, must be soluble in the fluids with which it is brought in contact. Hence the marks of tattooing, and the discoloration produced by silver nitrate taken internally, remain. Mercury may be absorbed even in the metallic state; and in that state may pass into and remain in the blood-vessels, or be deposited from them; and such substances as exceedingly finely-divided charcoal, when taken into the alimentary canal, have been found in the mesenteric veins; the insoluble materials of ointments may also be rubbed into the blood-vessels; but there are no facts to determine how these various substances effect their passage. Oil, minutely divided, as in an emulsion, will pass slowly into blood-vessels, as it will through a filter moistened with water; and, without doubt, fatty matters find their way into the blood-vessels as well as the lymph-vessels of the intestinal canal, although the latter seem to be specially intended for their absorption.

2. The less dense the fluid to be absorbed, the more speedy, as a general rule, is its absorption by the living blood-vessels. Hence the rapid absorption of water from the stomach; also of weak saline solutions; but with strong solutions, there appears less absorption into, than effusion from, the blood-vessels.

3. The absorption is the less rapid the fuller and tenser the blood-vessels are; and the tension may be so great as to hinder altogether the entrance of more fluid. Thus, if water is injected into a dog's veins to repletion, poison is absorbed very slowly; but when the tension of the vessels is diminished by bleeding, the poison acts quickly. So, when cupping-glasses are placed over a poisoned wound, they retard the absorption of the poison not only by diminishing the velocity of the circulation in the part, but by filling all its vessels too full to admit more.

On the same ground, absorption is the quicker the more rapid the circulation of the blood; not because the fluid to be absorbed is more quickly imbibed into the tissues, or mingled with the blood, but because as fast as it enters the blood, it is carried away from the part, and the blood being constantly renewed, is constantly as fit as at the first for the reception of the substance to be absorbed.

CHAPTER X.

ANIMAL HEAT.

THE Average Temperature of the human body in those *internal* parts which are most easily accessible, as the mouth and rectum, is from 98·5° to 99·5° F. (36·9°—37·4° C.). In different parts of the *external* surface of the human body the temperature varies only to the extent of two or three degrees (F.), when all are alike protected from cooling influences; and the difference which under these circumstances exists, depends chiefly upon the different degrees of blood-supply. In the arm-pit—the most convenient situation, under ordinary circumstances, for examination by the thermometer—the average temperature is 98·6° F. (36·9° C.). In different internal parts, the variation is one or two degrees; those parts and organs being warmest which contain most blood, and in which there occurs the greatest amount of chemical change, *e.g.*, the glands and the muscles; and the temperature is highest, of course, when they are most actively working: while those tissues which, subserving only a mechanical function, are the seat of least active circulation and chemical change, are the coolest. These differences of temperature, however, are actually but slight, on account of the provisions which exist for maintaining uniformity of temperature in different parts.

Circumstances causing Variations in Temperature.—The chief circumstances by which the temperature of the healthy body is influenced are the following:—*Age; Sex; Period of the day; Exercise; Climate and Season; Food and Drink.*

Age.—The average temperature of the new-born child is only about 1° F. (·54° C.) above that proper to the adult; and the difference becomes still more trifling during infancy and early childhood. The temperature falls to the extent of about ·2°—·5° F. from early infancy to puberty, and by about the same amount from puberty to fifty or sixty years of age. In old age the temperature again rises, and approaches that of infancy; but although this is the case, yet the power of resisting cold is less in them— exposure to a low temperature causing a greater reduction of heat than in young persons.

The same rapid diminution of temperature has been observed to occur in the new-born young of most carnivorous and rodent animals when they are removed from the parent, the temperature of the atmosphere being

between 50° and 53·5° F. (10°–12° C.); whereas while lying close to the
body of the mother, their temperature is only 2 or 3 degrees F. lower
than hers.   The same law applies to the young of birds.

Sex.—The average temperature of the female would appear to be very
slightly higher than that of the male.

Period of the Day.—The temperature undergoes a gradual alteration,
to the extent of about 1° to 1·5° F. (·54—·8° C.) in the course of the day
and night; the *minimum* being at night or in the early morning, the
maximum late in the afternoon.

Exercise.—*Active exercise* raises the temperature of the body from 1°
to 2° F. (·54°— 1.08° C.).   This may be partly ascribed to generally increased combustion-processes, and partly to the fact, that every muscular
contraction is attended by the development of one or two degrees of heat
in the acting muscle; and that the heat is increased according to the
number and rapidity of these contractions, and is quickly diffused by the
blood circulating from the heated muscles.   Possibly, also, some heat
may be generated in the various movements, stretchings, and recoilings
of the other tissues, as the arteries, whose elastic walls, alternately dilated
and contracted, may give out some heat, just as caoutchouc alternately
stretched and recoiling becomes hot.   But the heat thus developed cannot
be great.   The great apparent increase of heat during exercise depends,
in a great measure, on the increased circulation and quantity of blood,
and, therefore, greater heat, in parts of the body (as the skin, and especially the skin of the extremities), which, at the same time that they feel
more acutely than others any changes of temperature, are, under ordinary conditions, by some degrees colder than organs more centrally
situated.

Climate and Season.—The temperature of the human body is the same
in temperate and tropical climates.   (Johnson, Boileau, Furnell.)   In
summer the temperature of the body is a little higher than in winter; the
difference amounting to about a third of a degree F.   (Wunderlich.)

Food and Drink.   The effect of a meal upon the temperature of a body
is but small.   A very slight rise usually occurs.   Cold alcoholic drinks
depress the temperature somewhat (·5° to 1° F.).   Warm alcoholic
drinks, as well as warm tea and coffee, raise the temperature (about ·5° F.).

In disease the temperature of the body deviates from the normal standard to a greater extent than would be anticipated from the slight effect
of external conditions during health.   Thus, in some diseases, as pneu-
monia and typhus, it occasionally rises as high as 106° or 107° F. (41°.—
41·6° C.); and considerably higher temperatures have been noted.   In
Asiatic cholera, on the other hand, a thermometer placed in the mouth
may sometimes rise only to 77° or 79° F. (25°—26·2° C.).

The temperature maintained by Mammalia in an active state of life,

according to the tables of Tiedemann and Rudolphi, averages 101°
(38.3° C.). The extremes recorded by them were 96° and 106°, the former
in the narwhal, the latter in a bat (Vespertilio pipistrella). In Birds, the
average is as high as 107° (41·2° C.); the highest temperature, 111·25°
(46·2° C.);· being in the small species, the linnets, etc. Among Reptiles,
while the medium they were in was 75° (23·9° C.) their average temperature was 82·5° (31·2° C.). As a general rule, their temperature, though
it falls with that of the surrounding medium, is, ·in temperate media, two
or more degrees higher; and though it rises also with that of the medium,
yet at very high degrees it ceases to do so, and remains even lower than
that of the medium. Fish and invertebrata present, as a general rule, the
same temperature as the medium in which they live, whether that be high
or low; only among fish, the tunny tribe, with strong hearts and red
meat-like muscles, and more blood than the average of fish have, are
generally 7° (3·8° C.) warmer than the water around them.

The difference, therefore, between what are commonly called the warm
and the cold-blooded animals, is not one of absolutely higher or lower
temperature; for the animals which to us in a temperate climate feel cold
(being like the air or water, colder than the surface of our bodies), would
in an external temperature of 100° (37·8° C.) have nearly the same temperature and feel hot to us. The real difference is that what we call
warm-blooded animals (Birds and Mammalia), have a certain "permanent
heat in all atmospheres," while the temperature of the others, which we
call cold-blooded, is "variable with every atmosphere." (Hunter.)

The power of maintaining a uniform temperature, which Mammalia
and Birds possess, is combined with the want of power to endure such
changes of body temperature as are harmless to the other classes; and
when their power of resisting change of temperature ceases, they suffer
serious disturbance or die.

Sources and Mode of Production of Heat in the Body.—

The heat which is produced in the body arises from combustion, and is
due to the fact that the oxygen of the atmosphere taken into the system
is combined with the carbon and hydrogen of the tissues. Any changes
which occur in the protoplasm of the tissues, resulting in an exhibition
of their function, is attended by the evolution of heat and also by the production of carbonic acid and water; and the more active the changes,
the greater the heat produced and the greater the amount of the carbonic
acid and water formed. But in order that the protoplasm may perform
its function, the waste of its own tissue (destructive metabolism), must
be repaired by the supply of food material, and therefore for the production of heat it is necessary to supply food. In the tissues, therefore,
two processes are continually going on: the building up of the protoplasm
from the food (constructive metabolism), which is not accompanied by
the evolution of heat but possibly by the reverse, and the oxidation of the
protoplastic materials, resulting in the production of energy, by which
heat is produced and carbonic acid and water are evolved. Some heat
will also be generated in the combination of sulphur and phosphorus with
oxygen, but the amount. thus produced is but small.

It is not necessary to assume that the combustion processes, which ultimately issue in the production of carbonic acid and water, are as simple as the bare statement of the fact might seem to indicate. But complicated as the various stages of combustion may be, the ultimate result is as simple as in ordinary combustion outside the body, and the products are the same. The same amount of heat will be evolved in the union of any given quantities of carbon and oxygen, and of hydrogen and oxygen, whether the combination be rapid and direct, as in ordinary combustion, or slow and almost imperceptible, as in the changes which occur in the living body. And since the heat thus arising will be distributed wherever the blood is carried, every part of the body will be heated equally, or nearly so.

This theory, that the maintenance of the temperature of the living body depends on continual chemical change, chiefly by oxidation, of combustible materials existing in the tissues, has long been established by the demonstration that the quantity of carbon and hydrogen which, in a given time, unites in the body with oxygen, is sufficient to account for the amount of heat generated in the animal within the same time: an amount capable of maintaining the temperature of the body at from 98° —100° F. (36·8°—37·8° C.), notwithstanding a large loss by radiation and evaporation.

It should be remembered that heat may be introduced into the body by means of warm drinks and foods, and, again, that it is possible for the preliminary digestive changes to be accompanied by the evolution of heat.

Chief Heat-producing Tissues.—The chemical changes, which produce the body-heat appear to be especially active in certain tissues:— (1), In the *Muscles*, which form so large a part of the organism. The fact that the manifestation of muscular energy is always attended by the evolution of heat and the production of carbonic acid has been demonstrated by actual experiment; and when not actually in a condition of active contraction, a metabolism, not so active but still actual, goes on, which is accompanied by the manifestation of heat. The total amount set free by the muscles, therefore, must be very great; and it has been calculated that even neglecting the heat produced by the quiet metabolism of muscular tissue, the amount of heat generated by muscular activity supplies the principal part of the total heat produced within the body. (2), In the *Secreting glands*, and principally in the liver as being the largest and most active. It has been found by experiment that the blood leaving the glands is considerably warmer than that entering them. The metabolism in the glands is very active, and, as we have seen, the more active the metabolism the greater the heat produced. (3), In *the Brain;* the venous blood having a higher temperature than the arterial. It must be remembered, however, that although the organs above mentioned are the chief heat-producing parts of the body, all living tissues contribute their quota, and this in direct proportion to their activity. The blood itself is also the seat of metabolism, and, therefore, of the production of heat; but the share which it takes in this respect, apart from the tissues in which it circulates, is very inconsiderable.

- **Regulation of the Temperature of the Human Body.**—The average temperature of the body is maintained under different conditions of external circumstances by mechanisms which permit of (1) variation in the amount of heat got rid of, and (2) variations in the amount of heat produced or introduced into the body. In healthy warm-blooded animals the loss and gain of heat are so nearly balanced one by the other that, under all ordinary circumstances, a uniform temperature, within two or three degrees, is preserved.

I. Methods of Variation in the amount of Heat got rid of.—The loss of heat from the human body is principally regulated by the amount lost by radiation and conduction from its surface, and by means of the constant evaporation of water from the same part, and (2) to a much less degree from the air-passages; in each act of respiration, heat is lost to a greater or less extent according to the temperature of the atmosphere; unless indeed the temperature of the surrounding air exceed that of the blood. We must remember too that all food and drink which enter the body at a lower temperature than itself abstract a small measure of heat: while the urine and fæces which leave the body at about its own temperature are also means by which a small amount is lost.

(a.) *Loss of Heat from the Surface of the Body: the Skin.*—By far the most important loss of heat from the body,—probably 70 or 80 per cent. of the whole amount, is that which takes place by radiation, conduction, and evaporation from the skin. The means by which the skin is able to act as one of the most important organs for regulating the temperature of the blood, are—(1), that it offers a large surface for radiation, conduction, and evaporation; (2), that it contains a large amount of blood; (3), that the quantity of blood contained in it is the greater under those circumstances which demand a loss of heat from the body, and *vice versâ*. For the circumstance which directly determines the quantity of blood in the skin, is that which governs the supply of blood to all the tissues and organs of the body, namely, the power of the vaso-motor nerves to cause a greater or less tension of the muscular element in the walls of the arteries, and, in correspondence with this, a lessening or increase of the calibre of the vessels, accompanied by a less or greater current of blood. A warm or hot atmosphere so acts on the nerve fibres of the skin, as to lead them to cause in turn a relaxation of the muscular fibre of the blood-vessels; and, as a result, the skin becomes full-blooded, hot, and sweating; and much heat is lost. With a low temperature, on the other hand, the blood-vessels shrink, and in accordance with the consequently diminished blood-supply, the skin becomes pale, and cold, and dry; and no doubt a

similar effect may be produced through the vaso-motor centre in the
medulla and spinal cord.   Thus, by means of a self-regulating apparatus,
the skin becomes the most important of the means by which the temperature of the body is regulated.

In connection with loss of heat by the skin, reference has been made
to that which occurs both by radiation and conduction, and by evaporation; and the subject of animal heat has been considered almost solely
with regard to the ordinary case of man living in a medium colder than
his body, and therefore losing heat in all the ways mentioned.   The importance of the means, however, adopted, so to speak, by the skin for regulating the temperature of the body, will depend on the conditions by
which it is surrounded; an inverse proportion existing in most cases between the loss by radiation and conduction on the one hand, and by
evaporation on the other.   Indeed, the small loss of heat by evaporation
in cold climates may go far to compensate for the greater loss by radiation; as, on the other hand, the great amount of fluid evaporated in hot
air may remove nearly as much heat as is commonly lost by both radiation and evaporation in ordinary temperatures; and thus, it is possible
that the quantities of heat required for the maintenance of a uniform
proper temperature in various climates and seasons are not so different
as they, at first thought, seem.

Many examples may be given of *the power which the body possesses of
resisting the effects of a high temperature,* in virtue of evaporation from
the skin.   Blagden and others supported a temperature varying between
198°—211° F. (92°—100° C.) in dry air for several minutes; and in a
subsequent experiment he remained eight minutes in a temperature of
260° F. (126·5° C.).   "The workmen of Sir F. Chantrey were accustomed
to enter a furnace, in which his moulds were dried, whilst the floor was
red-hot and a thermometer in the air stood at 350° F. (177·8° C.); and
Chabert, the fire-king, was in the habit of entering an oven the temperature of which was from 400° to 600° F." (205°—315° C.)   (Carpenter.)
But such heats are not tolerable when the air is moist as well as hot,
so as to prevent evaporation from the body.   C. James states, that in the
vapor baths of Nero he was almost suffocated in a temperature of 112° F.
(44·5° C.), while in the caves of Testaccio, in which the air is dry, he
was but little incommoded by a temperature of 176° F. (80° C.).   In
the former, evaporation from the skin was impossible; in the latter it was
abundant, and the layer of vapor which would rise from all the surface
of the body would, by its very slowly conducting power, defend it for a
time from the full action of the external heat.

(The glandular apparatus, by which secretion of fluid from the skin is
effected, will be considered in the Section on the Skin.)

The ways by which the skin may be rendered more efficient as a cool-
ing-apparatus, by exposure, by baths, and by other means which man
instinctively adopts for lowering his temperature when necessary, are too
well known to need more than to be mentioned.

Although under any ordinary circumstances, the external application of cold only temporarily depresses the temperature to a slight extent, it is otherwise in cases of high temperature in fever. In these cases a tepid bath may reduce the temperature several degrees, and the effect so produced lasts· in some cases for many hours.

(b.) *Loss of Heat from the Lungs.*—As a means for lowering the temperature, the lungs and air-passages are very inferior to the skin; although, by giving heat to the air we breathe, they stand next to the skin in importance. As a *regulating* power, the inferiority is still more marked. The air which is expelled from the lungs leaves the body at about the temperature of the blood, and is always saturated with moisture. No inverse proportion, therefore, exists between the loss of heat by radiation and conduction on the one hand, and by evaporation on the other. The colder the air, for example, the greater will be the loss in all ways. Neither is the quantity of blood which is exposed to the cooling influence of the air diminished or increased, so far as is known, in accordance with any need in relation to temperature. It is true that by varying the number and depth of the respirations, the quantity of heat given off by the lungs may be made, to some extent, to vary also. But the respiratory passages, while they must be considered important means by which heat is lost, are altogether subordinate, in the power of regulating the temperature, to the skin.

(c.) *By Clothing.*—The influence of external coverings for the body must not be unnoticed. In warm-blooded animals, they are always adapted, among other purposes, to the maintenance of uniform temperature; and man adapts for himself such as are, for the same purpose, fitted to the various climates to which he is exposed. By their means, and by his command over food and fire, he maintains his temperature on all accessible parts of the surface of the earth.

II. **Methods of Variation in the amount of Heat produced.** —It may seem to have been assumed, in the foregoing pages. that the only regulating apparatus for temperature required by the human body is one that shall, more or less, produce a *cooling* effect; and as if the amount of heat produced were always, therefore, in excess of that which is required. Such an assumption would be incorrect. We have the power of regulating the production of heat, as well as its loss.

(a.) *By Regulating the Quantity and Quality of the Food taken.*—In food we have a means for elevating our temperature. It is the fuel, indeed, on which animal heat ultimately depends altogether. Thus, when more heat is wanted, we instinctively take more food, and take such kinds of it as are good for combustion; while every-day experience shows the different power of resisting cold possessed, respectively, by the well-fed and by the starved. In northern regions, again, and in the colder seasons of more southern climes, the quantity of food consumed is (speaking very generally) greater than that consumed by the same men or animals in opposite conditions of climate and season. And the food. which appears naturally adapted to the inhabitants of the coldest climates, such as the several fatty and oily substances, abounds in carbon and hydrogen, and is fitted to combine with the large quantities of oxygen which, breathing cold dense air, they absorb from their lungs.

(b.) *By Exercise.*—In exercise, we have an important means of raising the temperature of our bodies (p. 310).

(c.) *By Influence of the Nervous System.*—The influence of the nervous system in modifying the production of heat must be very important, as upon nervous influence depends the amount of the metabolism of the tissues. The experiments and observations which best illustrate it are those showing, first, that when the supply of nervous influence to a part is cut off, the temperature of that part falls below its ordinary degree; and, secondly, that when death is caused by severe injury to, or removal of, the nervous centres, the temperature of the body rapidly falls, even though artificial respiration be performed, the circulation maintained, and to all appearance the ordinary chemical changes of the body be completely effected. It has been repeatedly noticed, that after division of the nerves of a limb its temperature falls; and this diminution of heat has been remarked still more plainly in limbs deprived of nervous influence by paralysis.

With equal certainty, though less definitely, the influence of the nervous system on the production of heat, is shown in the rapid and momentary increase of temperature, sometimes general, at other times quite local, which is observed in states of nervous excitement; in the general increase of warmth of the body, sometimes amounting to perspiration, which is excited by passions of the mind; in the sudden rush of heat to the face, which is not a mere sensation; and in the equally rapid diminution of temperature in the depressing passions. But none of these instances suffice to prove that heat is generated by mere nervous action, independent of any chemical change; all are explicable, on the supposition that the nervous system alters, by its power of controlling the calibre of the blood-vessels, the quantity of blood supplied to a part; while any influence which the nervous system may have in the production of heat, apart from this influence on the blood-vessels, is an indirect one, and is derived from its power of causing such nutritive change in the tissues as may, by involving the necessity of chemical action, involve the production of heat.

Inhibitory heat-centre.—Whether a centre exists which regulates the production of heat in warm-blooded animals, is still undecided. Experiments have shown that exposure to cold at once increases the oxygen taken in, and the carbonic acid given out, indicating an increase in the activity of the metabolism of the tissues, but that in animals poisoned by urari, exposure to cold diminishes both the metabolism and the temperature, and warm-blooded animals then re-act to variations of the external temperature just in the same way as cold-blooded. These experiments seem to suggest that there is a centre, to which, under normal circumstances, the impression of cold is conveyed, and from which by efferent nerves impulses pass to the muscles, whereby an increased metabolism is induced, and so an increased amount of heat is generated. The centre is probably situated above the medulla. Thus in urarized animals, as the nerves to the muscles, the metabolism of which is so important in the production of heat, are paralyzed, efferent impulses from the centre cannot induce the necessary metabolism for the production of heat, even though afferent impulses from the skin, stimulated by the alteration of temperature, have conveyed to it the necessity of altering the amount of heat to be produced. The same effect is produced when the medulla is cut.

Influence of Extreme Heat and Cold.—In connection with the regulation of animal temperature, and its maintenance in health at the normal height, may be noted the result of circumstances too powerful, either in raising or lowering the heat of the body, to be controlled by the proper regulating apparatus. Walther found that rabbits and dogs, when tied to a board and exposed to a hot sun, reached a temperature of 114·8° F., and then died. Cases of sunstroke furnish us with several examples in the case of man; for it would seem that here death ensues chiefly or solely from elevation of the temperature. In many febrile diseases the immediate cause of death appears to be the elevation of the temperature to a point inconsistent with the continuance of life.

The effect of mere loss of bodily temperature in man is less well known than the effect of heat. From experiments by Walther, it appears that rabbits can be cooled down to 48° F. (8.9° C.), before they die, if artificial respiration be kept up. Cooled down to 64° F. (17.8° C.), they cannot recover unless external warmth be applied together with the employment of artificial respiration. Rabbits not cooled below 77° F. (25° C.) recover by external warmth alone.

CHAPTER XI.

Secretion is the process by which materials are separated from the blood, and from the organs in which they are formed, for the purpose either of serving some ulterior office in the economy, or of being discharged from the body as useless or injurious. In the former case, the separated materials are termed *secretions;* in the latter, they are termed *excretions.*

Most of the secretions consist of substances which, probably, do not pre-exist in the same form in the blood, but require special organs and a process of elaboration for their formation, *e.g.*, the liver for the formation of bile, the mammary gland for the formation of milk. The excretions, on the other hand, commonly or chiefly consist of substances which exist ready-formed in the blood, and are merely abstracted therefrom. If from any cause, such as extensive disease or extirpation of an excretory organ, the separation of an excretion is prevented, and an accumulation of it in the blood ensues, it frequently escapes through other organs, and may be detected in various fluids of the body. But this is never the case with secretions; at least with those that are most elaborated; for after the removal of the special organs by which any of them is elaborated, it is no longer formed. Cases sometimes occur in which the secretion continues to be formed by the natural organ, but not being able to escape toward the exterior, on account of some obstruction, is re-absorbed into the blood, and afterward discharged from it by exudation in other ways; but these are not instances of true vicarious secretion, and must not be thus regarded.

These circumstances, and their final destination, are, however, the only particulars in which secretions and excretions can be distinguished; for, in general, the structure of the parts engaged in eliminating excretions is as complex as that of the parts concerned in the formation of secretions. And since the differences of the two processes of separation, corresponding with those in the several purposes and destinations of the fluids, are not yet ascertained, it will be sufficient to speak in general terms of the process of separation or secretion.

Every secreting apparatus possesses, as essential parts of its structure, a simple and almost textureless membrane, named the *primary* or *base-*

ment-membrane; certain *cells;* and *blood-vessels.* These three structural elements are arranged together in various ways; but all the varieties ma**y** be classed under one or other of two principal divisions, namely, *membranes* and *glands.*

ORGANS AND TISSUES OF SECRETION.

The principal secreting membranes are (1) the Serous and Synovial membranes; (2) the Mucous membranes; (3) the Mammary gland; (4) the Lachrymal gland; and (5) the Skin.

(1) **Serous Membranes.**—The serous membranes are especially distinguished by the characters of the endothelium covering their free sur-

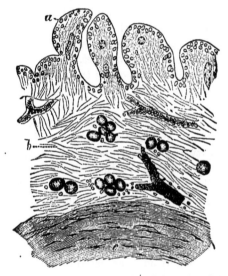

FIG. 219.—Section of synovial membrane. *a*, endothelial covering of elevations of the membrane; *b*, subserous tissue containing fat and blood-vessels; *c*, ligament covered by the synovial membrane. (Cadiat.)

face: it always consists of a single layer of polygonal cells. The ground substance of most serous membranes consists of connective-tissue corpuscles of various forms lying in the branching spaces which constitute the "lymph canalicular system" (p. 292), and interwoven with bundles of white fibrous tissue, and numerous delicate elastic fibrillæ, together with blood-vessels, nerves, and lymphatics. In relation to the process of secretion, the layer of connective tissue serves as a groundwork for the ramification of blood-vessels, lymphatics, and nerves. But in its usual form it is absent in some instances, as in the arachnoid covering the dura mater, and in the interior of the ventricles of the brain. The primary membrane and epithelium are always present, and are concerned in the formation of the fluid by which the free surface of tne membrane is moistened.

Serous membranes are of two principal kinds: 1*st.* Those which line visceral cavities,—the *arachnoid, pericardium, pleuræ, peritoneum,* and *tunicæ vaginales.* 2*nd.* The *synovial membranes* lining the joints, and the sheaths of tendons and ligaments, with which, also, are usually included the *synovial bursæ,* or *bursæ mucosæ,* whether these be subcutaneous, or situated beneath tendons that glide over bones.

The serous membranes form closed sacs, and exist wherever the free surfaces of viscera come into contact with each other or lie in cavities unattached to surrounding parts. The viscera invested by a serous membrane are, as it were, pressed into the shut sac which it forms, carrying before them a portion of the membrane, which serves as their investment. To the law that serous membranes form shut sacs, there is, in the human subject, one exception, viz.: the opening of the Fallopian tubes into the abdominal cavity,—an arrangement which exists in man and all Vertebrata, with the exception of a few fishes.

Functions.—The principal purpose of the serous and synovial membranes is to furnish a smooth, moist surface, to facilitate the movements of the invested organ, and to prevent the injurious effects of friction. This purpose is especially manifested in joints, in which free and extensive movements take place; and in the stomach and intestines, which, from the varying quantity and movements of their contents, are in almost constant motion upon one another and the walls of the abdomen.

Serous Fluid.—The fluid secreted from the free surface of the serous membranes is, in health, rarely more than sufficient to ensure the maintenance of their moisture. The opposed surfaces of each serous sac are at every point in contact with each other. After death, a larger quantity of fluid is usually found in each serous sac; but this, if not the product of manifest disease, is probably such as has transuded after death, or in the last hours of life. An excess of such fluid in any of the serous sacs constitutes dropsy of the sac.

The fluid naturally secreted by the serous membranes appears to be identical, in general and chemical characters, with the serum of the blood, or with very dilute liquor saguinis. It is of a pale yellow or straw color, slightly viscid, alkaline, and, on account of the presence of albumen, coagulable by heat. This similarity of the serous fluid to the liquid part of blood, and to the fluid with which most animal tissues are moistened, renders it probable that it is, in great measure, separated by simple transudation, through the walls of the blood-vessels. The probability is increased by the fact that, in jaundice, the fluid in the serous sacs is, equally with the serum of the blood, colored with the bile. But there is reason for supposing that the fluid of the cerebral ventricles and of the arachnoid sac are exceptions to this rule; for they differ from the fluids of the other serous sacs not only in being pellucid, colorless, and of much less specific gravity, but in that they seldom receive the tinge of bile when present in the blood, and are not colored by madder, or other similar substances introduced abundantly into the blood.

Synovial Fluid: Synovia.—It is also probable that the formation of synovial fluid is a process of more genuine and elaborate secretion, by means of the epithelial cells on the surface of the membrane, and especially of those which are accumulated on the edges and processes of the synovial fringes; for, in its peculiar density, viscidity, and abundance of albumin, synovia differs alike from the serum of blood and from the fluid of any of the serous cavities.

(2) **Mucous Membranes.**—The *mucous membranes* line all those passages by which internal parts communicate with the exterior, and by which either matters are eliminated from the body or foreign substances taken into it. They are soft and velvety, and extremely vascular. The external surfaces of mucous membranes are attached to various other tissues; in the tongue, for example, to muscle; on cartilaginous parts, to perichondrium; in the cells of the ethmoid bone, in the frontal and sphenoidal sinuses, as well as in the tympanum, to periosteum; in the intestinal canal, it is connected with a firm submucous membrane, which on its exterior gives attachment to the fibres of the muscular coat. The mucous membranes line certain principal tracts—Gastro-Pulmonary and Genito-Urinary; the former being subdivided into the Digestive and Respiratory tracts. 1. The *Digestive* tract commences in the cavity of the mouth, from which prolongations pass into the ducts of the salivary glands. From the mouth it passes through the fauces, pharynx, and œsophagus, to the stomach, and is thence continued along the whole tract of the intestinal canal to the termination of the rectum, being in its course arranged in the various folds and depressions already described, and prolonged into the ducts of the intestinal glands, the pancreas and liver, and into the gall-bladder. 2. The *Respiratory* tract includes the mucous membrane lining the cavity of the nose, and the various sinuses communicating with it, the lachrymal canal and sac, the conjunctiva of the eye and eyelids, and the prolongation which passes along the Eustachian tubes and lines the tympanum and the inner surface of the membrana tympani. Crossing the pharynx, and lining that part of it which is above the soft palate, the respiratory tract leads into the glottis, whence it is continued, through the larynx and trachea, to the bronchi and their divisions, which it lines as far as the branches of about $\frac{1}{30}$ of an inch in diameter, and continuous with it is a layer of delicate epithelial membrane which extends into the pulmonary cells. 3. The *Genito-urinary* tract, which lines the whole of the urinary passages, from their external orifice to the termination of the tubuli uriniferi of the kidneys, extends also into the organs of generation in both sexes, and into the ducts of the

glands connected with them; and in the female becomes continuous with the serous membrane of the abdomen at the fimbriæ of the Fallopian tubes.

Structure.—Along each of the above tracts, and in different portions of each of them, the mucous membrane presents certain structural peculiarities adapted to the functions which each part has to discharge; yet in some essential characters mucous membrane is the same, from whatever part it is obtained. In all the principal and larger parts of the several tracts, it presents, as just remarked, an external layer of epithelium, situated upon *basement-membrane*, and beneath this, a stratum of vascular tissue of variable thickness, containing lymphatic vessels and nerves which in different cases presents either outgrowths in the form of papillæ and villi, or depressions or involutions in the form of glands. But in the prolongations of the tracts, where they pass into gland-ducts, these constituents are reduced in the finest branches of the ducts to the epithelium, the primary or basement-membrane, and the capillary blood-vessels spread over the outer surface of the latter in a single layer.

The primary or basement-membrane is a thin transparent layer, simple, homogeneous, or composed of endothelial cells. In the minuter divisions of the mucous membranes, and in the ducts of glands, it is the layer continuous and correspondent with this basement-membrane that forms the proper walls of the tubes. The cells also which, lining the larger and coarser mucous membranes, constitute their epithelium, are continuous with, and often similar to those which, lining the gland-ducts, are called *gland-cells*. No certain distinction can be drawn between the epithelium-cells of mucous membranes and gland-cells. It thus appears, that the tissues essential to the production of a secretion are, in their simplest form, a membrane, having on one surface blood-vessels, and on the other a layer of cells, which may be called either epithelium-cells or gland-cells.

Mucous Fluid: Mucus.—From all mucous membranes there is secreted either from the surface or from certain special glands, or from both, a more or less viscid, greyish, or semi-transparent fluid, of alkaline reaction and high specific gravity, named *mucus*. It mixes imperfectly with water, but, rapidly absorbing liquid, it swells considerably when water is added. Under the microscope it is found to contain epithelium and leucocytes. It is found to be made up, chemically, of a nitrogenous principle called *mucin* which forms its chief bulk, of a little albumen, of salts chiefly chlorides and phosphates, and water with traces of fats and extractives.

Secreting Glands.—The structure of the elementary portions of a secreting apparatus, namely epithelium, simple membrane, and blood-vessels having been already described in this and previous chapters, we may proceed to consider the manner in which they are arranged to form the varieties of *secreting glands.*

The secreting glands are the organs to which the function of *secretion* is more especially ascribed; for they appear to be occupied with it alone. They present, amid manifold diversities of form and composition, a general plan of structure, by which they are distinguished from all other textures of the body; especially, all contain, and appear constructed with particular regard to, the arrangement of the cells, which, as already expressed, both line their tubes or cavities as an epithelium, and elaborate, as secreting cells, the substances to be discharged from them. Glands are provided also with lymphatic vessels and nerves. The distribution of the former is not peculiar, and need not be here considered. Nerve-fibres are distributed both to the blood-vessels of the gland and to its ducts; and, in some glands, to the secreting cells also (p. 229).

Varieties.—1. The *simple tubule*, or *tubular gland* (A, Fig. 220), examples of which are furnished by some mucous glands, the follicles of Lieberkühn (Fig. 186), and the tubular glands of the stomach. These appear to be simple tubular depressions of the mucous membrane, the wall of which is formed of primary membrane, and is lined with secreting cells arranged as an epithelium. To the same class may be referred the elongated and tortuous sudoriferous glands.

The compound *tubular glands* (D, Fig. 220) form another division. These consist of main gland-tubes, which divide and subdivide. Each gland may consist of the subdivisions of one or more main-tubes. The ultimate subdivisions of the tubes are generally highly convoluted. They are formed of a basement-membrane, lined by epithelium of various forms. The larger tubes may have an outside coating of fibrous, areolar, or muscular tissue. The kidney, testis, salivary glands, pancreas, Brunner's glands with the lachrymal and mammary glands, and some mucous glands are examples of this type, but present more or less marked variations among themselves.

2. The *aggregate* or *racemose glands*, in which a number of vesicles or *acini* are arranged in groups or lobules (C, Fig. 220). The Meibomian follicles are examples of this kind of gland.

These various organs differ from each other only in secondary points of structure; such as, chiefly, the arrangement of their excretory ducts, the grouping of the *acini* and lobules, their connection by areolar tissue, and supply of blood-vessels. The acini commonly appear to be formed by a kind of fusion of the walls of several vesicles, which thus combine to form one cavity lined or filled with secreting cells which also occupy recesses from the main cavity. The smallest branches of the gland-ducts sometimes open into the centres of these cavities; sometimes the acini are clustered round the extremities, or by the sides of the ducts: but, whatever secondary arrangement there may be, all have the same essential character of rounded groups of vesicles containing gland-cells, and opening by a common central cavity into minute ducts, which ducts in the large glands converge and unite to form larger and larger branches, and at length by one common trunk, open on a free surface of membrane.

Among these varieties of structure, all the secreting glands are alike in some essential points, besides those which they have in common with all truly secreting structures. They agree in presenting a large extent of secreting surface within a comparatively small space; in the circumstance

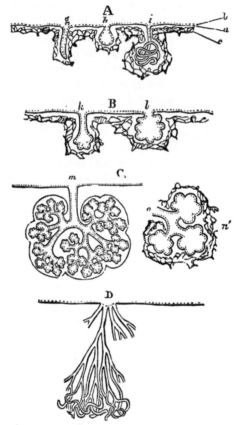

Fig. 220.—Plans of extension of secreting membrane by inversion or recession in form of cavities. A, simple glands, viz. *g*, straight tube; *h*, sac; *i*, coiled tube. B, multilocular crypts; *k*, of tubular form; *l*, saccular. C., racemose. or saccular compound gland; *m*, entire gland, showing branched duct and lobular structure; *n*, a lobule, detached with *o*, branch of duct proceeding from it. D, compound tubular gland. (Sharpey.)

that while one end of the gland-duct opens on a free surface, the opposite end is always closed, having no direct communication with blood-vessels, or any other canal; and in a uniform arrangement of capillary blood-vessels, ramifying and forming a network around the walls and in the interstices of the ducts and acini.

Process of Secretion.—In secretion two distinct processes are concerned which may be spoken of as, 1. *Physical,* and 2. *Chemical.*

1. *Physical processes.*—These are such as can be closely imitated in the laboratory, inasmuch as they consist in the operation of well-known physical laws: they are—

(*a*) Filtration. (*b*) Diffusion.

(*a*) *Filtration* is simply the passage of a fluid through a porous membrane under the influence of pressure. If two fluids be separated by a porous membrane, and the pressure on one side is greater than on the other, it is evident that in the absence of counteracting osmotic influences (see below) there will be a filtration through the membrane until the pressure on the two sides is equalized. Of course there may only be fluid on one side of the membrane, as, in the ordinary process of filtering through blotting-paper, and then the filtration will continue as long as the pressure (in this case, the weight of the fluid) is sufficient to force it through the pores of the filter. The necessary inequality of pressure may be obtained either by diminishing it on one side, as in the case of cupping; or increasing it on the other, as in the case of the increased blood-pressure and consequent increased flow of urine resulting from copious drinking. By filtration, not merely water, but various salts in solution, may transude from the blood-vessels. It seems probable that some fluids, such as the secretions of serous membranes, are simply exudations or oozings (filtration) from the blood-vessels, whose qualities are determined by those of the liquor sanguinis, while the quantities are liable to variation, and are chiefly dependent upon the blood-pressure.

(*b*) *Diffusion* is the passage of fluids through a moist animal membrane independent of pressure, and sometimes actually in opposition to it. There must always be in this process two fluids differing in composition, one or both possessing an affinity for the intervening membrane, and the fluids capable of being mixed one with the other; the osmotic current continuing in each direction (when both fluids have an affinity for the membrane) until the chemical composition of the fluid on each side of the septum becomes the same.

2. *Chemical processes.*—These constitute the process of *secretion* properly so called as distinguished from mere transudation spoken of above. In the *chemical* process of *secretion* various materials which do not exist as such in the blood are elaborated by the agency of the gland-cells from the blood, or, to speak more accurately, from the *plasma* which exudes from the blood-vessels into the interstices of the gland-textures.

The best evidence for this view is: 1*st*. That cells and nuclei are constituents of all glands, however diverse their outer forms and other characters, and are in all glands placed on the surface or in the cavity whence the secretion is poured. 2*nd*. That many secretions which are visible with the microscope may be seen in the cells of their glands before they are discharged. Thus, bile may be often discerned by its yellow tinge in the gland-cells of the liver; spermatozoids in the cells of the tubules of the testicles; granules of uric acid in those of the kidneys (of fish); fatty particles, like those of milk, in the cells of the mammary gland.

Secreting cells, like the cells or other elements of any other organ, appear to develop, grow, and attain their individual perfection by appropriating nutriment from the fluid exuded by adjacent blood-vessels and elaborating it, so that it shall form part of their own substance. In this perfected state, the cells subsist for some brief time, and when that period is over they appear to dissolve, wholly or in part, and yield their contents to the peculiar material of the secretion. And this appears to be the case in every part of the gland that contains the appropriate gland- • cells; therefore not in the extremities of the ducts or in the acini alone, but in great part of their length.

We have described elsewhere the changes which have been noticed from actual experiment in the cells of the salivary glands, pancreas, and peptic gland (pp. 235, 259, 265).

Discharge of Secretions from glands may either take place as soon as they are formed; or the secretion may be long retained within the gland or its ducts. The former is the case with the sweat glands. But the secretions of those glands whose activity of function is only occasional are usually retained in the cells in an undeveloped form during the periods of the gland's inaction. And there are glands which are like both these classes, such as the lachrymal, which constantly secrete small portions of fluid, and on occasions of greater excitement discharge it more abundantly.

When discharged into the ducts, the further course of secretions is affected partly by the pressure from behind; the fresh quantities of secretion propelling those that were formed before. In the larger ducts, its propulsion is assisted by the contraction of their walls. All the larger ducts, such as the ureter and common bile-duct, possess in their coats plain muscular fibres; they contract when irritated, and sometimes manifest peristaltic movements. Rhythmic contractions in the pancreatic and bile-ducts have been observed, and also in the ureters and vasa deferentia. It is probable that the contractile power extends along the ducts to a considerable distance within the substance of the glands whose secretions can be rapidly expelled. Saliva and milk, for instance, are sometimes ejected with much force; doubtless by the energetic and simultaneous contraction of many of the ducts of their respective glands.

Circumstances Influencing Secretion.—Amongst the principal conditions which influence secretion are (1) variations in the quantity of blood, (2) in the quantity of the peculiar materials for any secretion that it may contain, and (3) in conditions of the nerves of the glands.

(1.) *An increase in the quantity of blood traversing a gland,* as in nearly all the instances before quoted, coincides generally with an augmentation of its secretion. Thus, the mucous membrane of the stomach becomes florid when, on the introduction of food, its glands begin to secrete; the mammary gland becomes much more vascular during lactation; and all circumstances which give rise to an increase in the quantity of material secreted by an organ produce, coincidently, an increased supply of blood; but we have seen that a discharge of saliva may occur under extraordinary circumstances, without increase of blood-supply (p. 233), and so it may be inferred that this condition of increased blood-supply is not absolutely essential.

(2.) When the blood contains more than usual of the materials which the glands are designed to separate or elaborate. Thus, when an excess of nitrogenous waste is in the blood, whether from excessive exercise, or from destruction of one kidney, a healthy kidney will excrete more urea than it did before.

(3.) *Influence of the Nervous System on Secretion.*—The process of secretion is largely influenced by the condition of the nervous system. The exact mode in which the influence is exhibited must still be regarded as somewhat obscure. In part, it exerts its influence by increasing or diminishing the quantity of blood supplied to the secreting gland, in virtue of the power which it exercises over the contractility of the smaller blood-vessels; while it also has a more direct influence, as was demonstrated at length in the case of the submaxillary gland, upon the secreting cells themselves; this may be called *trophic* influence. Its influence over secretion, as well as over other functions of the body, may be excited by causes acting directly upon the nervous centres, upon the nerves going to the secreting organ, or upon the nerves of other parts. In the latter case, a reflex action is produced: thus the impression produced upon the nervous centres by the contact of food in the mouth, is reflected upon the nerves supplying the salivary glands, and produces, through these, a more abundant secretion of saliva (p. 232).

Through the nerves, various conditions of the brain also influence the secretions. Thus, the thought of food may be sufficient to excite an abundant flow of saliva. And, probably, it is the mental state which excites the abundant secretion of urine in hysterical paroxysms, as well as the perspirations and, occasionally, diarrhœa, which ensue under the influence of terror, and the tears excited by sorrow or excess of joy. The quality of a secretion may also be affected by the mind; as in the cases in which, through grief or passion, the secretion of milk is altered, and is sometimes so changed as to produce irritation in the alimentary canal of the child, or even death (Carpenter).

Relations between the Secretions.—The secretions of some of the glands seem to bear a certain relation or antagonism to each other, by which an increased activity of one is usually followed by diminished activity of one or more of the others; and a deranged condition of one is apt to entail a disordered state in the others. Such relations appear to exist among the various mucous membranes; and. the close relation be. tween the secretion of the kidney and that of the skin is a subject of con. stant observation.

THE MAMMARY GLANDS AND THEIR SECRETION:—MILK.

Structure.—The mammary glands are composed of large divisions or lobes, and these are again divisible into lobules,—the lobules being com-. posed of the convoluted subdivision of ducts (alveoli). The lobes and lobules are bound together by areolar tissue; penetrating between the lobes, and covering the general surface of the gland, with the exception of the nipple, is a considerable quantity of yellow fat, itself lobulated by

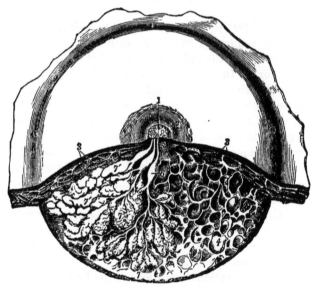

Fig. 221.—Dissection of the lower half of the female mamma during the period of lactation. ⅔.—In the left-hand side of the dissected part the glandular lobes are exposed and partially unraveled; and on the right-hand side, the glandular substance has been removed to show the reticular loculi of the connective-tissue in which the glandular lobules are placed: 1, upper part of the mamilla or nipple; 2, areola; 3, subcutaneous masses of fat; 4, reticular loculi of the connective-tissue which support the glandular substance and contain the fatty masses; 5. one of three lactiferous ducts shown passing toward the mamilla where they open; 6, one of the sinus lactei or reservoirs; 7, some of the glandular lobules which have been unraveled; 7', others massed together. (Luschka.)

sheaths and processes of tough areolar tissue (Fig. 221) connected both with the skin in front and the gland behind; the same bond of connection extending also from the under surface of the gland to the sheathing connective tissue of the great pectoral muscle on which it lies. The main ducts of the gland, fifteen to twenty in number, called the *lactiferous* or *galactophorous* ducts, are formed by the union of the smaller (lobular) ducts, and open by small separate orifices through the nipple. At the points of junction of lobular ducts to form lactiferous ducts, and just be-fore these enter the base of the nipple, the ducts are dilated (6, Fig. 221);

and, during lactation, the period of active secretion by the gland, the dilatations form reservoirs for the milk, which collects in them and distends them. The walls of the gland-ducts are formed of areolar and elastic with some muscular tissue, and are lined internally by short columnar and near the nipple by squamous epithelium. The alveoli consist of a membrana propria of flattened endothelial cells lined by low columnar epithelium, and are filled with fat globules.

The nipple, which contains the terminations of the lactiferous ducts, is composed also of areolar tissue, and contains unstriped muscular fibres. Blood-vessels are also freely supplied to it, so as to give it a species of erectile structure. On its surface are very sensitive papillæ; and around it is a small area or *areola* of pink or dark-tinted skin, on which are to be seen small projections formed by minute secreting glands.

Blood-vessels, nerves, and lymphatics are plentifully supplied to the mammary glands; the calibre of the blood-vessels, as well as the size of the glands, varying very greatly under certain conditions, especially those of pregnancy and lactation.

Changes in the Glands at certain Periods.—The minute changes which occur in the mammary gland during its periods of evolu-

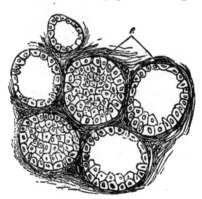

Fig. 222.—Section of mammary gland of rabbit near the end of pregnancy, showing six acini. *e*, epithelial cells of a polyhedral or short columnar form, with which the acini are packed. ' × 200. (Schofield.)

tion (pregnancy), and involution (when lactation has ceased), are the following:—

The most favorable period for observing the epithelium of the mammary gland fully developed is shortly before the end of pregnancy. At this period the acini which form the lobules of the gland, are found to be lined with a mosaic of polyhedral epithelial cells (Fig. 222), and supported by a connective tissue stroma.

The rapid formation of milk during *lactation* results from a fatty metamorphosis of the epithelial cells: "The secretion may be said to be produced by a transformation of the substance of successive generations of epithelial cells, and in the state of full activity this transformation is so complete that it may be called a deliquescence" (Creighton).

In the earlier days of lactation, epithelial cells partially transformed are discharged in the secretion: these are termed "colostrum corpuscles," but later on the cells are completely transformed before the secretion is discharged.

After the end of lactation, the mamma gradually returns to its original size (*involution*). The acini, in the early stages of involution, are lined with cells in all degrees of vacuolation (Fig. 223). As involution proceeds the acini diminish considerably in size, and at length, instead of a mosaic of lining epithelial cells (twenty to thirty in each acinus), we have five or six nuclei (some with no surrounding protoplasm) lying in an irregular heap

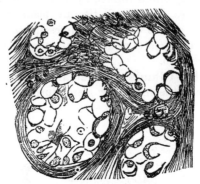

FIG. 223.—Section of mammary gland of ewe shortly after the end of lactation, showing parts of four acini, which contain numerous epithelial cells undergoing vacuolation *in situ;* they very closely resemble young fat-cells, and are in fact just like " Colostrum corpuscles." × 800. (Creighton.)

within the acinus. During the later stages of involution, large yellow granular cells are to be seen. As the acini diminish in size, the connective tissue and fatty matter between them increase, and in some animals, when the gland is completely inactive, it is found to consist of a thin film of glandular tissue overlying a thick cushion of fat. Many of the products of waste are carried off by the lymphatics.

During *pregnancy* the mammary glands undergo changes (*evolution*) which are readily observable. They enlarge, become harder and more distinctly lobulated: the veins on the surface become more prominent. The areola becomes enlarged and dusky, with projecting papillæ; the nipple too becomes more prominent, and milk can be squeezed from the orifices of the ducts. This is a very gradual process, which commences about the time of conception, and progresses steadily during the whole period of gestation. The acini enlarge, and a series of changes occur, exactly the reverse of those just described under the head of Involution.

THE MAMMARY SECRETION:—MILK.

Under the microscope, milk is found to contain a number of globules of various sizes (Fig. 224), the majority about $\frac{1}{10000}$ of an inch in diameter. They are composed of oily matter, probably coated by a fine layer of albuminous material, and are called *milk-globules;* while, accompanying these, are numerous minute particles, both oily and albuminous, which exhibit ordinary molecular movements. The milk which is secreted in the first few days after parturition, and which is called the *colostrum,* differs from ordinary milk in containing a larger quantity of solid matter; and under the microscope are to be seen certain granular

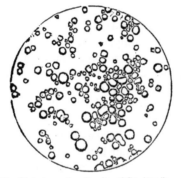

FIG. 224.—Globules and molecules of Cow's milk. × 400.

masses called *colostrum-corpuscles.* These, which appear to be small masses of albuminous and oily matter, are probably secreting cells of the gland, either in a state of fatty degeneration, or old cells which in their attempt at secretion under the new circumstances of active need of milk, are filled with oily matter; which, however, being unable to discharge, they are themselves shed bodily to make room for their successors. Colostrum-corpuscles have been seen to exhibit contractile movements and to squeeze out drops of oil from their interior (Stricker).

Chemical Composition.—Milk is in reality an emulsion consisting of numberless little globules of fat, coated with a thin layer of albuminous matter, floating in a large quantity of water which contains in solution casein, serum-albumin, milk-sugar (lactose), and several salts. Its percentage composition has been already mentioned, but may be here repeated. Its reaction is alkaline: its specific gravity about 1030.

TABLE OF THE CHEMICAL COMPOSITION OF MILK.

	Human.	Cows.
Water	890	858
Solids	110	142
	1000	1000
Proteids, including Casein and Serum-Albumin . . .	35	68
Fats or Butter	25	38
Sugar (with extractives) . .	48	30
Salts . . .	2	6
	110	142

When milk is allowed to stand, the fat globules, being the lightest portion, rise to the top, forming *cream*. If a little acetic acid be added to a drop of milk under the microscope, the albuminous film coating the oil drops is dissolved, and they run together into larger drops. The same result is produced by the process of *churning*, the effect of which is to break up the albuminous coating of the oil drops: they then coalesce to form *butter*.

Curdling of Milk.—If milk be allowed to stand for some time, its reaction becomes acid: in popular language it "turns sour." This change appears to be due to the conversion of the milk-sugar into lactic acid, which causes the precipitation of the casein (curdling): the curd contains the fat globules: the remaining fluid (whey) consists of water holding in solution albumin, milk-sugar and certain salts. The same effect is produced in the manufacture of cheese, which is really casein coagulated by the agency of rennet (p. 248). When milk is boiled, a scum of serum-albumin forms on the surface.

Curdling Ferments.—The effect of the ferments of the gastric, pancreatic, and intestinal juices in curdling milk (*curdling ferments*) has already been mentioned in the Chapter on Digestion.

The salts of milk are chlorides, sulphates, phosphates, and carbonates of potassium, sodium, calcium.

CHAPTER XII.

THE SKIN AND ITS FUNCTIONS.

THE skin serves—(1), as an external integument for the protection of the deeper tissues, and (2), as a sensitive organ in the exercise of touch; it is also (3), an important excretory, and (4), an absorbing organ; while it plays an important part in (5) the regulation of the temperature of the body.

Structure of the Skin.—The skin consists, principally, of a vascular tissue, named the *corium*, *derma*, or *cutis vera*, and an external covering of epithelium termed the *cuticle* or *epidermis*. Within and beneath the corium are imbedded several organs with special function, namely *sudoriferous* glands, *sebaceous* glands, and *hair follicles;* and on its surface are sensitive *papillæ*. The so-called appendages of the skin—the *hair* and *nails*—are modifications of the epidermis.

Epidermis.—The epidermis is composed of several strata of cells of various shapes, and closely resembles in its structure that which lines the mouth. The following four layers may be distinguished. 1. *Stratum corneum* (Fig. 225, *a*), consisting of many superposed layers of horny scales. The different thickness of the epidermis in different regions of the body is chiefly due to variations in the thickness of this layer; *e.g.*, on the horny parts of the palms of the hands and soles of the feet it is of great thickness. The stratum corneum of the buccal epithelium chiefly differs from that of the epidermis in the fact that nuclei are to be distinguished in some of the cells even of its most superficial layers.

2. *Stratum lucidum*, a bright homogeneous membrane consisting of squamous cells closely arranged, in some of which a nucleus can be seen.

3. *Stratum granulosum*, consisting of one layer of flattened cells which appear fusiform in vertical section: they are distinctly nucleated, and a number of granules extend from the nucleus to the margins of the cell.

4. *Stratum Malpighii* or *Rete mucosum*, which consists of many strata. The deepest cells, placed immediately above the cutis vera, are columnar with oval nuclei: this layer of columnar cells is succeeded by a number of layers of more or less polyhedral cells with spherical nuclei; the cells of the more superficial layers are considerably flattened. The deeper surface of the rete mucosum is accurately adapted to the papillæ of the true skin, being, as it were, moulded on them. It is very constant in thickness in all parts of the skin. The cells of the middle layers of the

stratum Malpighii are almost all connected by processes, and thus form
"prickle cells" (p. 21). The pigment of the skin, the varying quantity
of which causes the various tints observed in different individuals and
different races, is contained in the deeper cells of the rete mucosum; the
pigmented cells as they approach the free surface gradually losing their
color. Epidermis maintains its thickness in spite of the constant wear
and tear to which it is subjected. The columnar cells of the deepest
layer of the "rete mucosum" elongate, and their nuclei divide into two

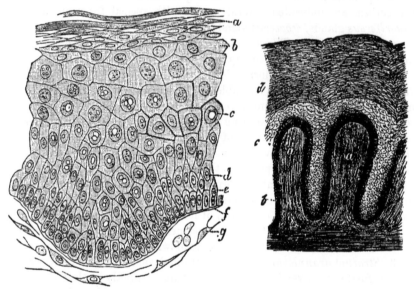

Fig. 225.—Vertical section of the epidermis of the prepuce. *a*, stratum corneum, of very few
layers, the stratum lucidum and stratum granulosum not being distinctly represented; *b, c, d*, and *e*,
the layers of the stratum Malpighii, a certain number of the cells in layers *d* and *e* showing signs of
segmentation; layer *c* consists chiefly of prickle or ridge and furrow cells; *f*, basement membrane;
g, cells in cutis vera. (Cadiat.)
Fig. 226.—Vertical section of skin of the negro. *a, a.* Cutaneous papillæ. *b.* Undermost and
dark-colored layer of oblong vertical epidermis-cells, *c.* Stratum Malpighii. *d.* Superficial layers,
including stratum corneum, stratum lucidum, and stratum granulosum, the last two not differentiated in figure. × 250. (Sharpey.)

(Fig. 225, *e*). Lastly the upper part of the cell divides from the lower;
thus from a long columnar cell are produced a polyhedral and a short
columnar cell: the latter elongates and the process is repeated. The
polyhedral cells thus formed are pushed up toward the free surface by the
production of fresh ones beneath them, and become flattened from pressure: they also become gradually horny by evaporation and transformation of their protoplasm into keratin, till at last by rubbing they are
detached as dry horny scales at the free surface. There is thus a constant production of fresh cells in the deeper layers, and a constant throwing off of old ones from the free surface. When these two processes are
accurately balanced, the epidermis maintains its thickness. When, by

intermittent pressure, a more active cell-growth is stimulated, the production of cells exceeds their waste and the epidermis increases in thickness, as we see in the horny hands of the laborer.

The thickness of the epidermis on different portions of the skin is directly proportioned to the friction, pressure, and other sources of injury to which it is exposed; for it serves as well to protect the sensitive and vascular cutis from injury from without, as to limit the evaporation of fluid from the blood-vessels. The adaptation of the epidermis to the latter purposes may be well shown by exposing to the air two dead hands or feet, of which one has its epidermis perfect, and the other is deprived of it; in a day, the skin of the latter will become brown, dry, and horn-like, while that of the former will almost retain its natural moisture.

Cutis vera.—The *corium* or *cutis*, which rests upon a layer of adipose and cellular tissue of varying thickness, is a dense and tough, but yielding and highly elastic structure, composed of fasciculi of fibro-cellular tissue, interwoven in all directions, and forming, by their interlacements, numerous spaces or areolæ. These areolæ are large in the deeper layers of the cutis, and are there usually filled with little masses of fat (Fig. 228): but, in the superficial parts, they are small or entirely obliterated. Plain muscular fibre is also abundantly present.

Papillæ.—The papillæ are conical elevations of the cutis vera, with a single or divided free extremity, more prominent and more densely set at some parts than at others (Figs. 227 and 230). The parts on which they are most abundant and most prominent, are the palmar surface of the

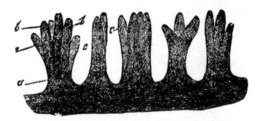

Fig. 227.—Compound papillæ from the palm of the hand; *a*, basis of a papilla; *b, b,* divisions or branches of the same; *c, c,* branches belonging to papillæ, of which the bases are hidden from view. × 60. (Kölliker.)

hands and fingers, and the soles of the feet—parts, therefore, in which the sense of touch is most acute. On these parts they are disposed in double rows, in parallel curved lines, separated from each other by depressions. Thus they may be seen easily on the palm, whereon each raised line is composed of a double row of papillæ, and is intersected by short transverse lines or furrows corresponding with the interspaces between the successive pairs of papillæ. Over other parts of the skin they are more or less thinly scattered, and are scarcely elevated above the surface. Their average length is about $\frac{1}{100}$ of an inch, and at their base they measure about $\frac{1}{250}$ of an inch in diameter. Each papilla is abundantly supplied with blood, receiving from the vascular plexus in the cutis one or more minute arterial twigs, which divide into capillary loops in its substance, and then reunite into a minute vein, which passes out at its base. The abundant supply of blood which the papillæ thus receive explains the turgescence or kind of erection which they undergo when the circulation through the skin is active. The majority, but not all, of

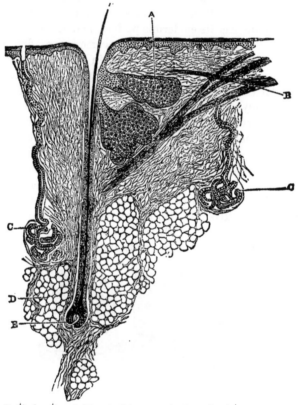

Fig. 228.—Vertical section of skin. A. Sebaceous gland opening into hair-follicle. B. Muscular fibres. C. Sudoriferous or sweat-gland. D. Subcutaneous fat. E. Fundus of hair-follicle, with hair-papillæ. (Klein and Noble Smith.)

the papillæ contain also one or more terminal nerve-fibres, from the ultimate ramifications of the cutaneous plexus, on which their exquisite sensibility depends.

Nerve-terminations.—In some parts, especially those in which the sense of touch is highly developed, as, for example, the palm of the hand and the lips, the nerve-fibres appear to terminate, in many of the papillæ, by one or more free ends in the substance of an oval-shaped body, occupying the principal part of the interior of the papillæ, and termed a *touch-*

corpuscle (Fig. 229). The nature of this body is obscure. Some regard it as little else than a mass of fibrous or connective tissue, surrounded by elastic fibres, and formed, according to Huxley, by an increased development of the primitive sheaths of the nerve-fibres, entering the papillæ. Others, however, believe that, instead of thus consisting of a homogeneous mass of connective tissue, they are special and peculiar bodies of laminated structure, directly concerned in the sense of touch. They do not occur in all the papillæ of the parts where they are found, and, as a rule, in the papillæ in which they are present there are no blood-vessels. Since

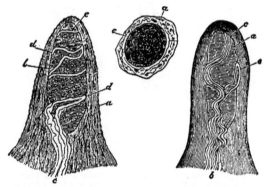

FIG. 229.—Papillæ from the skin of the hand, freed from the cuticle and exhibiting tactile corpuscles. A. Simple papilla with four nerve-fibres: *a*, tactile corpuscles; *b*, nerves. B. Papilla treated with acetic acid; *a*, cortical layer with cells and fine elastic filaments; *b*, tactile corpuscle with transverse nuclei; *c*, entering nerve with neurilemma or perineurium; *d*, nerve-fibres winding round the corpuscle. c, Papilla viewed from above so as to appear as a cross-section: *a*, cortical layer; *b*, nerve-fibre; *c*, sheath of the tactile corpuscle containing nuclei; *d*, core. × 350. (Kölliker.)

these peculiar bodies in which the nerve-fibres end are only met with in the papillæ of highly sensitive parts, it may be inferred that they are specially concerned in the sense of touch, yet their absence from the papillæ of other tactile parts shows that they are not essential to this sense.

Closely allied in structure to the touch-corpuscles, are some little bodies called *end-bulbs*, about $\frac{1}{600}$ inch in diameter (Krause). They are generally oval or spheroidal, and composed externally of a coat of connective tissue enclosing a softer matter, in which the extremity of a nerve terminates. These bodies have been found chiefly in the lips, tongue, palate, and the skin of the glans penis (Fig. 230).

Glands of the Skin.—The skin possesses glands of two kinds: (*a*) Sudoriferous, or Sweat Glands; (*b*) Sebaceous Glands.

(*a*) *Sudoriferous, or Sweat Glands.*—Each of these glands consists of a small lobular mass, formed of a coil of tubular gland-duct, surrounded by blood-vessels and embedded in the subcutaneous adipose tissue (Fig. 228, c.). From this mass, the duct ascends, for a short distance, in a spiral manner through the deeper part of the cutis, then passing straight,

and then sometimes again becoming spiral, it passes through the·cuticle
and opens by an oblique valve-like aperture.   In the parts where the epi.
dermis is thin the ducts themselves are thinner and more nearly straight
in their course (Fig. 228). The duct, which maintains nearly the same
diameter throughout, is lined with a layer of columnar epithelium (Fig.

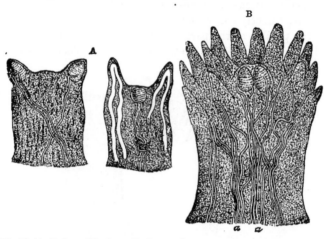

Fig. 230.—End-bulbs in papillæ (magnified) treated with acetic acid. A, from the lips: the white
loops in one of them are capillaries. B, from the tongue. Two end-bulbs seen in the midst of the
simple papillæ: *a, a*, nerves. (Kölliker.)

231) continuous with the epidermis; while the part which passes through
the epidermis is composed of the latter structure only; the cells which
immediately form the boundary of the canal in this part being somewhat
differently arranged from those of the adjacent cuticle.

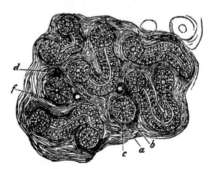

Fig. 231.—Glomeruli of sudoriferous gland, divided in various directions. *a*, sheath of the gland;
b, columnar epithelial lining of gland tube; *c*, lumen of tube; *d*, divided blood-vessel· *e* loose-con-
nective-tissue, forming a capsule to the gland. (Biesiadecki.)

The sudoriferous glands are abundantly distributed over the whole surface of the body; but are especially numerous, as well as very large, in
the skin of the palm of the hand, and of the sole of the foot.   The glands

by which the peculiar odorous matter of the axillæ is secreted form a nearly complete layer under the cutis, and are like the ordinary sudoriferous glands, except in being larger and having very short ducts.

· The peculiar bitter yellow substance secreted by the skin of the external auditory passage is named *cerumen,* and the glands themselves *ceruminous* glands; but they do not much differ in structure from the ordinary sudoriferous glands.

(*b*) *Sebaceous Glands.*—The sebaceous glands (Fig. 232), like the sudoriferous glands, are abundantly distributed over most parts of the body. They are most numerous in parts largely supplied with hair, as the scalp

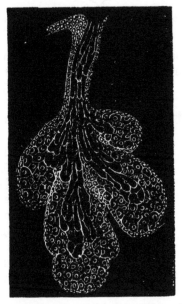

Fig. 232.—Sebaceous gland from human skin. (Klein and Noble Smith.)

and face, and are thickly distributed about the entrances of the various passages into the body, as the anus, nose, lips, and external ear. They are entirely absent from the palmar surface of the hand and the plantar surfaces of the feet. They are minutely lobulated glands composed of an aggregate of small tubes or sacculi filled with opaque white substances, like soft ointment. Minute capillary vessels overspread them; and their ducts open either on the surface of the skin, close to a hair, or, which is more usual, directly into the follicle of the hair. In the latter case, there are generally two or more glands to each hair (Fig. 228).

Hair.—A hair is produced by a peculiar growth and modification of the epidermis. Externally it is covered by a layer of fine scales closely imbricated, or overlapping like the tiles of a house, but with the free edges turned upward (Fig. 233, A). It is called the *cuticle* of the hair. Beneath this is a much thicker layer of elongated horny cells, closely packed together so as to resemble a fibrous structure. This, very com-

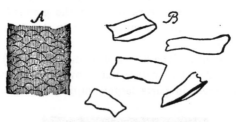

FIG. 233.—Surface of a white hair, magnified 160 diameters. The wave lines mark the upper or free edges of the cortical scales. *B*, separated scales, magnified 350 diameters. (Kölliker.)

monly, in the human subject, occupies the whole of the inside of the hair; but in some cases there is left a small central space filled by a substance

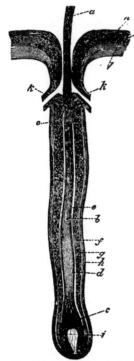

FIG. 234.—Medium-sized hair in its follicle. *a*, stem cut short; *b*, root; *c*, knob; *d*, hair cuticle; *e*, internal, and *f*, external root-sheath; *g, h*, dermic coat of follicle; *i*, papilla; *k, k*, ducts of sebaceous glands; *l*, corium; *m*, mucous layer of epidermis; *o*, upper limit of internal root-sheath. × 50. (Kölliker.) See also Fig. 235.

called the *medulla* or *pith*, composed of small collections of irregularly shaped cells, containing sometimes pigment granules or fat, but mostly air.

The follicle, in which the root of each hair is contained (Fig. 235), forms a tubular depression from the surface of the skin,—descending into the subcutaneous fat, generally to a greater depth than the sudoriferous glands, and at its deepest part enlarging in a bulbous form, and often curving from its previous rectilinear course. It is lined throughout by cells of epithelium, continuous with those of the epidermis, and its walls are formed of pellucid membrane, which commonly, in the follicles of the largest hairs, has the structure of vascular fibrous tissue. At the bottom of the follicle is a small papilla, or projection of true skin, and it is by the production and out-growth of epidermal cells from the surface of this papilla that the hair is formed. The inner wall of the follicle is lined by epidermal cells continuous with those covering the general surface of the skin; as if indeed the follicle had been formed by a simple thrusting in of the surface of the integument (Fig. 234). This epidermal lining of the hair follicle, or *root-sheath* of the hair, is composed of two layers, the inner one of which is so moulded on the imbricated scaly cuticle of the hair, that its inner surface becomes imbricated also, but of course in the opposite direction. When a hair is pulled out, the inner layer of the *root-sheath* and part of the outer layer also are commonly pulled out with it.

Nails.—A *nail*, like a hair, is a peculiar arrangement of epidermal cells, the undermost of which, like those of the general surface of the integument, are rounded or elongated, while the superficial are flattened,

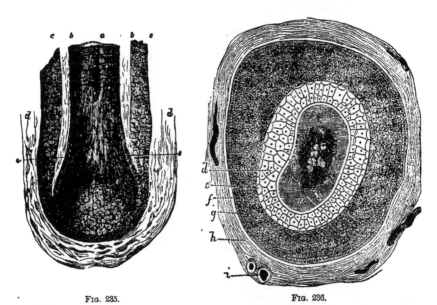

FIG. 235.   FIG. 236.

FIG. 235.—Magnified view of the root of a hair. *a*, stem or shaft of hair cut across; *b*, inner, and *c*, outer layer of the epidermal lining of the hair-follicle, called also the inner and outer root-sheath; *d*, dermal or external coat of the hair-follicle, shown in part; *e*, imbricated scales about to form a cortical layer on the surface of the hair. The adjacent cuticle of the root-sheath is not represented, and the papilla is hidden in the lower part of the knob where that is represented lighter. (Kohlrausch.)

FIG. 236.—Transverse section of a hair and hair-follicle made below the opening of the sebaceous gland. *a*, medulla, or pith of the hair; *b*, fibrous layer or cortex; *c*, cuticle; *d*, Huxley's layer, *e*, Henle's layer of internal root-sheath; *f* and *g*, layers of external root-sheath, outside of *g* is a light layer, or "glassy membrane," which is equivalent to the basement membrane; *h*, fibrous coat of hair sac; *i*, vessels. (Cadiat.)

and of more horny consistence. That specially modified portion of the corium, or true skin, by which the nail is secreted, is called the *matrix*.

The back edge of the nail, or the *root* as it is termed, is received into a shallow crescentic groove in the *matrix*, while the front part is free and projects beyond the extremity of the digit. The intermediate portion of the nail rests by its broad under-surface on the front part of the matrix, which is here called the *bed* of the nail. This part of the matrix is not uniformly smooth on the surface, but is raised in the form of longitudinal and nearly parallel ridges or laminæ, on which are moulded the epidermal cells of which the nail is made up (Fig. 237).

The growth of the nail, like that of the hair, or of the epidermis generally, is effected by a constant production of cells from beneath and behind, to take the place of those which are worn or cut away. Inasmuch, however, as the posterior edge of the nail, from its being lodged in a groove of the skin, cannot grow backward, on additions being made to it, so easily as it can pass in the opposite direction, any growth at its hinder part pushes the whole forward. At the same time fresh cells are added to its under surface, and thus each portion of the nail becomes gradually thicker as it moves to the front, until, projecting beyond the

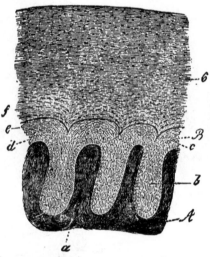

FIG. 287.—Vertical transverse section through a small portion of the nail and matrix largely magnified. *A*, corium of the nail-bed, raised into ridges or laminæ *a*, fitting in between corresponding laminæ *b*, of the nail. *B*, Malpighian, and *C*, horny layer of nail; *d*, deepest and vertical cells; *e*, upper flattened cells of Malpighian layer. (Kölliker.)

surface of the matrix, it can receive no fresh addition from beneath, and is simply moved forward by the growth at its root, to be at last worn away or cut off.

FUNCTIONS OF THE SKIN.

(1.) By means of its toughness, flexibility and elasticity, the skin is eminently qualified to serve as the general integument of the body, for defending the internal parts from external violence, and readily yielding and adapting itself to their various movements and changes of position.

(2.) The skin is the chief organ of the sense of touch. Its whole surface is extremely sensitive; but its *tactile* properties are due more especially to the abundant papillæ with which it is studded. (See Chapter on Special Senses.)

Although destined especially for the sense of touch, the papillæ are not so placed as to come into direct contact with external objects; but like the rest of the surface of the skin, are covered by one or more layers of epithelium, forming the cuticle or epidermis. The papillæ adhere very intimately to the cuticle, which is thickest in the spaces between them, but tolerably level on its outer surface: hence, when stripped off from the cutis, as after maceration, its internal surface presents a series of pits and elevations corresponding to the papillæ and their interspaces, of which it thus forms a kind of mould. Besides affording by its impermeability a check to undue evaporation from the skin, and providing the sensitive cutis with a protecting investment, the cuticle is of service in relation to the sense of touch. For by being thickest in the spaces, between the papillæ, and only thinly spread over the summits of these processes, it may serve to subdivide the sentient surface of the skin into a number of isolated points, each of which is capable of receiving a distinct. impression from an external body. By covering the papillæ it renders the sensation produced by external bodies more obtuse, and in this manner also is subservient to touch: for unless the very sensitive papillæ were thus defended, the contact of substances would give rise to pain, instead of the ordinary impressions of touch. This is shown in the extreme sensitiveness and loss of tactile power in a part of the skin when deprived of its epidermis. If the cuticle is very thick, however, as on the heel, touch becomes imperfect, or is lost.

(3.) **The Secretion of Sebaceous Glands, and Hair-follicles.**— The secretion of the *sebaceous* glands and *hair-follicles* (for their products cannot be separated) consists of cast-off epithelium-cells, with nuclei and granules, together with an oily matter, extractive matter, and stearin; in certain parts, also, it is mixed with a peculiar odorous principle, which contains caproic, butyric, and rutic acids. It is, perhaps, nearly similar in composition to the unctuous coating, or *vernix caseosa*, which is formed on the body of the fœtus while in the uterus, and which contains large quantities of ordinary fat. Its purpose seems to be that of keeping the skin moist and supple, and, by its oily nature, of both hindering the evaporation from the surface, and guarding the skin from the effects of the long-continued action of moisture. But while it thus serves local purposes, its removal from the body entitles it to be reckoned among the excretions of the skin; though the share it has in the purifying of the blood cannot be discerned.

(4.) **The Excretion of the Skin : the Sweat.**—The fluid secreted by the *sudoriferous* glands is usually formed so gradually, that the watery portion of it escapes by evaporation as fast as it reaches the surface. But, during strong exercise, exposure to great external warmth, in some diseases, and when evaporation is prevented, the secretion becomes more sensible, and collects on the skin in the form of drops of fluid.

The *perspiration* of the skin, as the term is sometimes employed in physiology, includes all that portion of the secretions and exudations from the skin which passes off by evaporation; the *sweat* includes that which may be collected only in drops of fluid on the surface of the skin. The two terms are, however, most often used synonymously; and for distinc· tion, the former is called *insensible* perspiration; the latter *sensible* per· spiration. The fluids are the same, except that the sweat is commonly mingled with various substances lying on the surface of the skin. The contents of the sweat are, in part, matters capable of assuming the form of vapor, such as carbonic acid and water, and in part, other matters which are deposited on the skin, and mixed with the sebaceous secretion.

Table of the Chemical Composition of Sweat.

Water 995	
Solids:—	
Organic Acids (formic, acetic, butyric, pro- ⎱ ·9 pionic, caproic, caprylic) . . . ⎰	
Salts, chiefly sodium chloride . . . 1·8	
Neutral fats and cholesterin . . . . ·7	
Extractives (including urea), with epithelium 1·6 5	
1000	

Of these several substances, however, only the carbonic acid and water need particular consideration.

Watery Vapor.—The quantity of *watery vapor* excreted from the skin is on an average between 1¼ and 2 lb. daily. This subject has been estimated very carefully by Lavoisier and Sequin. The latter chemist enclosed his body in an air-tight bag, with a mouth-piece. The bag being closed by a strong band above, and the mouth-piece adjusted and gummed to the skin around the mouth, he was weighed, and then re-mained quiet for several hours, after which time he was again weighed. The difference in the two weights indicated the amount of loss by pul-monary exhalation. Having taken off the air-tight dress, he was imme-diately weighed again, and a fourth time after a certain interval. The difference between the two weights last ascertained gave the amount of the cutaneous and pulmonary exhalation together; by subtracting from this the loss by pulmonary exhalation alone, while he was in the air-tight dress, he ascertained the amount of cutaneous transpiration. During a state of rest, the average loss by cutaneous and pulmonary exhalation in a minute, is eighteen grains,—the minimum eleven grains, the maximum thirty-two grains; and of the eighteen grains, eleven pass off by the skin, and seven by the lungs.

The quantity of watery vapor lost by transpiration is of course influ-enced by all external circumstances which affect the exhalation from other evaporating surfaces, such as the temperature, the hygrometric state, and the stillness of the atmosphere. But, of the variations to which it is subject under the influence of these conditions, no calculation has been exactly made.

Carbonic Acid.—The quantity of *carbonic acid* exhaled by the skin on an average is about $\frac{1}{150}$ to $\frac{1}{200}$ of that furnished by the pulmonary respiration.

The cutaneous exhalation is most abundant in the lower classes of animals, more particularly the naked Amphibia, as frogs and toads, whose skin is thin and moist, and readily permits an interchange of gases between the blood circulating in it and the surrounding atmosphere. Bischoff found that, after the lungs of frogs had been tied and cut out, about a quarter of a cubic inch of carbonic acid gas was exhaled by the skin in eight hours. And this quantity is very large, when it is remembered that a full-sized frog will generate only about half a cubic inch of carbonic acid by his lungs and skin together in six hours. (Milne-Edwards and Müller.)

The importance of the respiratory function of the skin, which was once thought to be proved by the speedy death of animals whose skins, after removal of the hair, were covered with an impermeable varnish, has been shown by further observations to have no foundation in fact; the immediate cause of death in such cases being the loss of temperature. A varnished animal is said to have suffered no harm when surrounded by cotton wadding, and to have died when the wadding was removed.

Influence of the Nervous System on Excretion.—Any increase in the amount of sweat secreted is usually accompanied by dilatation of the cutaneous vessels. It is, however, probable that the secretion is like the other secretions, *e.g.*, the saliva, under the direct action of a special nervous apparatus, in that various nerves contain fibres which act directly upon the cells of the sweat glands in the same way that the chorda tympani contains fibres which act directly upon the salivary cells. The nerve fibres which induce sweating may act independently of the vaso-motor fibres, whether vaso-dilator or vaso-constrictor. The local apparatus is under control of the central nervous system—sweat centres probably existing both in the medulla and spinal cord—and may be reflexly as well as directly excited. This will explain the fact that sweat occurs not only when the skin is red, but also when it is pale, and the cutaneous circulation languid, as in the sweat which accompanies syncope or fainting, or which immediately precedes death.

(5.) **Absorption by the Skin.**—Absorption by the skin has been already mentioned, as an instance in which that process is most actively accomplished. Metallic preparations rubbed into the skin have the same action as when given internally, only in a less degree. Mercury applied in this manner exerts its specific influence upon syphilis, and excites salivation; potassio-tartrate of antimony may excite vomiting, or an eruption extending over the whole body; and arsenic may produce poisonous effects. Vegetable matters, also, if soluble, or already in solution, give rise to their peculiar effects, as cathartics, narcotics, and the like, when rubbed into the skin. The effect of rubbing is probably to convey the particles of the matter into the orifices of the glands, whence they are more readily absorbed than they would be through the epidermis. When simply left in contact with the skin, substances, unless in a fluid state, are seldom absorbed.

It has long been a contested question whether the skin covered with the epidermis has the power of absorbing water; and it is a point the more difficult to determine because the skin loses water by evaporation. But, from the result of many experiments, it may now be regarded as a well-ascertained fact that such absorption really occurs. The absorption of water by the surface of the body may take place in the lower animals very rapidly. Not only frogs, which have a thin skin, but lizards, in which the cuticle is thicker than in man, after having lost weight by being kept for some time in a dry atmosphere, were found to recover both their weight and plumpness very rapidly when immersed in water. When merely the tail, posterior extremities, and posterior part of the body of the lizard were immersed, the water absorbed was distributed throughout the system. And a like absorption through the skin, though to a less extent, may take place also in man.

In severe cases of dysphagia, when not even fluids can be taken into the stomach, immersion in a bath of warm water or of milk and water may assuage the thirst; and it has been found in such cases that the weight of the body is increased by the immersion. Sailors also, when destitute of fresh water, find their urgent thirst allayed by soaking their clothes in salt water and wearing them in that state; but these effects are in part due to the hindrance to the evaporation of water from the skin.

(6.) **Regulation of Temperature.**—For an account of this important function of the skin, see Chapter on Animal Heat.

CHAPTER XIII.

THE KIDNEYS AND THE EXCRETION OF URINE.

The Kidneys are two in number, and are situated deeply in the lumbar region of the abdomen, on either side of the spinal column, behind the peritoneum. They correspond in position to the last two dorsal and two upper lumbar vertebræ; the right being slightly lower than the left in consequence of the position of the liver on the right side of the abdomen. They are characteristic in shape, about 4 inches long, 2¼ inches broad, and 1½ inch thick. The weight of each kidney is about 4½ oz.

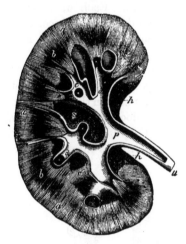

Fig. 238.—Plan of a longitudinal section through the pelvis and substance of the right kidney. ½; a, the cortical substance; b, b, broad part of the pyramids of Malpighii; c, c, the divisions of the pelvis named calyces, laid open; c', one of those unopened; d, summit of the pyramids of papillæ projecting into calyces; e, e, section of the narrow part of two pyramids near the calyces; p, pelvis or enlarged divisions of the ureter within the kidney; u, the ureter; s, the sinus; h, the hilus.

Structure of the Kidneys.—The kidney is covered by a rather tough fibrous capsule, which is slightly attached by its inner surface to the proper substance of the organ by means of very fine fibres of areolar tissue and minute blood-vessels. From the healthy kidney, therefore, it may be easily torn off without injury to the subjacent cortical portion of the organ. At the *hilus* or notch of the kidney, it becomes continuous with the external coat of the upper and dilated part of the ureter (Fig. 238).

On making a section lengthwise through the kidney (Fig. 238) the main part of its substance is seen to be composed of two chief portions, called respectively the *cortical* and the *medullary* portion, the latter being also sometimes called the *pyramidal* portion, from the fact of its being composed of about a dozen conical bundles of urine-tube, each bundle being called a pyramid. The upper part of the duct of the organ, or the *ureter*, is dilated into what is called the *pelvis* of the kidney; and this, again, after separating into two or three principal divisions, is finally sub-divided into still smaller portions, varying in number from about 8 to 12, or even more, and called *calyces*. Each of these little calyces or cups, which are often arranged in a double row, receives the pointed extremity or *papilla* of a *pyramid*. Sometimes, however, more than one papilla is received by a *calyx*.

The kidney is a compound *tubular* gland, and both its cortical and medullary portions are composed essentially of secreting tubes, the *tubuli uriniferi*, which, by one extremity, in the *cortical* portion, end commonly in little saccules containing blood-vessels, called *Malpighian bodies*, and, by the other, open through the *papillæ* into the pelvis of the kidney, and thus discharge the urine which flows through them.

Fig. 239.—A. Portion of a secreting tubule from the cortical substance of the kidney. B. The epithelial or gland-cells. × 700 times.

In the pyramids the tubes are chiefly straight—dividing and diverging as they ascend through these into the cortical portion; while in the latter region they spread out more irregularly, and become much branched and convoluted.

Tubuli Uriniferi.—The tubuli uriniferi (Fig. 239) are composed of a nearly homogeneous membrane, and are lined internally by epithelium. They vary considerably in size in different parts of their course, but are, on an average, about $\frac{1}{500}$ of an inch in diameter, and are found to be made up of several distinct sections which differ from one another very markedly, both in situation and structure. According to Klein, the following segments may be made out: (1) *The Malpighian corpuscle* (Figs.

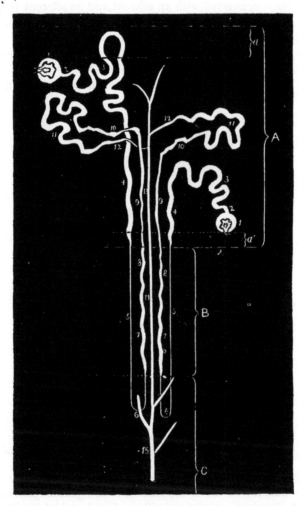

Fig. 240.— A Diagram of the sections of uriniferous tubes. A, Cortex limited externally by the capsule; *a*, subcapsular layer not containing Malpighian corpuscles; *a'*, inner stratum of cortex, also without Malpighian capsules; B, Boundary layer; C, Papillary part next the boundary layer; 1, Bowman's capsule of Malpighian corpuscle; 2, neck of capsule; 3. proximal convoluted tubule; 4, spiral tubule of Schachowa; 5, descending limb of Henle's loop; 6, the loop proper; 7, thick part of the ascending limb; 8, spiral part of ascending limb; 9, narrow ascending limb in the medullary ray; 10, the irregular tubule; 11, the intercalated section of Schweigger-Seidel, or the distal convoluted tubule; 12, the curved collecting tubule; 13, the straight collecting tubule of the medullary ray; 14, the collecting tube of the boundary layer; 15, the large collecting tube of the papillary part which, joining with similar tubes, forms the duct. (Klein and Noble Smith.)

240, 241), composed of a hyaline membrana propria, thickened by a varying amount of fibrous tissue, and lined by flattened nucleated epithelial plates.   This capsule is the dilated extremity of the uriniferous tubule, and contains within it a glomerulus of convoluted capillary blood-vessels supported by connective tissue, and covered by flattened epithelial plates. The glomerulus is connected with an efferent and an afferent vessel.   (2) The constricted *neck* of the capsule (Fig. 240, 2), lined in a similar man-ner, connects it with (3) The *Proximal convoluted tubule,* which forms several distinct curves and is lined with short columnar cells, which vary somewhat in size.   The tube next passes almost vertically downward, forming (4) The *Spiral tubule,* which is of much the same diameter, and

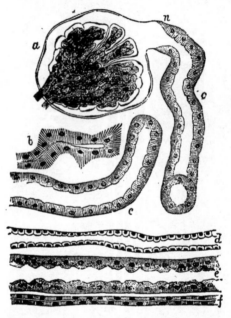

Fig. 241.—From a vertical section through the kidney of a dog—the capsule of which is supposed to be on the right.   *a* The capillaries of the Malpighian corpuscle—viz., the glomerulus, are arranged in lobules; *n*, neck of capsule; *c*, convoluted tubes cut in various directions: *b*, irregular tubule; *d*, *e*, and *f*, are straight tubes running toward capsules forming a so-called *medullary ray; d,* collecting tube; *e,* spiral tube; *f,* narrow section of ascending limb.   × 380.   (Klein and Noble Smith.)

is lined in the same way as the convoluted portion.   So far the tube has been contained in the cortex of the kidney, it now passes vertically down-ward through the most external part (boundary layer) of the Malpighian pyramid into the more internal part (papillary layer), where it curves up sharply, forming altogether the (5 and 6) *Loop of Henle,* which is a very narrow tube lined with flattened nucleated cells.   Passing vertically up-ward just as the tube reaches the boundary layer (7) it suddenly enlarges and becomes lined with polyhedral cells.   (8) About midway in the boun-dary layer the tube again narrows, forming the *ascending spiral of Henle's loop,* but is still lined with polyhedral cells.   At the point where the tube enters the cortex (9) the ascending limb narrows, but the diameter varies considerably; here and there the cells are more flattened, but both in this as in (8) the cells are in many places very angular, branched, and imbricated. It then joins (10) the *"irregular tubule,"* which has a very irregular and angular outline, and is lined with angular and imbri. cated cells. The tube next becomes convoluted, (11) forming the *distal convoluted tube or intercalated section of Schweigger-Seidel,* which is identical in all respects with the proximal convoluted tube (12 and 13). The curved and straight collecting tubes, the former entering the latter at right angles, and the latter passing vertically downward, are lined with polyhedral, or spindle-shaped, or flattened, or angular cells. The straight collecting tube now enters the boundary layer (14), and passes on to the

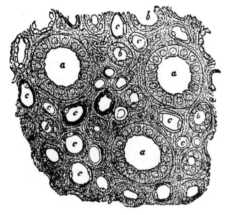

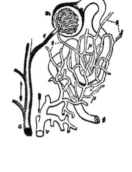

FIG. 242.   FIG. 243.

FIG. 242.—Transverse section of a renal papilla; *a*, larger tubes or papillary ducts; *b*, smaller tubes of Henle; *c*, blood-vessels, distinguished by their flatter epithelium; *d*, nuclei of the stroma. (Kölliker.) × 300.

FIG. 243.—Diagram showing the relation of the Malpighian body to the uriniferous ducts and blood-vessels. *a*, one of the interlobular arteries; *a'*, afferent artery passing into the glomerulus; *c*, capsule of the Malpighian body, forming the termination of and continuous with *t*, the uriniferous tube; *e'*, *e'*, efferent vessels which subdivide in the plexus *p*, surrounding the tube, and finally terminate in the branch of the renal vein *e* (after Bowman).

papillary layer, and, joining with other collecting tubes, form larger tubes, which finally open at the apex of the papilla. These collecting tubes are lined with transparent nucleated columnar or cubical cells (14, 15, 16).

The cells of the tubules with the exception of Henle's loop and all parts of the collecting tubules, are, as a rule, possessed of the intra-nuclear as well as of the intra-cellular network of fibres, of which the vertical rods are most conspicuous parts.

Heidenhain observed that indigo-sulphate of sodium, and other pigments injected into the jugular vein of an animal, were apparently excreted by the cells which possessed these rods, and therefore concluded that the pigment passes through the cells, rods, and nucleus themselves.

Klein, however, believes that the pigment passes through the intercellular substances, and not through the cells.

In some places, it is stated that a distinct membrane of flattened cells can be made out lining the lumen of the tubes (*centrotubular membrane*).

Blood-vessels of Kidneys.—In connection with the general distribution of blood-vessels to the kidney, the *Malpighian Corpuscles* may be further considered. They (Fig. 243) are found only in the cortical part of the kidney, and are confined to the central part, which, however, makes up about seven-eighths of the whole cortex. On a section of the organ, some of them are just visible to the naked eye as minute red points; others are too small to be thus seen. Their average diameter is about $\frac{1}{120}$ of an inch. Each of them is composed, as we have seen above, of the dilated extremity of a urinary tube, or Malpighian *capsule*, enclosing a tuft of blood-vessels.

The renal artery divides into several branches, which, passing in at the hilus of the kidney, and covered by a fine sheath of areolar tissue derived from the capsule, enter the substance of the organ in the intervals between the papillæ, chiefly at the junction between the cortex and the boundary layer. The chief branches then pass almost horizontally, giving off smaller branches upward to the cortex and downward to the medulla. The former are for the most part straight, they pass almost vertically to the surface of the kidney, giving off laterally in all directions longer or shorter branches, which supply the afferent arteries to the Malpighian bodies.

The small *afferent* artery (Figs. 243 and 245) which enters the Malpighian corpuscle, breaks up as before mentioned in the interior into a dense and convoluted and looped capillary plexus, which is ultimately gathered up again into a single small *efferent* vessel, comparable to a minute vein, which leaves the Malpighian capsule just by the point at which the afferent artery enters it. On leaving, it does not immediately join other small veins as might have been expected, but again breaking up into a network of capillary vessels, is distributed on the exterior of the tubule, from whose dilated end it had just emerged. After this second breaking up it is finally collected into a small vein, which, by union with others like it, helps to form the radicles of the renal vein. Thus, in the kidney, the blood entering by the renal artery traverses *two* sets of capillaries before emerging by the renal vein, an arrangement which may be compared to the *portal* system in miniature.

The tuft of vessels in the course of development is, as . were, thrust into the dilated extremity of the urinary tubule, which finally completely invests it just as the pleura invests the lungs or the tunica vaginalis the testicle. Thus the Malpighian capsule is lined by a parietal layer of squamous cells and a visceral or reflected layer immediately covering the vascular tuft (Fig. 241), and sometimes dipping down into its interstices.

This reflected layer of epithelium is readily seen in young subjects, but cannot always be demonstrated in the adult. (See Figs. 244 and 245.)

The vessels which enter the medullary layer break up into smaller arterioles, which pass through the boundary layer and proceed in a straight course between the tubules of the papillary layer, giving off on their way branches, which form a fine arterial meshwork around the tubes, and end in a similar plexus, from which the venous radicles arise.

Besides the small *afferent* arteries of the Malpighian bodies, there are, of course, others, which are distributed in the ordinary manner, for nutrition's sake, to the different parts of the organ; and in the pyramids, be-

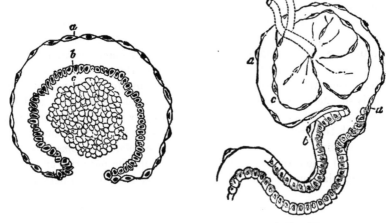

FIG. 244.   FIG. 245.

FIG. 244.—Transverse section of a developing Malpighian capsule and tuft (human) × 300. From a fœtus at about the fourth month; *a*, flattened cells growing to form the capsule; *b*, more rounded cells, continuous with the above, reflected round *c*, and finally enveloping it; *c*, mass of embryonic cells which will later become developed into blood-vessels. (W. Pye.)

FIG. 245.—Epithelial elements of a Malpighian capsule and tuft, with the commencement of a urinary tubule showing the afferent and efferent vessel; *a*, layer of tessellated epithelium forming the capsule; *b*, similar, but rather larger epithelial cells, placed in the walls of the tube; *c*, cells, covering the vessels of the capillary tuft; *d*, commencement of the tubule, somewhat narrower than the rest of it. (W. Pye.)

tween the tubes, there are numerous straight vessels, the *vasta recta*, supposed by some observers to be branches of *vasa efferentia* from Malpighian bodies, and therefore comparable to the venous plexus around the tubules in the *cortical* portion, while others think that they arise directly from small branches of the renal arteries.

Between the tubes, vessels, etc., which make up the substance of the kidney, there exists, in small quantity, a fine matrix of areolar tissue.

Nerves.—The nerves of the kidney are derived from the renal plexus.

Structure of the Ureters.—The duct of the kidney, or *ureter*, is a tube about the size of a goose-quill, and from a foot to sixteen inches in length, which, continuous above with the pelvis of the kidney, ends below by perforating obliquely the walls of the bladder, and opening on its internal surface. It is constructed of three principal coats (*a*) an outer, tough, *fibrous and elastic* coat; (*b*) a middle, *muscular coat*, of which the fibres are unstriped, and arranged in three layers—the fibres of the central layer being circular, and those of the other two longitudinal in direction; and (*c*) an internal *mucous* lining continuous with that of the pelvis of the kidney above, and of the urinary bladder below. The epithelium of all these parts (Fig. 246) is alike stratified and of a somewhat peculiar form; the cells on the free surface of the mucous membrane being usually spheroidal or polyhedral with one or more spherical or oval nuclei; while beneath these are pear-shaped cells, of which the broad ends are directed toward the free surface, fitting in beneath the cells of the first row, and the apices are prolonged into processes of various lengths, among which, again, the deepest cells of the epithelium are found spheroidal, irregularly oval, spindle-shaped or conical.

Structure of Urinary Bladder.—The urinary bladder, which forms a receptacle for the temporary lodgment of the urine in the intervals of its expulsion from the body, is more or less pyriform, its widest part, which is situate above and behind, being termed the *fundus:* and the narrow constricted portion in front and below, by which it becomes continuous with the urethra, being called its *cervix* or *neck*. It is constructed of four principal coats,—*serous, muscular, areolar* or *submucous*, and *mucous*. (*a*) The *serous* coat, which covers only the posterior and upper half of the bladder, has the same structure as that of the peritoneum,

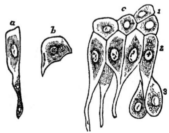

Fig. 246.—Epithelium of the bladder; *a*, one of the cells of the first row; *b*, a cell of the second row; *c*, cells *in situ*, of first, second, and deepest layers. (Obersteiner.)

with which it is continuous. (*b*) The fibres of the *muscular* coat, which are unstriped, are arranged in three principal layers, of which the external and internal (Ellis) have a general longitudinal, and the middle layer a circular direction. The latter are especially developed around the *cervix* of the organ, and are described as forming a *sphincter vesicæ*. The muscular fibres of the bladder, like those of the stomach, are arranged not in simple circles, but in figure-of-8 loops. (*c*) The *areolar* or *submucous* coat is constructed of connective tissue with a large proportion of elastic fibres. (*d*) The *mucous* membrane, which is rugose in the contracted state of the organ, does not differ in essential structure from mucous membranes in general. Its epithelium is stratified and closely resembles that of the pelvis of the kidney and the ureter (Fig. 246).

The mucous membrane is provided with mucous glands, which are more numerous near the neck of the bladder.

The bladder is well provided with blood and lymph vessels, and with nerves. The latter are branches from the sacral plexus (spinal) and hypogastric plexus (sympathetic). A few ganglion-cells are found, here and there, in the course of the nerve-fibres.

THE EXCRETION OF THE KIDNEY :—THE URINE.

Physical Properties.—Healthy urine is a perfectly transparent, amber-colored liquid, with a peculiar, but not disagreeable odor, a bitterish taste, and slight acid reaction. Its specific gravity varies from 1015 to 1025. On standing for a short time, a little mucous appears in it as a flocculent cloud.

Chemical Composition.—The urine consists of water, holding in solution certain organic and saline matters as its ordinary constituents, and occasionally various matters taken into the stomach as food—salts, coloring matter, and the like.

Table of the Chemical Composition of the Urine (modified from Becquerel).

Water	967
Urea	14.230
Other nitrogenous crystalline bodies—	
Uric acid, principally in the form of alkaline urates, a trace only free.	
Kreatinin, xanthin, hypoxanthin.	10·635
Hippuric acid, leucin, tyrosin, taurin, cystin, etc., all in small amounts and not constant.	
Mucus and pigment.	
Salts :—	
Inorganic—	
Principally sulphates, phosphates, and chlorides of sodium, and potassium, with phosphates of magnesium and calcium, traces of silicates and of chlorides.	8·135
Organic—	
Lactates, hippurates, acetates and formates, which only appear occasionally.	
Sugar	a trace sometimes.
Gases (nitrogen and carbonic acid principally).	
	1000

Reaction of the Urine—The normal reaction of the urine is slightly acid. This acidity is due to acid phosphate of sodium, and is less marked after meals. The urine contains no appreciable amount of free acid, as it gives no precipitate with sodium hyposulphite. After standing for some time the acidity increases from a kind of fermentation, due in all probability to the presence of mucus, and acid urates or free uric acid is deposited. After a time, varying in length according to the temperature, the reaction becomes strongly alkaline from the change of urea into ammonium carbonate—while at the same time a strong ammoniacal and fœtid odor appears, with deposits of triple phosphates and alkaline urates. As this does not occur unless the urine is exposed to the air, or, at least, until air has had access to it, it is probable that the decomposition is due to atmospheric germs.

Reaction of Urine in different classes of Animals.—In most herbivorous animals the urine is alkaline and turbid. The difference depends, not on any peculiarity in the mode of secretion, but on the differences in the food on which the two classes subsist: for when carnivorous animals, such as dogs, are restricted to a vegetable diet, their urine becomes pale, turbid, and alkaline, like that of an herbivorous animal, but resumes its former acidity on the return to an animal diet; while the urine voided by herbivorous animals, *e.g.*, rabbits, fed for some time exclusively upon animal substances, presents the acid reaction and other qualities of the urine of Carnivora, its ordinary alkalinity being restored only on the substitution of a vegetable for the animal diet. Human urine is not usually rendered alkaline by vegetable diet, but it becomes so after the free use of alkaline medicines, or of the alkaline salts with carbonic or vegetable acids; for these latter are changed into alkaline carbonates previous to elimination by the kidneys.

AVERAGE QUANTITY OF THE CHIEF CONSTITUENTS OF THE URINE EXCRETED IN 24 HOURS BY HEALTHY MALE ADULTS (PARKES).

Water	52· fluid ounces.
Urea	512·4 grains.
Uric acid	8·5 "
Hippuric acid, uncertain probably 10 to 15·	"
Sulphuric acid	31·11 "
Phosphoric acid	45· "
Potassium, Sodium, Ammonium Chlorides and free Chlorine . . . . } 323·25	"
Lime	3·5
Magnesia	3·
Mucus	7·
Extractives { Kreatinin Pigment Xanthin Hypoxanthin Resinous matter, etc. } . . 154·	"

Variations in Quantity of Constituents.—From these proportions, however, most of the constituents are, even in health, liable to variations.

The variations of the water in different seasons, and according to the quantity of drink and exercise, have already been mentioned. The water of the urine is also liable to be influenced by the condition of the nervous system, being sometimes greatly increased in hysteria, and some other nervous affections; and at other times diminished. In some diseases it is enormously increased; and its increase may be either attended with an augmented quantity of solid matter, as in ordinary diabetes, or may be nearly the sole change, as in the affection termed diabetes insipidus. In other diseases, e.g., the various forms of albuminuria, the quantity may be considerably diminished. A febrile condition almost always diminishes the quantity of water; and a like diminution is caused by any affection which draws off a large quantity of fluid from the body through any other channel than that of the kidneys, e.g., the bowels or the skin.

Method of estimating the Solids.—A useful rule for approximately estimating the total solids in any given specimen of healthy urine is to multiply the last two figures representing the specific gravity by 2·33. Thus, in urine of sp. gr. 1025, 2·33×25=58·25 grains of solids, are contained in 1000 grains of the urine. In using this method it must be remembered that the limits of error are much wider in diseased than in healthy urine.

Variations in the Specific Gravity.—The specific gravity of the human urine is about 1020. Probably no other animal fluid presents so many varieties in density within twenty-four hours as the urine does; for the relative quantity of water and of solid constituents of which it is composed is materially influenced by the condition and occupation of the body during the time at which it is secreted, by the length of time which has elapsed since the last meal, and by several other accidental circumstances. The existence of these causes of difference in the composition of the urine has led to the secretion being described under the three heads of *urina sanguinis, urina potus,* and *urina cibi.* The first of these names signifies the urine, or that part of it which is secreted from the blood at times in which neither food nor drink has been recently taken, and is applied especially to the urine which is evacuated in the morning before breakfast. The term *urina potus* indicates the urine secreted shortly after the introduction of any considerable quantity of fluid into the body: and the *urina cibi,* the portions secreted during the period immediately succeeding a meal of solid food. The last kind contains a larger quantity of solid matter than either of the others; the first or second, being largely diluted with water, possesses a comparatively low specific gravity. Of these three kinds, the morning urine is the best calculated for analysis in health, since it represents the simple secretion unmixed with the elements of food or drink; if it be not used, the whole of the urine passed during a period of twenty-four hours should be taken.

In accordance with the various circumstances above-mentioned, the specific gravity of the urine may, consistently with health, range widely on both sides of the usual average. The average healthy range may be stated at from 1015 in the winter to 1025 in the summer; but variations of diet and exercise, and many other circumstances, may make even greater differences than these. In disease, the variation may be greater; sometimes descending, in albuminuria, to 1004, and frequently ascending in diabetes, when the urine is loaded with sugar, to 1050, or even to 1060.

Quantity.—The total quantity of urine passed in twenty-four hours is affected by numerous circumstances. On taking the mean of many observations by several experimenters, the average quantity voided in twenty-four hours by healthy male adults from 20 to 40 years of age has been found to amount to about 52¼ fluid ounces (1⅓ to 2 litres).

Abnormal Constituents.—In disease, or after the ingestion of special foods, various abnormal substances occur in urine, of which the following may be mentioned—serum-albumin, globulin, ferments (apparently present in health also), blood, sugar, bile acids, and pigments, fats, oxalates, various salts taken as medicine, and other matters, as bacteria and renal casts.

THE SOLIDS OF THE URINE.

Urea (CH_4N_2O).—Urea is the principal solid constituent of the urine, forming nearly one-half of the whole quantity of solid matter. It is also the most important ingredient, since it is the chief substance by which the nitrogen of decomposed tissue and superfluous food is excreted

from the body. For its removal, the secretion of urine seems especially provided; and by its retention in the blood the most pernicious effects are produced.

Properties.—Urea, like the other solid constituents of the urine, exists in a state of solution. But it may be procured in the solid state, and then appears in the form of delicate silvery acicular crystals, which, under the microscope, appear as four-sided prisms (Fig. 247). It is obtained in this state by evaporating urine carefully to the

FIG. 247.—Crystals of Urea.

consistence of honey, acting on the inspissated mass with four parts of alcohol, then evaporating the alcoholic solution, and purifying the residue by repeated solution in water or alcohol, and finally allowing it to crystallize. It readily combines with some acids, like a weak base; and may thus be conveniently procured in the form of crystals of nitrate or oxalate of urea.

Urea is colorless when pure; when impure, yellow or brown: without smell, and of a cooling nitre-like taste; has neither an acid nor an alkaline reaction, and deliquesces in a moist and warm atmosphere. At 59° F. (15° C.) it requires for its solution less than its weight of water; it is dissolved in all proportions by boiling water; but it requires five times its weight of cold alcohol for its solution. It is insoluble in ether. At 248° F. (120° C.) it melts without undergoing decomposition; at a still higher temperature ebullition takes place, and carbonate of ammonium sublimes; the melting mass gradually acquires a pulpy consistence; and if the heat is carefully regulated, leaves a grey-white powder, cyanic acid.

Chemical Nature of Urea.—The chemical nature of urea is explained elsewhere,[1] but it will be as well to mention here that urea is isomeric with ammonium cyanate, and that it was first artificially produced from this substance. Thus:—Ammonium cyanate $(NH_4.CNO)$ = urea (CH_4N_2O). The action of heat upon urea in evolving ammonium carbonate and leaving cyanic acid, is thus explained. A similar decomposition of the urea with development of ammonium carbonate ensues spontaneously when urine is kept for some days after being voided, and explains the ammoniacal odor then evolved (p. 356). The urea is sometimes decomposed before it leaves the bladder, when the mucous membrane is diseased, and the mucus secreted by it is both more abundant, and, probably, more prone to act as a ferment; although the decomposition does not often occur unless atmospheric germs have had access to the urine.

Variations in the Quantity of Urea.—The quantity of urea excreted is, like that of the urine itself, subject to considerable variation. For a healthy adult 500 grains (about 32·5 grms.) per diem may be taken as rather a high average. Its percentage in healthy urine is 1·5 to 2·5. It is materially influenced by diet, being greater when animal food is exclusively used, less when the diet is mixed, and least of all with a vegetable diet. As a rule, men excrete a larger quantity than women, and persons in the middle periods of life a larger quantity than infants or old people. The quantity of urea excreted by children, relatively to their body-weight, is much greater than in adults. Thus the quantity of urea excreted per kilogram of weight was, in a child, 0·8 grm.: in an adult only 0·4 grm. Regarded in this way, the excretion of carbonic acid gives similar result, the proportion in the child and adult being as 82 : 34.

The quantity of urea does not necessarily increase and decrease with that of the urine, though on the whole it would seem that whenever the amount of urine is much augmented, the quantity of urea also is usually increased; and it appears that the quantity of urea, as of urine, may be especially increased by drinking large quantities of water. In various

---

[1] Appendix.

diseases the quantity is reduced considerably below the healthy standard, while in other affections it is above it.

Estimation of Urea.—A convenient apparatus for estimating the quantity of urea in a given sample of urine is that devised by Russell and West.

Urea contains nearly half its weight of nitrogen; hence this gas may be taken as a measure of the urea. A small quantity of urine is mixed with a large excess of solution of sodium hypobromite, which completely decomposes the urea, liberating all the nitrogen in a gaseous form: a gentle heat promotes the reaction. The percentage of urea can of course be readily calculated from the volume of nitrogen evolved from a measured quantity of the urine, but this calculation is avoided by graduating the tube in which the nitrogen is collected with numbers which indicate the corresponding percentage of urea. $CON_2H_4 + 3NaBrO + 2NaHO = 3NaBr + 3H_2O + Na_2CO_3 + N_2$.

Uric Acid ($C_5H_4N_4O_3$).—This substance, which was formerly termed lithic acid, on account of its existence in many forms of urinary calculi, is rarely absent from the urine of man or animals, though in the feline tribe it seems to be sometimes entirely replaced by urea. The proportionate quantity of uric acid varies considerably in different animals. In man, and Mammalia generally, especially the Herbivora, it is comparatively small. . In the whole tribe of birds, and of serpents, on the other hand, the quantity is very large, greatly exceeding that of the urea. In the urine of granivorous birds, indeed, urea is rarely if ever found, its place being entirely supplied by uric acid.

Variations in Quantity.—The quantity of uric acid, like that of urea, in human urine, is increased by the use of animal food, and decreased by the use of food free from nitrogen, or by an exclusively vegetable diet. In most febrile diseases, and in plethora, it is formed in unnaturally large quantities; and in gout it is deposited in, and around, joints, in the form of urate of soda, of which the so-called chalk-stones of this disease are principally composed. The average amount secreted in twenty-four hours is 8·5 grains (rather more than half a gramme).

Condition of Uric Acid in the Urine.—The condition in which uric acid exists in solution in the urine has formed the subject of some discussion, because of its difficult solubility in water. It is found chiefly in the form of urate of sodium, produced by the uric acid as soon as it is formed, combining with part of the base of the alkaline sodium phosphate of the blood. Hippuric acid, which exists in human urine also, acts upon the alkaline phosphate in the same way, and increases still more the quantity of acid phosphate, on the presence of which it is probable that a part of the natural acidity of the urine depends. It is scarcely possible to say whether the union of uric acid with the base sodium and probably ammonium, takes place in the blood, or in the act of secretion in the kidney:

the latter is the more likely opinion; but the quantity of either uric acid or urates in the blood is probably too small to allow of this question being solved.

Owing to its existence in combination in healthy urine, uric acid for examination must generally be precipitated from its bases by a stronger acid. Frequently, however, when excreted in excess, it is deposited in a crystalline form (Fig. 248), mixed with large quantities of ammonium or sodium urate. In such cases it may be procured for microscopic examination by gently warming the portion of urine containing the sediment; this dissolves urate of ammonium and sodium, while the comparatively insoluble crystals of uric acid subside to the bottom.

The most common form in which uric acid is deposited in urine, is that of a brownish or yellowish powdery substance, consisting of granules of

FIG. 248.—Various forms of uric acid crystals.   FIG. 249.—Crystals of hippuric acid.

ammonium—or sodium urate. When deposited in crystals, it is most frequently in rhombic or diamond-shaped laminæ, but other forms are not uncommon (Fig. 248). When deposited from the urine, the crystals are generally more or less deeply colored, from being combined with the coloring principles of the urine.

There are two chief tests for uric acid besides the microscopic evidence of its crystalline structure: (1) *The Murexide test,* which consists of evaporating to dryness a mixture of strong nitric acid and uric acid in a water bath. This leaves a yellowish-red residue of *Alloxan* ($C_4H_2N_2O_4$) and urea, and this, on addition of ammonium hydrate, gives a beautiful purple (ammonium purpurate, $C_8H_4(NH_4)N_5O_6$), deepened on addition of caustic potash. (2) *Schiff's test.* Dissolve the uric acid in sodium carbonate solution, and drop some of it on a filter paper moistened with silver nitrate, a black spot appears, which corresponds to the reduction of silver by the uric acid.

Hippuric Acid ($C_9H_9NO_3$) has long been known to exist in the urine of herbivorous animals in combination with soda. It also exists naturally in the urine of man, in quantity equal or rather exceeding that of the uric acid.

Pigments.—The coloring matters of the urine are: (1) *Uro-bilin,* a substance connected with the coloring matters of the blood and bile (p. 275); it is especially seen in febrile urine and exists normally, but to less amount; it is of a yellowish-red color; (2) *Uro-chrome,* which on exposure undergoes oxidation, and becomes *Uro-erythrin,* the former being yellowish and the latter sandy red; and (3) *Indican* is occasionally present.

Indican is not itself pigmentary, though by its decomposition indigo blue and indigo red are produced. Its presence can usually be detected by adding to a small quantity of urine an equal bulk of strong hydrochloric acid, and gently heating the solution; on the addition of two or three drops of strong nitric acid a delicate purplish tint is developed, and indigo blue and red crystals separate out.

Mucus.—*Mucus* in the urine consists principally of the epithelial débris of the mucous surface of the urinary passages. Particles of epithelium, in greater or less abundance, may be detected in most samples of urine, especially if it has remained at rest for some time and the lower strata are then examined (Fig. 250). As urine cools, the mucous is some-

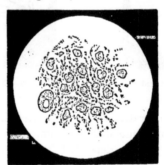

Fig. 250.—Mucus deposited from urine.

times seen suspended in it as a delicate opaque cloud, but generally it falls. In inflammatory affections of the urinary passages, especially of the bladder, mucus in large quantities is poured forth, and speedily undergoes decomposition. The presence of the decomposing mucus excites (as already stated, p. 356) chemical changes in the urea, whereby ammonia, or carbonate of ammonium, is formed, which, combining with the excess of acid in the super-phosphates in the urine, produces insoluble neutral or alkaline phosphates of calcium and magnesium, and phosphate of ammonium and magnesium. These mixing with the mucus, constitute the peculiar white, viscid, mortar-like substance which collects upon the mucous surface of the bladder, and is often passed with the urine, forming a thick tenacious sediment.

Extractives.—Besides mucus and coloring matter, urine contains a considerable quantity of nitrogenous compounds, usually described under the generic name of *extractives*. Of these, the chief are: (1) *Kreatinin* ($C_4H_7N_3O$) a substance derived, probably, from the metamorphosis of muscular tissue, crystallizing in colorless oblique rhombic prisms; a fairly definite amount of this substance, about 15 grains (1 grm.), appears in the urine daily, so that it must be looked upon as a normal constituent; it is increased on an increase of the nitrogenous constituents of the food; (2) *Xanthin* ($C_5N_4H_4O_2$), an amorphous powder soluble in hot water; (3) *Hypo-xanthin*, or sarkin ($C_5N_4H_4O$); (4) *Oxaluric acid* ($C_3H_4N_2O_4$), in combination with ammonium; (5) *Allantoin* ($C_4H_6N_4O_3$), in the urine of the new-born child. All these extractives are chiefly interesting as being closely connected with urea, and mostly yielding that substance on oxidation. Leucin and tyrosin can scarcely be looked upon as normal constituents of urine.

Saline Matter.—The *sulphuric acid* in the urine is combined chiefly or entirely with sodium or potassium; forming salts which are taken in very small quantity with the food, and are scarcely found in other fluids or tissues of the body; for the sulphates commonly enumerated among the constituents of the ashes of the tissues and fluids are for the most part, or entirely, produced by the changes that take place in the burning. Only about one-third of the sulphuric acid found in the urine is derived directly from the food (Parkes). Hence the greater part of the sulphuric acid which the sulphates in the urine contain, must be formed in the blood, or in the act of secretion of urine; the sulphur of which the acid is formed being probably derived from the decomposing nitrogenous tissues, the other elements of which are resolved into urea and uric acid. It may be in part derived also from the sulphur-holding *taurin* and *cystin*, which can be found in the liver, lungs, and other parts of the body, but not generally in the excretions; and which, therefore, must be broken up. The oxygen is supplied through the lungs, and the heat generated during combination with the sulphur, is one of the subordinate means by which the animal temperature is maintained.

Besides the sulphur in these salts, some also appears to be in the urine, uncombined with oxygen; for after all the sulphates have been removed from urine, sulphuric acid may be formed by drying and burning it with nitre. From three to five grains of sulphur are thus daily excreted. The combination in which it exists is uncertain: possibly it is in some compound analogous to cystin or cystic oxide (p. 365). Sulphuric acid also exists normally in the urine in combination with phenol (C_6H_6O) as phenol sulphuric acid or its corresponding salts, with sodium, etc.

The *phosphoric acid* in the urine is combined partly with the alkalies, partly with the alkaline earths—about four or five times as much with the former as with the latter. In blood, saliva, and other alkaline fluids of the body, phosphates exist in the form of alkaline, neutral, or acid salts. In the urine they are acid salts, viz., the sodium, ammonium, calcium, and magnesium phosphates, the excess of acid being (Liebig) due to the appropriation of the alkali with which the phosphoric acid in the blood is combined, by the several new acids which are formed or discharged at the kidneys, namely, the uric, hippuric, and sulphuric acids, all of which are neutralized with soda.

The phosphates are taken largely in both vegetable and animal food; some thus taken are excreted at once; others, after being transformed and incorporated with the tissues. Calcium phosphate forms the principal earthy constituent of bone, and from the decomposition of the osseous tissue the urine derives a large quantity of this salt. The decomposition of other tissues also, but especially of the brain and nerve-substance, furnishes large supplies of phosphorus to the urine, which

Fig. 251.—Urinary sediment of triple phosphates (large prismatic crystals) and urate of ammonium, from urine which had undergone alkaline fermentation.

phosphorus is supposed, like the sulphur, to be united with oxygen, and then combined with bases. This quantity is, however, liable to considerable variation. Any undue exercise of the brain, and all circumstances producing nervous exhaustion, increase it. The earthy phosphates are more abundant after meals, whether on animal or vegetable food, and are diminished after long fasting. The alkaline phosphates are increased after animal food, diminished after vegetable food. Exercise increases the alkaline, but not the earthy phosphates (Bence Jones). Phosphorus uncombined with oxygen appears, like sulphur, to be excreted in the urine (Ronalds). When the urine undergoes alkaline fermentation, phosphates are deposited in the form of a *urinary sediment*, consisting chiefly of ammonio-magnesium phosphate (triple phosphate) (Fig. 251). This compound does not, as such, exist in healthy urine. The ammonia is chiefly or wholly derived from the decomposition of urea (p. 359).

The *chlorine* of the urine occurs chiefly in combination with sodium, but slightly also with ammonium, and perhaps potassium. As the chlorides exist largely in food and in most of the animal fluids, their occurrence in the urine is easily understood.

Cystin (C$_6$H$_7$NSO$_2$) (Fig. 252) is an occasional constituent of urine. It resembles taurin in containing a large quantity of sulphur—more than 25 per cent. It does not exist in healthy urine.

Another common morbid constituent of the urine is *oxalic acid*, which is frequently deposited in combination with calcium (Fig. 253) as a

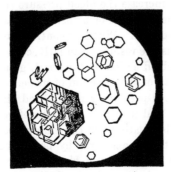

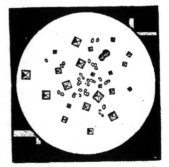

FIG. 252.—Crystals of cystin.   FIG. 253.—Crystals of calcium oxalate.

urinary sediment. Like cystin, but much more commonly, it is the chief constituent of certain calculi.

Of the other abnormal constituents of the urine mentioned it will be unnecessary to speak at length in this work.

Gases.—A small quantity of gas is naturally present in the urine in a state of solution. It consists of carbonic acid (chiefly) and nitrogen and a small quantity of oxygen.

THE METHOD OF THE EXCRETION OF URINE.

The excretion of the urine by the kidney is believed to consist of two more or less distinct processes—viz., (1.) *of filtration,* by which the water and the ready-formed salts are eliminated; and (2.) *of true secretion,* by which certain substances forming the chief and more important part of the urinary solids are removed from the blood. This division of function corresponds more or less to the division in the functions of other glands of which we have already treated. It will be as well to consider them separately.

(1.) **Of Filtration.**—This part of the renal function is performed within the Malpighian corpuscles by the renal glomeruli. By it not only the water is strained off, but also certain other constituents of the urine, *e.g.,* sodium chloride, are separated. The amount of the fluid filtered off depends almost entirely upon the blood-pressure in the glomeruli.

The greater the blood-pressure in the arterial system generally, and consequently in the renal arteries, the greater, *cæteris paribus,* will be the blood-pressure in the glomeruli, and the greater the quantity of urine separated; but even without increase of the general blood-pressure, if the renal arteries be locally dilated, the pressure in the glomeruli will be increased and with it the secretion of urine. On the other hand, if the local blood-pressure be diminished, the amount of fluid will be lessened. All the numerous causes, therefore, which increase the blood-pressure (p. 152) will, as a rule, secondarily increase the secretion of urine. Of these the heart's action is amongst the most important. When its contractions are increased in force, increased diuresis is the result. Similarly, causes which lower the blood-pressure, *e.g.,* enfeebled action of the heart, great loss of blood, etc., will diminish the activity of the secretion of urine.

The close connection between the blood-pressure generally and the nervous system has been before considered, and it will be clear, therefore, that the amount of urine secreted depends greatly upon the influence of the nervous system. Thus, division of the spinal cord, by producing general vascular dilatation, causes a great diminution of blood-pressure, and so diminishes the amount of water passed; since the local dilatation in the renal arteries is not sufficient to counteract the general diminution of pressure. Stimulation of the cut cord produces, strangely enough, the same results—*i.e.,* a diminution in the amount of the urine passed, but in a different way, viz., by constricting the arteries generally, and, among others, the renal arteries; the diminution of blood-pressure resulting from the local resistance in the renal arteries being more potent to diminish blood-pressure in the glomeruli than the general increase of blood-pressure is to increase it. Section of the renal nerves or of any others which produce *local* dilatation without greatly diminishing the general blood-pressure will cause an increase in the quantity of fluid passed.

The fact that in summer or in hot weather the urine is diminished may be attributed partly to the copious elimination of water by the skin in the form of sweat which occurs in summer, as contrasted with the greatly diminished functional activity of the skin in winter, but also to the dilated condition of the vessels of the skin causing a decrease in the general blood-pressure. Thus we see that in regard to the elimination of water from the system, the skin and kidneys perform similar functions, and are capable to some extent of acting vicariously, one for the other. Their relative activities are inversely proportional to each other.

The intimate connection between the condition of the kidney and the blood-pressure has been exceedingly well shown by the introduction of an instrument called the *Oncometer,* recently introduced by Roy, which is a modification of the plethysmograph (Fig. 138). By means of this apparatus any alteration in the volume of the kidney is communicated to an apparatus (oncograph) capable of recording graphically, with a writing lever, such variations. It has been found that the kidney is extremely sensitive to any alteration in the general blood-pressure, every fall in the general blood-pressure being accompanied by a decrease in the volume of the kidney, and every rise, unless produced by considerable constriction of the peripheral vessels, including those of the kidney, being accompanied by a corresponding increase of volume. Increase of volume is followed by an increase in the amount of urine secreted, and decrease of volume by a decrease in the secretion. In addition, however, to the response of the kidney to alterations in the general blood-pressure, it has been further observed that certain substances, when injected into the blood, will also produce an increase in volume of the kidney, and consequent increased flow of urine, without affecting the general blood-pressure—such bodies as sodium acetate and other diuretics. These observations appear to prove that local dilatation of the renal vessels may be produced by alterations in the blood upon a local nervous mechanism, as the effect is produced when all of the renal nerves have been divided. The alterations are not only produced by the addition of drugs, but also by the introduction of comparatively small quantities of water or saline solution. To this alteration of the blood acting upon the renal vessels (either directly or) through a local vaso-motor mechanism, and not to any great alteration in the general blood-pressure, must we attribute the effect of meals, etc., observed by Roberts. "The renal excretion is increased after meals and diminished during fasting and sleep. The increase began within the first hour after breakfast, and continued during the succeeding two or three hours; then a diminution set in, and continued until an hour or two after dinner. The effect of dinner did not appear until two or three hours after the meal; and it reached its maximum about the fourth hour. From this period the excretion steadily decreased until bedtime. During sleep it sank still lower, and reached its minimum—being not more than one-third of the quantity excreted during the hours of digestion." The increased amount of urine passed after drinking large quantities of fluid probably depends upon the diluted condition of the blood thereby induced.

The following table[1] will help to explain the dependence of the filtration function upon the blood-pressure and the nervous system:—

Table of the Relation of the Secretion of Urine to Arterial Pressure.

A. Secretion of Urine may be increased—

   a. By increasing the general blood-pressure, by
     1. Increase of force or frequency of heart-beat.
     2. Constriction of small arteries of areas other than the kidney.

---

[1] Modified from M. Foster.

b. *By relaxation of the renal artery without compensating relaxa-*
   tion elsewhere, by
   1. Division of the renal nerves (causing polyuria).
   2. "   "   " and afterward stimulating cord
      below medulla (causing greater polyuria).
   3. Division of the splanchnic nerves; but polyuria is less than
      in 1 or 2, as these nerves are distributed to a wider area,
      the dilatation of the renal artery is accompanied by dila-
      tation of other vessels, and therefore with a somewhat
      diminished general blood supply.
   4. Puncture of the floor of fourth ventricle or mechanical irri-
      tation of the superior cervical ganglion of the sympathetic,
      possibly from dilatation of the renal arteries.

B. Secretion of urine may be diminished—

a. *By diminishing the general blood-pressure,* by
   1. Diminishing the force or frequency of the heart-beats.
   2. Dilatation of capillary areas other than the kidney.
   3. Division of spinal cord below medulla, which causes dilata-
      tion of general abdominal area, and urine generally ceases
      being secreted.
b. *By increasing the blood-pressure,* by stimulation of spinal cord
   below medulla, the constriction of the renal artery not being
   compensated for by the increase of general blood-pressure.
c. *By constriction of the renal artery,* by stimulating the renal or
   splanchnic nerves, or by stimulating the spinal cord.

Although it is convenient to call the processes which go on in the renal
glomeruli, filtration, there is reason to believe that they are not absolutely
mechanical, as the term might seem to imply, since, when the epithelium
of the Malpighian capsule has been, as it were, put out of order by ligature of the renal artery, on removal of the ligature, the urine has been
found temporarily to contain albumen, indicating that a selective power
resides in the healthy epithelium, which allows a certain constituent part
of the blood to be filtered off and not others.

(2.) **Of True Secretion.**—That there is a second part in the process
of the excretion of urine, which is true secretion, is suggested by the
structure of the tubuli uriniferi, and the idea is supported by various
experiments. It will be remembered that the convoluted portions of the
tubules are lined with epithelium, which bears a close resemblance to the
secretory epithelium of other glands, whereas the Malpighian capsules
and portions of the loops of Henle are lined simply by endothelium. The
two functions are, then, suggested by the differences of epithelium, and
also by the fact that the blood supply is different, since the convoluted
tubes are surrounded by capillary vessels derived from the breaking up of
the efferent vessels of the Malpighian tufts. The theory first suggested
by Bowman (1842), and still generally accepted, of the function of the

two parts of the tubules, is that the cells of the convoluted tubes, by a process of true secretion, separate from the blood substances such as urea, whereas from the glomeruli are separated the water and the inorganic salts. Another theory suggested by Ludwig (1844) is that in the glomeruli is filtered off from the blood all the constituents of the urine in a very diluted condition. When this passes along the tortuous uriniferous tube, part of the water is re-absorbed into the vessels surrounding them, leaving the urine in a more concentrated condition—retaining all its proper constituents. This osmosis is promoted by the high specific gravity of the blood in the capillaries surrounding the convoluted tubes, but the return of the urea and similar substances is prevented by the secretory epithelium of the tubules. Ludwig's theory, however plausible, must, we think, give way to the first theory, which is more strongly supported by direct experiment.

By using the kidney of the newt, which has two distinct vascular supplies, one from the renal artery to the glomeruli, and the other from the renal portal vein to the convoluted tubes, Nussbaum has shown that certain substances, e.g., peptones, sugar, when injected into the blood, are eliminated by the glomeruli, and so are not got rid of when the renal arteries are tied; whereas certain other substances, e.g., urea, when injected into the blood, are eliminated by the convoluted tubes, even when the renal arteries have been tied. This evidence is very direct that urea is excreted by the convoluted tubes.

Heidenhain also has shown by experiment that if a substance (sodium sulphindigotate), which ordinarily produces blue urine, be injected into the blood after section of the medulla which causes lowering of the blood-pressure in the renal glomeruli, that when the kidney is examined, the cells of the convoluted tubules (and of these alone) are stained with the substance, which is also found in the lumen of the tubules. This appears to show that under ordinary circumstances the pigment at any rate is eliminated by the cells of the convoluted tubules, and that when by diminishing the blood-pressure, the filtration of urine ceases, the pigment remains in the convoluted tubes, and is not, as it is under ordinary circumstances, swept away from them by the flushing of them which ordinarily takes place with the watery part of urine derived from the glomeruli. It therefore is probable that the cells, if they excrete the pigment, excrete urea and other substances also. But urea acts somewhat differently to the pigment, as when it is injected into the blood of an animal in which the medulla has been divided and the secretion of urine stopped, a copious secretion of urine results, which is not the case when the pigment is used instead under similar conditions. The flow of urine, independent of the general blood-pressure, might be supposed to be due to the action of the altered blood upon some local vaso-motor mechanism; and, indeed, the local blood-pressure is directly affected in this way, but there is reason for believing that part of the increase of the secretion is due to the direct stimulation of the cells by the urea contained in the blood.

To sum up, then, the relation of the two functions: (1.) The process of filtration, by which the chief part if not the whole of the *fluid* is eliminated, together with certain inorganic salts, and possibly other solids, is directly dependent upon blood-pressure, is accomplished by the renal glomeruli, and is accompanied by a free discharge of solids from the tubules. (2.) The process of secretion proper, by which urea and the principal urinary solids are eliminated, is only indirectly, if at all, dependent upon blood-pressure, and is accomplished by the cells of the convoluted tubes. It is sometimes accompanied by the elimination of copious fluid, produced by the chemical stimulation of the epithelium of the same tubules.

Sources of the Nitrogenous Urinary Solids.

Urea.—In speaking of the method of the secretion of urine, it was assumed that the part played by the cells of the uriniferous tubules was that of mere separation of the constituents of the urine which existed ready-formed in the blood: there is considerable evidence to favor this assumption. What may be called the specially characteristic solid of the urine, *i.e.*, urea (as well as most of the other solids), may be detected in the blood, and in other parts of the body, *e.g.*, the humors of the eye (Millon), even while the functions of the kidneys are unimpaired; but when from any cause, especially extensive disease or extirpation of the kidneys, the separation of urine is imperfect, the urea is found largely in the blood and in most other fluids of the body.

It must, therefore, be clear that the urea is for the most part made somewhere else than in the kidneys, and simply brought to them by the blood for elimination. It is not absolutely proved, however, that all the urea is formed away from these organs, and it is possible that a small quantity is actually secreted by the cells of the tubules. The sources of the urea, which is brought to the kidneys for excretion, are stated to be two.

(1.) *From the splitting up of the Elements of the Nitrogenous Food.*— The origin of urea from this source is shown by the increase which ensues on substituting an animal or highly nitrogenous for a vegetable diet; in the much larger amount—nearly double—excreted by Carnivora than Herbivora, independent of exercise; and in its diminution to about one-half during starvation, or during the exclusion of non-nitrogenous principles of food. Part, at any rate, of the increased amount of urea which appears in the urine soon after a full meal of proteid material may be attributed to the production of a considerable amount of leucin and tyrosin as by-products of pancreatic digestion. These substances are carried by the portal vein to the liver, and it is there that the change in all probability takes place; as when the functions of the organ are gravely interfered with, as in the case of acute yellow atrophy, the amount of urea is distinctly diminished, and its place appears to be taken by leucin and tyrosin. It has been found by experiment, too, that if these substances be introduced into the alimentary canal, the introduction is followed by a corresponding increase in the amount of urea, but not by the presence of the bodies themselves in the urine.

(2.) *From the Nitrogenous metabolism of the Tissues.*—This second origin of urea is shown by the fact that it continues to be excreted, though in smaller quantity than usual, when all nitrogenous substances are strictly excluded from the food, as when the diet consists for several days of sugar, starch, gum, oil, and similar non-nitrogenous substances (Lehmann). It is excreted also, even though no food at all be taken for a considerable time; thus it is found in the urine of reptiles which have fasted for months; and in the urine of a madman who had fasted eighteen days, Lassaigne found both urea and all the components of healthy urine.

Turning to the muscles, however, as the most actively metabolic tissue, we find as a result of their activity not urea, but *kreatin;* and although it may be supposed that some of this latter body appears naturally as *kreatinin*, yet it is not in sufficient quantity to represent the large amount of it formed by the muscles, and, indeed, by others of the tissues. It is assumed that kreatin therefore is the nitrogenous antecedent of urea; where its conversion into urea takes place is doubtful, but very likely the liver, and possibly the spleen, may be the seats of the change. It may be, however, that part—but if so, a small part—reaches the kidneys without previous change, leaving it to the cells of the renal tubules to complete the action. In speaking of kreatin as the antecedent of urea, it should be recollected that other nitrogenous products, such as xanthin ($C_6H_4N_4O_2$), appear in conjunction with it, and that these may also be converted into urea.

It was formerly taken for granted that the quantity of urea in the urine is greatly increased by *active exercise;* but numerous observers have failed to detect more than a slight increase under such circumstances; and our notions concerning the relation of this excretory product to the destruction of muscular fibre, consequent on the exercise of the latter, have undergone considerable modification. There is no doubt, of course, that like all parts of the body, the muscles have but a limited term of existence, and are being constantly although very slowly renewed, at the same time that a part of the products of their disintegration appears in the urine in the form of urea. But the waste is not so fast as it was formerly supposed to be; and the theory that the amount of work done by the muscle is expressed by the quantity of urea excreted in the urine must without doubt be given up.

Uric Acid.—Uric acid probably arises much in the same way as urea, either from the disintegration of albuminous tissues, or from the food. The relation which uric acid and urea bear to each other is, however, still obscure: but uric acid is said to be a less advanced stage of the oxidation of the products of proteid metabolism. The fact that they often exist together in the same urine, makes it seem probable that they have different origins; but the entire replacement of either by the other, as of urea by uric acid in the urine of birds, serpents, and many insects, and of uric acid by urea, in the urine of the feline tribe of Mammalia, shows that either alone may take the place of the two. At any rate, although it is true that one molecule of uric acid is capable of splitting up into two molecules of urea and one of mes-oxalic acid, there is no evidence for believing that uric acid is an antecedent of urea in the nitrogenous metabolism of the body. Some experiments seem to show that uric acid ·is formed in the kidney.

Hippuric Acid ($C_9H_9NO_3$).—Hippuric acid is closely allied to benzoic acid; and this substance when introduced into the system, is excreted by the kidneys as hippuric acid (Ure). Its source is not satisfactorily determined: in part it is probably derived from some constituents of vegetable diet, though man has no hippuric acid in his food, nor, commonly, any benzoic acid that might be converted into it; in part from the natural disintegration of tissues, independent of vegetable food, for Weismann constantly found an appreciable quantity, even when living on an exclusively animal diet. Hippuric acid arises from the union of benzoic acid with glycin ($C_2H_5NO_2 + C_7H_6O_2 = C_9H_9NO_3 + H_2O$), which union may take place in the kidneys themselves. as well as in the liver.

Extractives.—The source of the extractives of the urine is probably in chief part the disintegration of the nitrogenous tissues, but we are unable to say whether these nitrogenous bodies are merely accidental, having resisted further decomposition into urea, or whether they are the representatives of the decomposition of special tissues, or of special forms of metabolism of the tissues. There is, however, one exception, and this is in the case of kreatinin; there is great reason for believing that the amount of this body which appears in the urine is derived from the metabolism of the nitrogenous food, as when this is diminished, it diminishes, and when stopped, it no longer appears in the urine.

THE PASSAGE OF URINE INTO THE BLADDER.

As each portion of urine is secreted it propels that which is already in the tubes onward into the pelvis of the kidney. Thence through the ureter the urine passes into the bladder, into which its rate and mode of entrance has been watched in cases of *ectopia vesicæ*, *i.e.*, of such fissures in the anterior or lower part of the walls of the abdomen, and of the front wall of the bladder, as expose to view its hinder wall together with the orifices of the ureters. The urine does not enter the bladder at any regular rate, nor is there a synchronism in its movement through the two ureters. During fasting, two or three drops enter the bladder every minute, each drop as it enters first raising up the little papilla on which, in these cases, the ureter opens, and then passing slowly through its orifice, which at once again closes like a sphincter. In the recumbent posture, the urine collects for a little time in the ureters, then flows gently, and, if the body be raised, runs from them in a stream till they are empty. Its flow is increased in deep inspiration, or straining, and in active exercise, and in fifteen or twenty minutes after a meal (Erichsen). The urine collecting is prevented from regurgitation into the ureters by the mode in which these pass through the walls of the bladder, namely, by their lying for between half and three-quarters of an inch between the muscular and mucous coats before they turn rather abruptly forward, and open through the latter into the interior of the bladder.

Micturition.—The contraction of the muscular walls of the bladder may by itself expel the urine with little or no help from other muscles, when the sphincter of the organ is relaxed. In so far, however, as it is a *voluntary* act, micturition is performed by means of the abdominal and other expiratory muscles which, in their contraction, press on the abdominal viscera, the diaphragm being fixed, and cause the expulsion of the contents of the bladder. The muscular coat of the bladder co-operates, in micturition, by reflex *involuntary* action, with the abdominal muscles; and the act is completed by the *accelerator urinæ*, which, as its name implies, quickens the stream, and expels the last drops of urine from the urethra. The act, so far as it is not directed by volition, is under the control of a nervous *centre* in the lumbar spinal cord, through which, as in the case of the similar centre for defæcation (p. 288), the various muscles concerned are harmonized in their action. It is well known that the act may be reflexly induced, *e.g.*, in children who suffer from intestinal worms, or other such irritation. Generally the afferent impulse which calls into action the desire to micturate is excited by over distention of the bladder, or even by a few drops of urine passing into the urethra.

END OF VOL. I.

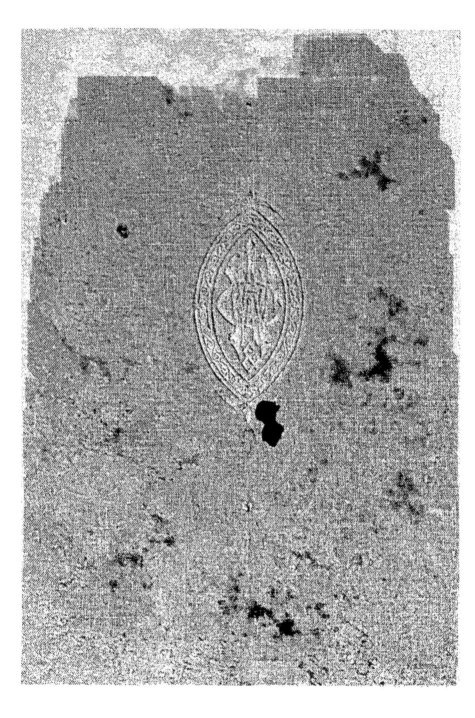

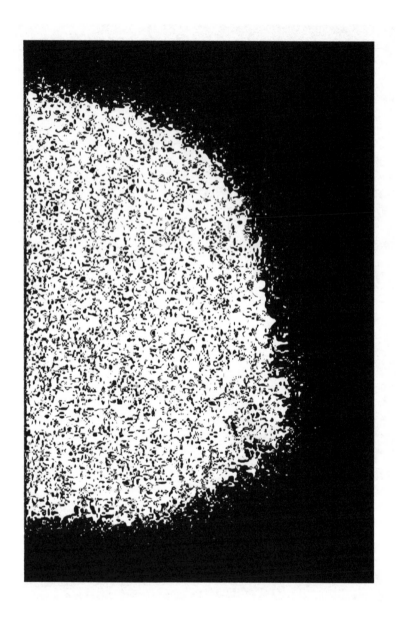

Lightning Source UK Ltd.
Milton Keynes UK
UKHW021449030219
336610UK00006B/369/P